CHIRAL SULFUR REAGENTS

Applications in Asymmetric and Stereoselective Synthesis

NEW DIRECTIONS in ORGANIC and BIOLOGICAL CHEMISTRY

Series Editor: C.W. Rees, CBE, FRS
Imperial College of Science, Technology and Medicine, London, UK

Advisory Editor: Alan R. Katritzky, FRS
University of Florida, Gainesville, Florida

Published and Forthcoming Titles

Chirality and the Biological Activity of Drugs
Roger J. Crossley

Enzyme-Assisted Organic Synthesis
Manfred Schneider and Stefano Servi

***C*-Glycoside Synthesis**
Maarten Postema

Organozinc Reagents in Organic Synthesis
Ender Erdik

Activated Metals in Organic Synthesis
Pedro Cintas

Capillary Electrophoresis: Theory and Practice
Patrick Camilleri

Cyclization Reactions
C. Thebtaranonth and Y. Thebtaranonth

Mannich Bases: Chemistry and Uses
Maurilio Tramontini and Luigi Angiolini

Vicarious Nucleophilic Substitution and Related Processes in Organic Synthesis
Mieczyslaw Makosza

Aromatic Fluorination
James H. Clark, David Wails, and Tony W. Bastock

Lewis Acids and Selectivity in Organic Synthesis
M. Santelli and J.-M. Pons

Dianion Chemistry in Organic Synthesis
Charles M. Thompson

Asymmetric Synthetic Methodology
David J. Ager and Michael B. East

Synthesis Using Vilsmeier Reagents
C. M. Marson and P. R. Giles

The Anomeric Effect
Eusebio Juaristi and Gabriel Cuevas

Chiral Sulfur Reagents
M. Mikolajczyk, J. Drabowicz, and P. Kielbasiński

Chemical Approaches to the Systhesis of Peptides and Proteins
Paul Lloyd-Williams, Fernanado Albericio, and Ernest Giralt

Concerted Organic Mechanisms
Andrew Williams

CHIRAL SULFUR REAGENTS

Applications in Asymmetric and Stereoselective Synthesis

Marian Mikołajczk
Jósef Drabowicz
Piotr Kiełbasiński
Center of Molecular and Macromolecular Studies
Polish Academy of Sciences
Lódź, Poland

CRC Press
Boca Raton New York

Library of Congress Cataloging-in-Publication Data

Mikołajczyk, Marian.
Chiral sulfur reagents : applications in asymmetric and stereoselective synthesis / Marian Mikołajczyk, Józef Drabowicz, and Piotr Kiełbasiński.
p. cm. -- (New directions in organic and biological chemistry)
Includes bibliographical references (p. –) and index.
ISBN 0-8493-9120-2 (alk. paper)
1. Organosulphur compounds. 2. Chirality. 3. Asymmetric synthesis. 4. Stereochemistry. I. Drabowicz, Józef. II. Kiełbasiński, Piotr. III. Title. IV. Series.
[DNLM: 1. Hepatitis B virus. QW 710 G289h]
QD305.S3M5 1996
547′.060459--dc20

96-6162
CIP

Direct all inquiries to CRC Press LLC, 2000 Corporate Blvd., N.W., Boca Raton, Florida 33431.

International Standard Book Number 0-8493-9120-2
Library of Congress Card Number 96-6162
Printed in the United States of America 1 2 3 4 5 6 7 8 9 0
Printed on acid-free paper

Preface

Although the first chiral organosulfur compounds were obtained at the beginning of this century, they have received more attention since the early 1960s. Initially, chiral sulfur compounds served as model compounds in studies on the mechanism and stereochemistry of nucleophilic substitution at sulfur's center. Quite soon, however, it was recognized that chiral sulfur compounds are of great value in asymmetric synthesis, since many reactions may be efficiently stereocontrolled by chiral sulfur auxiliaries which later are easily removable under mild conditions by reductive or eliminative methods. As a result, there has been a literature explosion in this field. On the one hand, over the last three decades more than 40 different classes of chiral sulfur compounds have been described in the chemical literature and a large number of useful procedures for the synthesis of enantiomerically pure sulfur compounds have been developed, especially of tri- and tetracoördinated sulfur structures. On the other hand, every year literature records dozens and dozens of diverse sulfur-mediated asymmetric syntheses applied in both academic and industrial laboratories for obtaining desirable chiral materials like natural products, drugs, or agrochemicals.

Many of these developments have received treatment in recent books and monographs on sulfur chemistry and organic synthesis, but no comprehensive text covering the whole area is available. The aim of this book, therefore, is to take a somewhat broader view, encompassing as many as possible of the chiral sulfur reagents, their preparation, and application in asymmetric and stereoselective synthesis. It does not seek to be totally comprehensive because compilation of such a text would entail a task of daunting proportions. Therefore, we apologize that many interesting contributions to this field which have been published could not be cited in this book.

As this is a practical book, the greater emphasis has been usually placed on describing the modern methodologies and procedures furnishing compounds with full or high enantiomeric purity. For the same reasons, some selected experimental data and examples of experimental procedures have been included which we felt were liable to be of greatest use both to students working for their first degree as well as to research chemists. However, some space has also been devoted to mechanistic aspects of the discussed asymmetric reactions.

Though the main purpose of this book is to demonstrate the great potential of enantiomerically pure sulfur reagents in transmitting chirality to other centers, the results obtained with racemic compounds are also discussed, particularly in cases where a high diastereoselectivity was observed. Such results can be easily transferred to enantiopure sulfur compounds, the only problem being the effective synthesis of the latter. Furthermore, several rather unsuccessful results (low extent of asymmetric induction, unsatisfactory yields) have also been mentioned to warn and prevent the reader from undertaking efforts which had already been made by others.

Finally, we believe that the scientific technological importance of chiral sulfur compounds justifies the conclusion that the subject of this book will be of interest to many chemists in many countries. We can only hope that our treatment of it has been adequate.

M. Mikołajczyk
J. Drabowicz
P. Kiełbasiński

Contents

Preface ... iii

1 Structure-Chirality Relationship in Organic Sulfur Compounds ... 1

References ... 5

2 Chiral Sulfinic Acid Derivatives ... 7
- 2.1 Sulfinic Esters ... 7
  - 2.1.1 Diastereomeric Sulfinic Esters ... 7
  - 2.1.2 Enantiomeric Sulfinic Esters ... 13
  - 2.1.3 Synthetic Application of Optically Active Sulfinate Esters ... 16
  - 2.1.4 Examples of Experimental Procedures ... 26
- 2.2 Chiral Sulfinamides and N-Alkylidenesulfinamides ... 28
  - 2.2.1 Diastereomeric Sulfinamides ... 28
  - 2.2.2 Enantiomeric Sulfinamides ... 30
  - 2.2.3 Enantiomeric N-Alkylidenesulfinamides ... 31
  - 2.2.4 Synthetic Application of Chiral Sulfinamides and N-Alkylidenesulfinamides ... 33
  - 2.2.5 Examples of Experimental Procedures ... 42

References ... 44

3 Chiral Sulfoxides ... 47
- 3.1 α-Sulfinyl Carbanions ... 47

References (3.1) ... 56

- 3.2 Dialkyl and Alkyl Aryl Sulfoxides ... 57
  - 3.2.1 Alkylation ... 57
    - 3.2.1.1 Synthetic Applications ... 59
  - 3.2.2 Michael Addition to α,β-Unsaturated Carbonyl Compounds ... 60
  - 3.2.3 Hydroxyalkylation ... 61
    - 3.2.3.1 Synthetic Applications ... 64
  - 3.2.4 Aminoalkylation ... 70
  - 3.2.5 Acylation ... 73
  - 3.2.6 Examples of Experimental Procedures ... 74

References (3.2) ... 77

- 3.3 Allyl Sulfoxides ... 78
  - 3.3.1 Asymmetric Allylic Sulfoxide-Sulfenate Rearrangements ... 78
    - 3.3.1.1 Chirality Transfer from Carbon to Sulfur ... 78
    - 3.3.1.2 Chirality Transfer from Sulfur to Carbon: Asymmetric Synthesis of Chiral Allyl Alcohols ... 80
  - 3.3.2 Reactions of Allyl Sulfoxide Carbanions ... 82
    - 3.3.2.1 Conjugate Addition of Chiral Allyl Sulfoxide Anions ... 83

3.3.2.2 Asymmetric Synthesis of (+)-Hirsutene 87
3.3.2.3 Asymmetric Synthesis of Enantiomers of 12,13-Epoxytrichothec-9-ene 88
3.3.2.4 Asymmetric Synthesis of (+)-Pentalene 89
3.3.3 Examples of Experimental Procedures 89

References (3.3) 90

3.4 Alkyl and Arylsulfenylmethyl Sulfoxides 91
3.4.1 Synthesis 91
3.4.2 Michael Addition 92
3.4.3 Hydroxyalkylation 93
3.4.4 Acylation 94
3.4.5 Examples of Experimental Procedures 95

References (3.4) 96

3.5 α-Sulfinyl-, α-Sulfonyl-, and α-Sulfoximino Sulfoxides 98
3.5.1 α-Sulfinyl Sulfoxides 98
3.5.2 α-Sulfonyl Sulfoxides 101
3.5.3 α-Sulfoximino Sulfoxides 104
3.5.4 Examples of Experimental Procedures 104

References (3.5) 105

3.6 α-Halogeno Sulfoxides 106
3.6.1 Examples of Experimental Procedures 114

References (3.6) 115

3.7 α-Phosphoryl Sulfoxides and α-Sulfinyl Phosphonium Salts 116
3.7.1 Synthesis of Chiral α-Phosphoryl Sulfoxides 116
3.7.2 Synthesis of Chiral α-Sulfinyl Phosphonium Salts and Ylides 119
3.7.3 α-Phosphoryl Sulfoxides and α-Sulfinyl Phosphonium Ylides as Key Reagents in the Synthesis of Chiral α,β-Unsaturated Sulfoxides 120
3.7.4 Asymmetric Reactions of α-Phosphoryl Sulfoxides 124
3.7.5 Examples of Experimental Procedures 127

References (3.7) 128

3.8 β-Oxosulfoxides and Related Derivatives 129
3.8.1 α-Sulfinylcarboxylates 129
3.8.1.1 Synthesis 129
3.8.1.2 Reactions of the Carbanion of α-Sulfinylcarboxylates with Electrophiles 131
3.8.2 β-Oxosulfoxides 135
3.8.2.1 Asymmetric Synthesis of Both Enantiomers of 4-Hydroxy-2-Cyclohexanone 137
3.8.2.2 Synthesis of Optically Active Epoxides 137
3.8.2.3 Asymmetric Synthesis of Protected α-Hydroxyaldehydes 138
3.8.2.4 Stereoselective Synthesis of (R)-3-Benzoyloxy-2-Butanone 138

3.8.2.5 Synthesis of Chiral 1,3-Diols 138
3.8.2.6 Reduction of Fluoroalkyl-Oxosulfoxides 139
3.8.3 β-Thioxosulfoxides 139
3.8.4 β-Iminosulfoxides and Analogs 140
3.8.5 Examples of Experimental Procedures 141

References (3.8) 143

3.9 Acyclic α,β-Unsaturated Sulfoxides 144
3.9.1 Electrophilic Addition 144
3.9.2 Michael Addition of Carbon Nucleophiles 145
3.9.3 Michael Addition of Oxygen Nucleophiles 148
3.9.3.1 Enantioselective Synthesis of Each Enantiomer of a Sex-Pheromone of an Olive Fly 148
3.9.3.2 Synthesis of the Chroman Ring of α-Tocopherol (Vitamin E) 150
3.9.3.3 Synthesis of (+) and (–)-(*cis*-6-Methyltetrahydropyran-2-yl)acetic Acids 152
3.9.4 Conjugate Addition of Nitrogen Nucleophiles 152
3.9.4.1 Total Synthesis of (+)-(R)-Carnegine and (+)- and (–)-Sedamine 154
3.9.4.2 Total Synthesis of (+)-(R)-Canadine 156
3.9.5 Conjugate Addition of Silicon Nucleophiles 157
3.9.6 Diels-Alder Cycloaddition 158
3.9.6.1 Enantiodivergent Synthesis of Fused Bicyclo [2.2.1] Heptane Lactones; Enantioselective Synthesis of (–)-Boschnialactone 166
3.9.6.2 Synthesis of the Ohno's Intermediate for the Preparation of Carbocyclic Nucleosides 168
3.9.6.3 Stereoselective Syntheses of Some Bicyclic Alkaloids 168
3.9.6.4 Reactions of Sulfinyl Dienophiles with Dane's Diene; Synthesis of Steroid Precursors 169
3.9.7 Miscellaneous Reactions of α,β-Unsaturated Sulfoxides 170
3.9.7.1 Additive Pummerer Rearrangement 169
3.9.7.2 Asymmetric Cyclopropanation 170
3.9.7.3 Asymmetric Radical Cyclizations 173
3.9.8 Cycloadditions of Sulfinyl Dienes 174
3.9.9 Examples of Experimental Procedures 175

References (3.9) 178

3.10 Cycloalkenone Sulfoxides 180
3.10.1 Synthesis of 2-Sulfinyl-2-Cycloalkenones 180
3.10.2 Synthesis of 2-Sulfinyl-2-Alkenolides 180
3.10.3 Conjugate Addition of Nucleophiles to Sulfinyl Cycloalkenones 182
3.10.4 Conjugate Addition of Nucleophiles to Sulfinyl Alkenolides 185
3.10.5 Applications of the Conjugate Addition of Nucleophiles to 2-Sulfinyl-2-Cycloalkenones and 2-Sulfinyl-2-Alkenolides in the Synthesis of Natural Products: Selected Examples 186
3.10.5.1 Formal Total Synthesis of 11-Oxoequilenin 186
3.10.5.2 Synthesis of (+)-α-Cuparenone 186
3.10.5.3 Synthesis of Natural (–)-Methyl Jasmonate 187

3.10.5.4 Synthesis of (+)-Estrone Methyl Ether 187
3.10.5.5 Synthesis of (+)-A Factor 188
3.10.5.6 Synthesis of Aphidicolin 188
3.10.5.7 Preparation of Chiral 2-Substituted Chroman-4-Ones 189
3.10.6 Asymmetric Radical Reaction 190
3.10.7 Examples of Experimental Procedures 191

References (3.10) 193

4 Chiral Sulfilimines and Sulfoximines 195
4.1 Synthesis of Chiral Sulfilimines 195
4.2 Synthesis of Chiral Sulfoximines 198
4.3 Chiral Sulfilimines and Sulfoximines in Asymmetric Synthesis 206
4.4 Examples of Experimental Procedures 233

References 239

5 Chiral "Onium" Derivatives of Sulfur and Chiral Sulfur Ylides 241
5.1 Preparation of Chiral Sulfur Onium Salts and Ylides 242
5.2 Synthetic Application of Chiral Sulfur "Onium" Salts and Ylides 251
5.3 Examples of Experimental Procedures 264

References 266

Index 267

Chapter 1

Structure-Chirality Relationship in Organic Sulfur Compounds

Sulfur forms a variety of organic compounds showing different structural and stereochemical properties. A very useful criterion to classify all the organosulfur compounds is the number (N) of ligands on sulfur.[1-3] Based on this criterion sulfur compounds are divided into six classes with ligand numbers N = 1 to 6. A limited number of monocoordinate (N = 1) sulfur compounds of linear structure are not interesting from the point of view of chirality at sulfur, since they are achiral compounds. Similarly, sulfur in dicoordinate (N = 2) compounds having angular structures cannot be a center of chirality. The only exception is the sulfenamide **1** which, due to the partial double bond character of the sulfur-nitrogen bond, shows, like allenes and carbodiimides, an axial chirality and may exist in two enantiomeric forms. This stereochemical feature of sulfenamides was first demonstrated by Kost and Raban[4] and then widely investigated.

1 ⇌ ent - **1** (1.1)

Sulfur compounds with the ligand number N = 3 may adopt a trigonal planar or trigonal pyramidal structure. However, planar arrangement of three different substituents around sulfur is not a sufficient condition for chirality at sulfur. On the contrary, pyramidal sulfur compounds, of the general structure **2**, containing three different ligands and the lone electron pair occupying the fourth position of a distorted tetrahedron are chiral and configurationally stable. This is in contrast to isoelectronic amines and carbonium ions. A higher, configurational stability of this class of sulfur compounds is due to the higher amount of *s*-character and the longer bond lengths about the central sulfur atom.

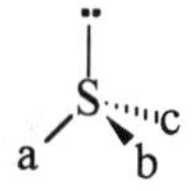

2 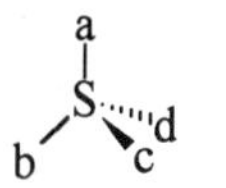 **3**

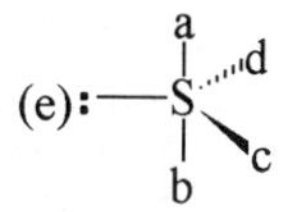

4 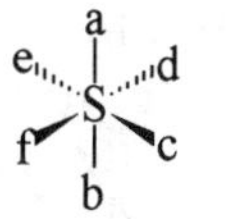 **5**

To the class of chiral tricoordinate sulfur compounds belong various sulfonium salts, $R^1R^2R^3S^+A^-$, sulfinyl compounds, $R^1R^2S{=}O$, and iminosulfinyl compounds, $R^1R^2S{=}NR$, which, in the majority of cases, are configurationally stable and have been obtained in optically active states. It is interesting to note that the first resolution of chiral sulfur compounds was reported in 1900 by Pope and Peachey[5] and by Smiles,[6] who resolved the sulfonium salts **6** and **7** via diastereomeric salts with chiral acids.

$Me(Et)S^{+}-CH_2CO_2H$

6

$Me(Et)S^{+}-CH_2C(O)Ph$

7

The isolation, around 1950, of many chiral sulfinyl compounds such as **8**, **9**, **10**, and **11** from natural sources resulted in intense activity in the preparation and study of new chiral tricoordinate sulfur structures.[7,8]

$CH_2{=}CHCH_2S(O)SCH_2CH{=}CH_2$

8

$MeS(O)CH{=}CHCH_2CH_2NCS$

9

$CH_2{=}CHCH_2S(O)CH_2CH(NH_2)CO_2H$

10

$MeS(O)CH_2CH(NH_2)CO_2H$

11

A great number of chiral tricoordinate sulfur compounds may be derived from sulfinic acids **12** which, however, are themselves effectively achiral. This is due to a fast proton exchange between two enantiomeric forms of **12** via the achiral sulfinic acid anion (Equation 1.2).

12 ⇌ [R–S(O)O⁻] H⁺ ⇌ ent - **12**

[^{16}O, ^{18}O]-**12**, R = p-Tol

13, R = t-Bu, Adamantyl, Triptycenyl

(1.2)

Replacement of one of the two ^{16}O oxygen atoms in **12** by ^{18}O or by sulfur leads to chiral structures of sulfinic acid [^{16}O, ^{18}O]-**12** and thiosulfinic acid **13**, respectively.[9] The former was recently obtained in optically active form by stereoselective synthesis.[10]

Other sulfinic acid derivatives such as sulfinates **14**, thiosulfinates **15**, sulfinamides **16**, sulfoxides **17**, and sulfoxonium salts **18** as well as their imino-analogs are chiral and have been obtained in enantiomeric forms.[2]

14, X = O
15, X = S
16, X = NR

17

18, X = O, S, NR
XR^1 = alkyl, aryl

Another series of chiral tricoordinate sulfur compounds may be derived from achiral sulfurous acid **19** by replacement of the hydroxy groups by suitable substituents. Thus, sulfites, amidosulfites, and amidothiosulfites of the general structure **20** belong to this class of compounds. Although these chiral compounds have been obtained in enantiomeric or diastereomeric forms, they are not in common use.

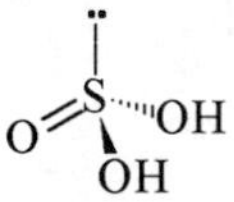

19

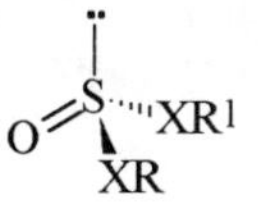

20, X = O, S, NR

The structure of sulfur compounds with four ligands (N = 4) and without a "stereochemically active" nonbonding electron pair is tetrahedral. If the four ligands are different as in **3**, such tetrahedral, tetracoordinate sulfur compounds are chiral and can be resolved or prepared in enantiomeric forms. Here, the situation is quite similar to the tetrahedral, sp^3-carbon compounds if one neglects the different bond order between some substituents and sulfur.

The most common chiral, tetracoordinate sulfur compounds are sulfoximines **21** which formally arise from achiral, unsymmetrical sulfones by replacement of one of the two oxygen atoms by the imino nitrogen. Consequently, replacement of both oxygen atoms in unsymmetrical sulfones by different imino groups leads to other chiral, tetracoordinate structures, namely sulfodiimides **22**. Of interest is that chiral, optically active sulfonimidoyl chlorides **23** have also been obtained. Due to the presence of a good leaving group, these chlorides are excellent substrates for nucleophilic substitution reaction and afford in a highly stereoselective way the corresponding esters and amides **24**.[2,11]

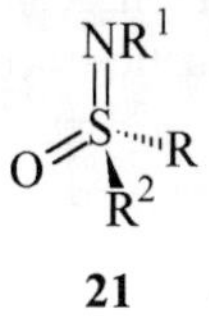

21 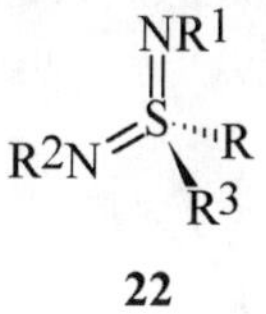 **22** 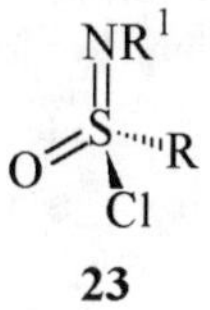 **23**

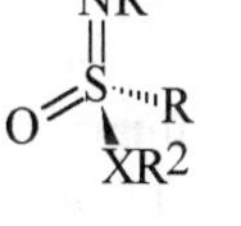

24, X = O, NH

As in the case of sulfinic acids, the chirality at sulfur in unsymmetrical sulfones may be generated by isotopic substitution. The first example of such a chiral sulfone, i.e., (–)-benzyl p-tolyl [^{16}O, ^{18}O]-sulfone **25** was described by Stirling[12] as early as 1963. A few years later, Sabol and Andersen[13] prepared diastereomerically pure (–)-menthyl [^{16}O, ^{18}O]-phenylmethanesulfonate **26**. The list of the tetracoordinate sulfur compounds chiral by virtue of isotopic oxygen substitution was recently extended by the synthesis of the lithium salt of chiral [^{16}O, ^{18}O]-p-toluenethiosulfonic acid **27**.[10]

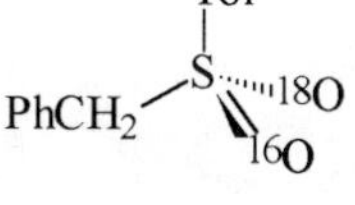

25 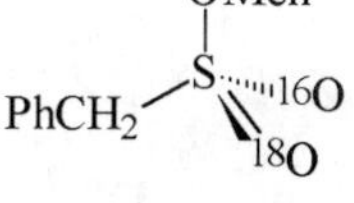 **26** 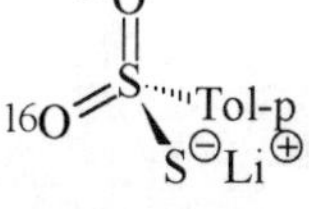 **27**

Oxosulfonium salts of the general formula **28** belong to the group of tetrahedral tetracoordinate sulfur compounds with four different ligands (N = 4). Therefore, they are chiral at sulfur and can exist in enantiomeric forms. Thus far, however, methylethylphenyloxosulfonium perchlorate **29** is the only compound of this type whose enantiomers have been isolated.[14]

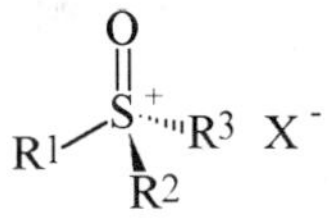

28 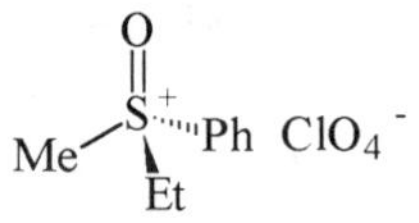 **29**

Tetracoordinate sulfur compounds (N = 4), which contain the lone electron pair as a phantom ligand, as well as pentacoordinate sulfur compounds (N = 5) possess a trigonal bipyramidal structure exemplified by **4**. A common name, sulfurane, is generally accepted for this type of high-coordinated compounds.[15]

In the present brief account at least three significant features of sulfuranes should be mentioned. The first concerns the positions of substituents in a trigonal bipyramidal structure. In such a structure two substituents occupy apical positions while the remaining three are placed in a basal plane in equatorial positions. This nonequivalence of ligand positions is preserved even in the case of the same substituents connected with the central sulfur atom. The tendency of a substituent to occupy an apical position is defined as apicophilicity. In the first approximation apicophilicity is related to electronegativity, i.e., the more electronegative the ligand the greater its apicophilicity. Ring strain, steric bulk, and electronic effects are the other factors affecting the apicophilicity order in sulfuranes.

Secondly, the most interesting phenomenon observed in sulfuranes as well as in other valency-shell expanded compounds is the internal ligand reorganization changing the relative positions of ligands in a trigonal bipyramidal structure. This process is commonly called pseudorotation. A single process of pseudorotation according to the Berry mechanism is visualized below (Equation 1.3).

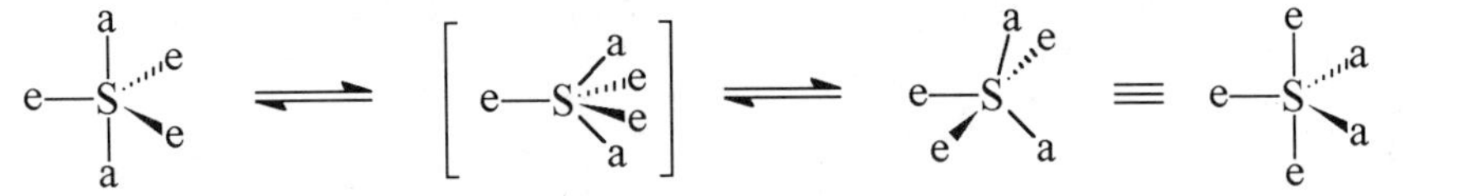

(1.3)

Since the energy required for pseudorotation is usually very low, this process may have an important influence on the stereochemical properties of sulfuranes.

Finally, it should be emphasized that sulfuranes may be chiral. However, the number of optically active isomers is dependent on the nature of substituents connected with sulfur, their apicophilicity, and the energy for pseudorotation. In this context, it is interesting to note that acyclic sulfuranes with five different ligands like **4** should exist in twenty isomeric chiral forms. However, in the case of sulfuranes, all structures containing at least three different ligands can be chiral. Thus, the sulfurane structure **30** is chiral in contrast to the more symmetrical **31**, which is achiral. Moreover, considering the topological properties of such trigonal bipyramidal molecules, it should be pointed out that, after incorporation of cyclic ligands into this structure, chirality may still appear in the more symmetrical spiro system **32**.

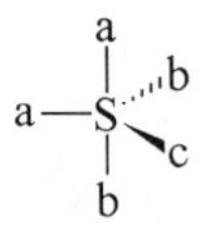

30 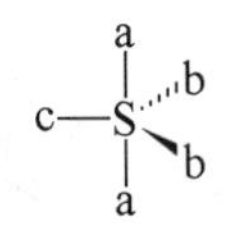 **31** 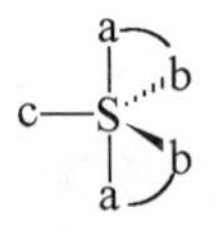 **32**

The first example of an optically active sulfurane was the dextrorotatory chlorosulfurane **33** prepared by Martin and Balthazor.[16] All other sulfuranes **34–38**, which have been till now prepared[17,18] as optically active species, belong to the group of spiro derivatives and are shown below.

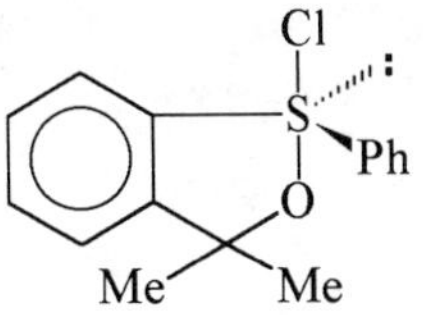

For hexacoordinate sulfur compounds (N = 6) the octahedral arrangement of ligands is characteristic as pictured in **5**. In spite of the fact that such compounds are known their stereochemistry has yet to begin. Till now, no optically active compounds of this type have been described in the literature.

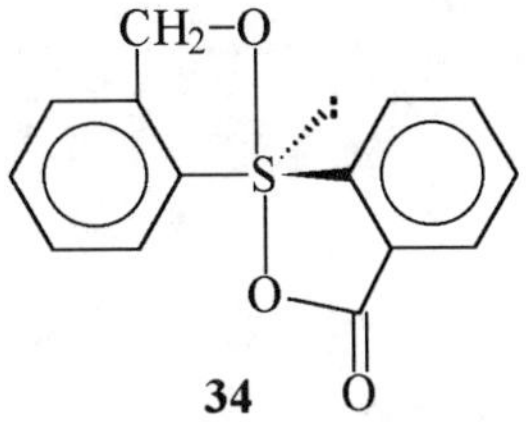

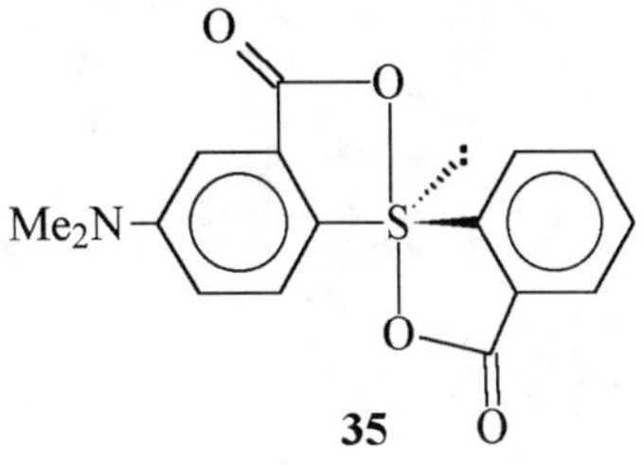

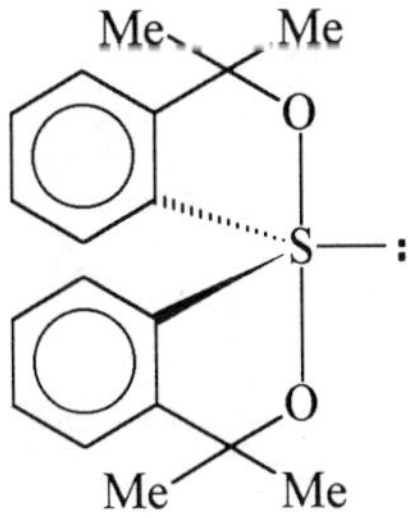

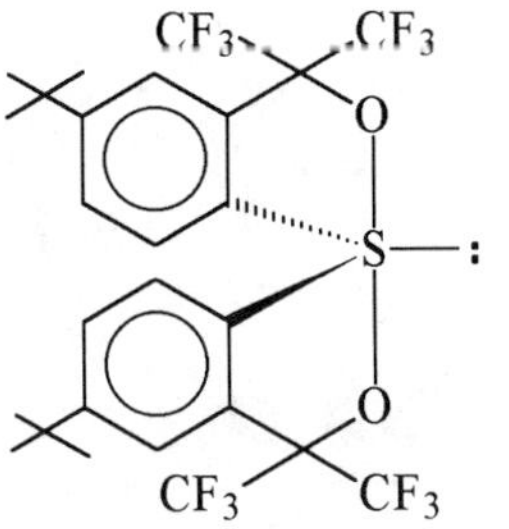

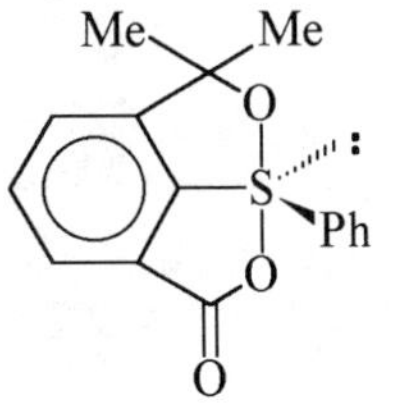

REFERENCES

1. Laur, P.H., in *Sulfur in Organic and Inorganic Chemistry*, Vol. 3, Senning, A., Ed., Marcel Dekker, New York, 1972.
2. Mikołajczyk, M., Drabowicz, J., *Top. Stereochem.*, 13, 333, 1982.
3. Oae, S., *Organic Sulfur Chemistry: Structure and Mechanism*, CRC Press, Boca Raton, FL, 1991, chap. 3, p.67.
4. Kost, D., Raban, M., in *The Chemistry of Sulfenic Acids and Their Derivatives*, Patai, S., Ed., John Wiley & Sons, Chichester, 1990, pp 23–82.
5. Pope, W.J., Peachey, S.J., *J. Chem. Soc.*, 77, 1072, 1900.
6. Smiles, S., *J. Chem. Soc.*, 77, 1174, 1900.
7. Challenger, F., *Aspects of the Organic Chemistry of Sulfur*, Butterworths, London, 1959.
8. Kjaer, A., *Pure Appl. Chem.*, 49, 137, 1977.
9. Mikołajczyk, M., Łyżwa, P., Drabowicz, J., *Angew. Chem.*, Int. Ed. Engl., 28, 97, 1989.
10. Drabowicz, J., Łyźwa, P., Bujnicki, B., Mikołajczyk, M., *Phosphorus, Sulfur and Silicon*, 95–96, 293, 1994.
11. Nudelman, A., *The Chemistry of Optically Active Sulfur Compounds*, Gordon and Breach, New York, 1984.
12. Stirling, C.J.M., *J. Chem. Soc.*, 5741, 1963.
13. Sabol, M., Andersen, K.K., *J. Am. Chem. Soc.*, 91, 3603, 1969.
14. Kobayashi, M., Kamiyama, K., Minato, H., Oishi, Y., Takada, Y., Hattori, Y., *J. Chem. Soc. D.*, 1577, 1971; Kamiyama, K., Minato, H., Kobayashi, M., *Bull. Chem. Soc. Jpn.*, 46, 3895, 1973.
15. Drabowicz, J., Łyźwa, P., Mikołajczyk, M., in *Supplement S: the Chemistry of Sulfur-Containing Functional Groups*, Patai, S. and Rappoport, Z., Eds., John Wiley & Sons, Chichester, 1993, pp. 799–956.
16. Martin, J. C., Balthazor, T. M., *J. Am. Chem. Soc.*, 99, 152, 1977.
17. Kapovits, I., *Phosphorus, Sulfur and Silicon*, 58, 39, 1991.
18. Drabowicz, J., Martin, J. C., *Tetrahedron: Asymmetry*, 4, 297, 1993.

Chapter 2

Chiral Sulfinic Acid Derivatives

2.1 SULFINIC ESTERS

Sulfinic esters of the general structure R-S(O)OR[1] belong to the oldest group of chiral organosulfur derivatives prepared as optically active species.[1] Depending on the nature of a substituent R^1, they can be obtained as enantiomers (if R^1 is achiral) or as diastereomeric mixtures (if R^1 contains at least one chiral center). A high stability of the pyramidal structure around the central sulfur atom[1,2] allows the synthesis of a rich family of stable, optically active sulfinic esters. Their importance as substrates in the synthesis of optically active sulfinyl derivatives and in establishing the absolute configuration of three- and tetracoordinated sulfur compounds is well recognized.[1,3] A presentation of synthetic routes reported for the preparation of the most important diastereomeric and enantiomeric sulfinic esters as well as their synthetic applications will follow these introductory remarks. A comprehensive review on the synthesis of sulfinates can be found in Reference 3.

2.1.1 Diastereomeric Sulfinic Esters

The oldest and up to now the most common procedure for the preparation of diastereomeric sulfinic esters involves condensation of sulfinyl chlorides with the appropriately selected enantiomerically or diastereomerically pure alcohols carried out in the presence of an organic or inorganic base. This method was for the first time used by Phillips[4] for obtaining O-menthyl p-toluenesulfinate **1a**. Thus, in the reaction between p-toluenesulfinyl chloride **2a** and (–)-(1R,2S,5R)-menthol in the presence of pyridine a mixture of the diastereomeric O-menthyl p-toluenesulfinates **1a** was formed (Scheme 2.1.1), from which Phillips was able to isolate a pure, solid, diastereomer (–)-(S)-O-menthyl p-toluenesulfinate **1a** by crystallization from acetone.

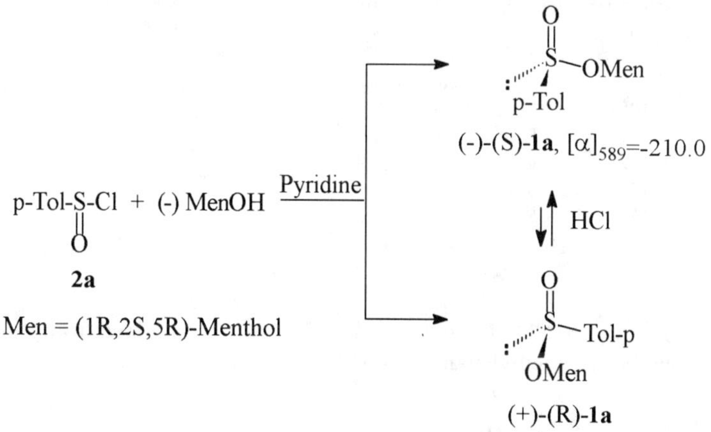

p-Tol-S(O)-Cl + (-) MenOH —Pyridine→

2a

Men = (1R,2S,5R)-Menthol

(-)-(S)-**1a**, $[\alpha]_{589}$=-210.0

HCl

(+)-(R)-**1a**

Scheme 2.1.1

Because of the importance of (–)-(S)-**1a** as a substrate in the synthesis of other optically active sulfinyl derivatives, a considerable effort has been devoted to improve its synthesis. The most important improvement is based on the observation reported by Herbrandson and Dickerson[5] as early as 1959 that the addition of hydrogen chloride gas to a mixture of diastereomeric **1a** causes epimerization of the liquid diastereomer (+)-(R)-**1a** to its crystalline epimer (–)-(S)-**1a**. Mioskowski and Solladie[6], by modifying the conditions of Herbrandson and Dickerson under which diastereomeric sulfinates **1a** undergo epimerization in favor of the less soluble (–)-(S)-**1a** isomer, were able to isolate it in 90% yield. In our laboratory,[7a] the above-discussed isomerization procedures did not always give fully reproducible results. It has been found that the more consistent results of epimerization of the sulfinates **1a** are observed when the solid diastereomer, once formed, is dissolved in the mother liquid and the crystallization process is repeated. Another modification of the reaction between p-toluenesulfinyl chloride and (–)-menthol, which allows the isolation of the solid (–)-(S)-**1a** diastereomer in a high yield, involves a very rapid addition of the reaction components.[8]

It is obvious that the use of (+)-(1S,2R,5S)-menthol leads to the formation of (+)-(R)-O-menthyl p-toluenesulfinate. Both sulfinates **1a** with the opposite configuration at the sulfinyl sulfur atom are now commercially available.

The Phillips approach to the synthesis of diastereomeric O-menthyl sulfinates is general in scope and a great number of other diastereomeric sulfinates **1b–k** were prepared in a similar way starting from the appropriate sulfinyl chlorides **2b–2k** (Scheme 2.1.2).

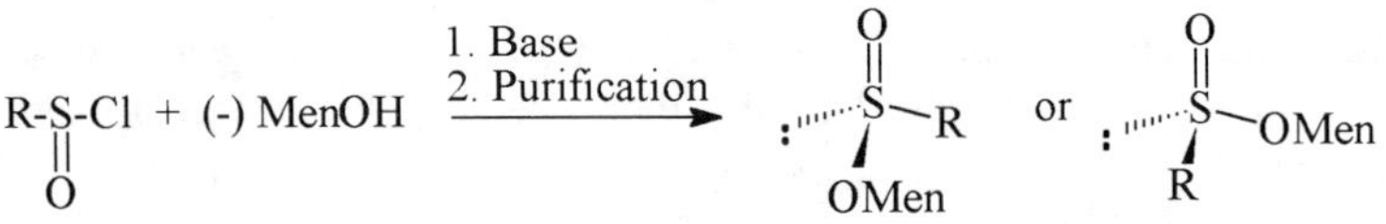

R-S(=O)-Cl + (-) MenOH —(1. Base; 2. Purification)→ :S(=O)(R)(OMen) or :S(=O)(OMen)(R)

2 (+)-(R)-**1b-k** (-)-(S)-**1b-k**

Sulfinyl chloride **2**		Sulfinic ester **1**				
No	R	No	$[\alpha]_{589}$	Abs.conf.	de	Ref.
b	Ph	**b**	–206.1	S	100	9
c	p-$MeOC_6H_4$	**c**	–189.1	S	100	10
d	p-ClC_6H_4	**d**	–181.1	S	100	2,11
e	p-IC_6H_4	**e**	–145.8	S	100	12
f	1-$C_{10}H_7$	**f**	–433.2	S	100	10a
g	$PhCH_2$	**g**	+105.0	R	90.5	13
g	$PhCH_2$	**g**	+123.0	R	100	14
h	Me	**h**	–99.1	R	13	10b
i	n-Bu	**i**	+50.0	R	47	12
k	2-MeO-$C_{10}H_6$	**k**	–183.0	S	100	18

Scheme 2.1.2

It should be noted that (–)-(S)-O-menthyl benzenesulfinate-**1b** was also prepared from benzenesulfinyl chloride **2b** and 1-menthoxytrimethylsilane **3** in 91% yield (Equation 2.1.1).[15]

$$\mathbf{2b} + (-)\ \mathrm{MenOSiMe_3}\ (\mathbf{3}) \xrightarrow[-\mathrm{Me_3SiCl}]{} (-)\text{-}(S)\text{-}\mathbf{1b} \quad [\alpha]_D = -195.3 \tag{2.1.1}$$

The *in situ* reduction of commonly available sulfonyl chlorides **4** with trimethyl phosphite in the presence of (–)-menthol was found to be a simple method for the preparation of diastereomeric O-menthyl sulfinates **1**, especially those for which there are no readily available sulfinyl chloride precursors.[16] According to authors, the transiently formed sulfinyl chlorides **2** are intercepted by menthol present in the reaction mixture to produce the sulfinates **1**. A variety of O-menthyl sulfinates **1**, which were prepared according to this procedure in up to 2:1 S/R diastereoselectivity, are collected in Scheme 2.1.3.

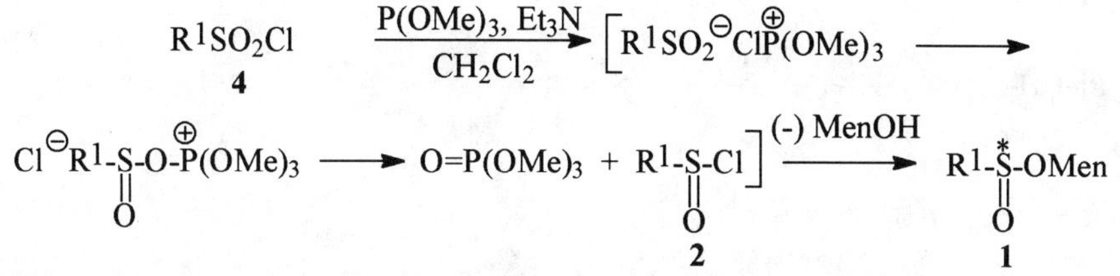

$$R^1SO_2Cl\ (\mathbf{4}) \xrightarrow[CH_2Cl_2]{P(OMe)_3,\ Et_3N} [R^1SO_2^{\ominus}\ ClP^{\oplus}(OMe)_3] \longrightarrow$$

$$Cl^{\ominus}\ R^1\text{-}S(=O)\text{-}O\text{-}P^{\oplus}(OMe)_3 \longrightarrow O{=}P(OMe)_3 + [R^1\text{-}S(=O)\text{-}Cl]\ (\mathbf{2}) \xrightarrow{(-)\ MenOH} R^1\text{-}S^*(=O)\text{-}OMen\ (\mathbf{1})$$

Sulfonyl chloride 4		Sulfinic ester 1			
No	R^1	No	R^1	Yield [%]	S:R ratio
a	p-Tol	**a**	p-Tol	90	1.4:1
b	p-$MeOC_6H_4$	**c**	p-$MeOC_6H_4$	89	1.3:1
c	p-ClC_6H_4	**d**	p-ClC_6H_4	92	1.6:1
d	2-$C_{10}H_7$	**l**	2-$C_{10}H_7$	96	1.4:1
e	2,4,6-$(iPr)_3C_6H_2$	**m**	2,4,6-$(iPr)_3C_6H_2$	36	[a]
f	p-(t-Bu)C_6H_4	**n**	p-(t-Bu)C_6H_4	48	1.6:1
g	o-$MeO_2CC_6H_4$	**o**	o-$MeO_2CC_6H_4$	[a]	1.6:1
h	2,4,6-$Me_3C_6H_2$	**p**	2,4,6-$Me_3C_6H_2$	70	1.5:1
i	8-Quinolyl	**q**	8-Quinolyl	52	1.9:1
j	2-Thienyl	**r**	2-Thienyl	92	1.8:1

[a] Not given.

Scheme 2.1.3

O-Menthyl alkanesulfinates prepared by condensation of alkanesulfinyl chlorides with laevo- and dextrorotatory menthol are liquids and, therefore, are not available in a diastereomerically pure state. To overcome this problem, a few other optically active alcohols were used instead of (–)-menthol as chiral reaction components. For example, from (–)-cholesterol **5** and methanesulfinyl chloride **2h** diastereomerically pure O-cholesteryl methanesulfinates (–)-(S)-**6** and (+)-(R)-**6** were obtained by fractional crystallization (Scheme 2.1.4), albeit in low yields.[17]

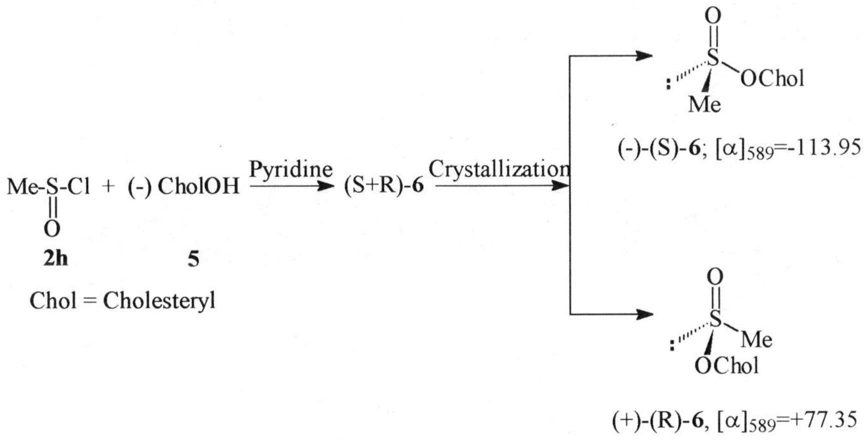

$$Me\text{-}S(=O)\text{-}Cl\ (\mathbf{2h}) + (-)\ CholOH\ (\mathbf{5}) \xrightarrow{Pyridine} (S{+}R)\text{-}\mathbf{6} \xrightarrow{Crystallization}$$

Chol = Cholesteryl

(-)-(S)-**6**; $[\alpha]_{589}$=-113.95

(+)-(R)-**6**, $[\alpha]_{589}$=+77.35

Scheme 2.1.4

Similarly, diastereomerically pure methane- and t-butanesulfinates (–)-(S)-**7a** and (–)-(S)-**7b** derived from L-N-methylephedrine **8** (EphOH) were isolated, although in poor chemical yields, by flash chromatography of the crude products formed in the reaction between the corresponding sulfinyl chlorides **2h** or **2j** and **8** (Equation 2.1.2).[7b]

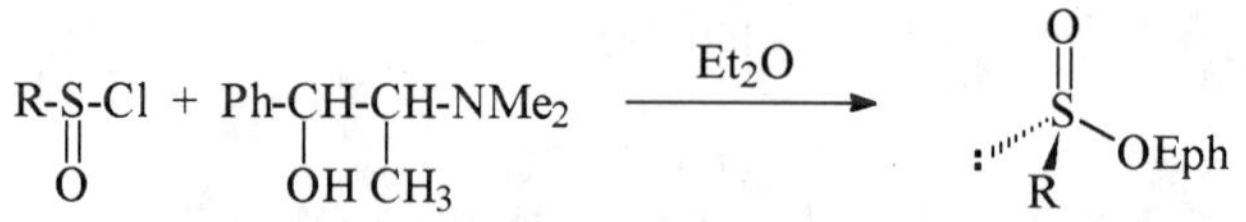

$$\text{R-S(=O)-Cl} + \text{Ph-CH(OH)-CH(CH}_3\text{)-NMe}_2 \xrightarrow{Et_2O} \text{:S(=O)(R)OEph}$$

2h,j (-)-**8** (Eph OH)

h; R=Me
j; R=t-Bu

R=Me; (-)-(S)-**7a**, $[\alpha]_{589}$=-167.0
R=t-Bu; (-)-(S)-**7b**, $[\alpha]_{589}$=-152

(2.1.2)

In the pool of enantiomerically pure chiral alcohols, sugar derivatives have been employed with good results in the preparation of diastereomerically pure sulfinic esters. The best results have been obtained with diacetone-D-glucose (DAG) **9a** — commercially available, sugar-derived secondary alcohol — and with dicyclohexylidene-D-glucose (DCG) **9b**. The first examples of arenesulfinates **10a–f** and **11a–e** derived from these sugars were described as early as 1982 by Ridley and Small.[19] However, only the (+)-(R)-mesitylenesulfinate **11e** was isolated as a crystalline pure diastereomer. Very recently, Alcudia and co-workers[20] have carried out very systematic studies on the reaction between sulfinyl chlorides and DAG and developed a general asymmetric synthesis of both epimerically pure alkane- and arenesulfinates **10** (see Scheme 2.1.5 and Table 2.1.1).

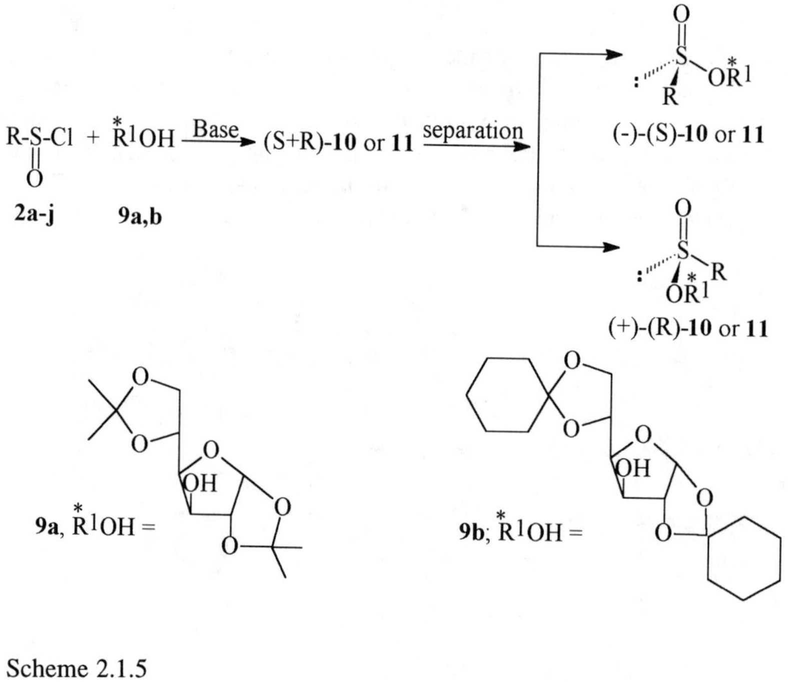

$$\text{R-S(=O)-Cl} + \overset{*}{R}{}^1\text{OH} \xrightarrow{\text{Base}} \text{(S+R)-}\mathbf{10}\text{ or }\mathbf{11} \xrightarrow{\text{separation}} \text{(-)-(S)-}\mathbf{10}\text{ or }\mathbf{11} + \text{(+)-(R)-}\mathbf{10}\text{ or }\mathbf{11}$$

2a-j **9a,b**

(-)-(S)-**10** or **11**: :S(=O)(R)$\overset{*}{O}R^1$

(+)-(R)-**10** or **11**: :S(=O)(R)$\overset{*}{O}R^1$

9a, $\overset{*}{R}{}^1$OH =

9b; $\overset{*}{R}{}^1$OH =

Scheme 2.1.5

Of interest is that the chirality at sulfur in the final sulfinate ester **10** is determined by the base used. Thus, sulfinates having the S-configuration are formed as the major isomers (de 89–95%) when iPr_2NEt is used, whereas sulfinates with the R-configuration are predominantly produced when pyridine is used as a condensing agent. The sulfinates **10b–f** can be easily purified by flash

Table 2.1.1 Reaction of DAG-**9a** or DCG-**9b** with Sulfinyl Chlorides **2** (See Scheme 2.1.5)

Sulfinyl chloride 2				**Sulfinic Ester 10/11**					
No	**R**	**Base**	**Solvent**	**No**	**R**	**R¹**	**R/S ratio**	**Yield [%]**	**Ref.**
2a	p-Tol	Py	THF	**10a**	p-Tol	**9a**	86/14	84	20
2a	p-Tol	iPr_2NEt	Toluene	**10a**	p-Tol	**9a**	6/94	87	20
2b	Ph			**11a**	Ph	**9a**			19
2c	$p\text{-}MeOC_6H_4$			**11b**	$pMeOC_6H_4$	**9b**			19
2d	$p\text{-}ClC_6H_4$			**11c**	$p\text{-}ClC_6H_4$	**9b**			19
2f	$1\text{-}C_{10}H_7$			**11d**	$1\text{-}C_{10}H_7$	**9b**			19
2h	Me	Py	THF	**10b**	Me	**9a**	93/7	87	20b
2h	Me	iPr_2NEt	Toluene	**10b**	Me	**9a**	2/98	90	20b
2k	Et	Py	THF	**10c**	Et	**9a**	86/14	85	20b
2k	Et	iPr_2NEt	Toluene	**10c**	Et	**9a**	2/98	90	20b
2m	n-Pr	Py	THF	**10d**	n-Pr	**9a**	85/15	75	20b
2n	i-Pr	iPr_2NEt	Toluene	**10e**	i-Pr	**9a**	2/98	50	20b
2j	t-Bu	NEt_3	Toluene	**10f**	t-Bu	**9a**	14/86	74	20c
2j	t-Bu	Py	THF	**10f**	t-Bu	**9a**	92/8	50	20c

chromatography. The only exception is the p-toluenesulfinate **10a**, where the two epimers are not sufficiently resolved for a convenient chromatographic separation. However, due to the good crystallizing properties of the (S)-p-toluenesulfinate **10a**, it can be obtained in a diastereomerically pure state by recrystallization from hexane.

Another procedure for the efficient preparation of diastereomerically pure sulfinate esters is based on the use of trans-2-phenylcyclohexanol **12** as a chiral auxiliary.[21] Reaction of this alcohol with an excess of alkane- and arenesulfinyl chlorides **1a, 2h, 2n, 2o** affords the corresponding sulfinate esters **13** (Scheme 2.1.6) in a good yield and with considerably better kinetic selectivity than that observed with menthol.[12] The diastereomers can be readily separated by chromatography and in all four examples shown in Scheme 2.1.6 the S-diastereomer is crystalline. Therefore, separation is also possible by recrystallization. For instance, the major diastereomer of **13a** was obtained in 98% de after two crystallizations with a recovery of 62%.

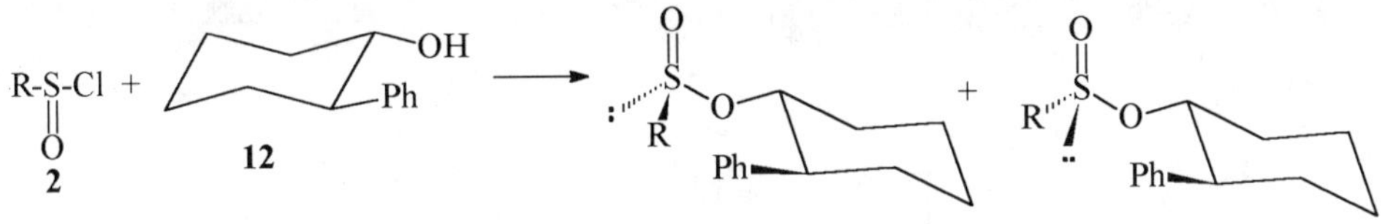

R-S(O)-Cl **2** + **12** → **(S)-13a-d** + **(R)-13a-d**

Sulfinyl chloride **2**		Sulfinic ester **13**				
					$[\alpha]_D$ (acetone)	
No	R	No	Yield [%]	S:R ratio	for S-**13**	for R-**13**
a	p-Tol	**a**	92	10:1	+82	+176.5
h	Me	**b**	80	9:2	+25.0	+157.0
n	i-Pr	**c**	100	9:2	+51.7	+196.6
o	2-Nph	**d**	84	65:1	+104.2	+124.5

Scheme 2.1.6

Another general approach to the preparation of diastereomeric sulfinates with a high value of diastereomeric purity is based on the reaction between organometallic reagents and chiral sulfurous acid derivatives such as acyclic chlorosulfites and cyclic sulfites. Thus, treatment of a nearly equimolar mixture of two diastereomeric chlorosulfites **14a** and **14b**, prepared *in situ* from the chiral trans-2-phenylcyclohexanol **12** and thionyl chloride, with an equivalent amount of dialkylzinc reagents **15** afforded high levels of conversion of both chlorosulfites to almost single diastereomers of the corresponding sulfinate esters **13b,c,e** (Scheme 2.1.7).[22]

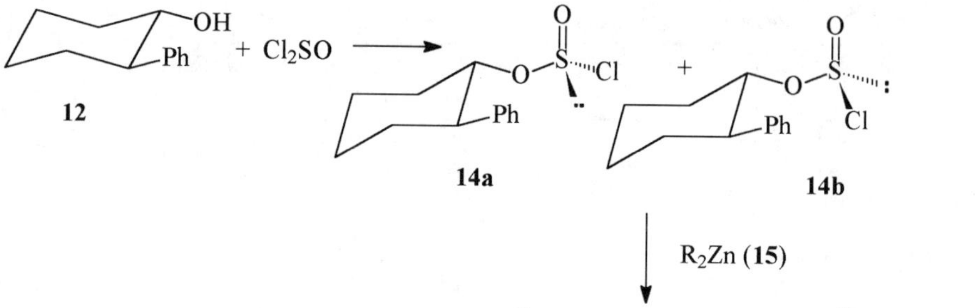

No	R	Yield [%]	R:S ratio
13b	Me	58	98:2
13c	i-Pr		10:1
13e	Et	100	19:1

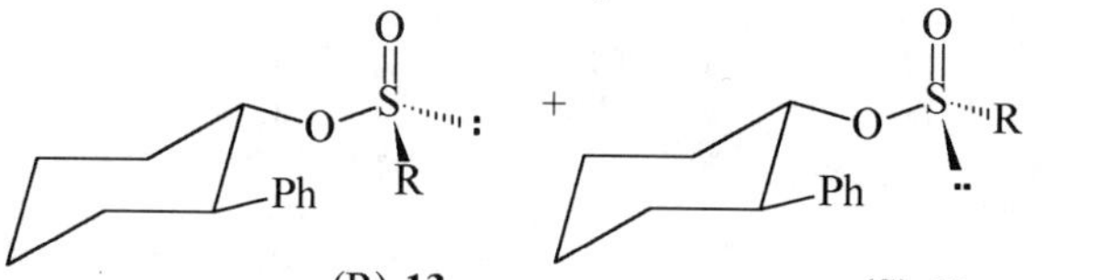

Scheme 2.1.7

On the other hand, the same mixture of chlorosulfites **14a,b** gave upon treatment with arylorganometallics the sulfinates **13a,b** and **13f** (R=Ph) with diastereomer ratios similar to those of the starting chlorosulfites (between 1:1 to 1:2). It should be noted, however, that the procedures which use trans-2-phenylcyclohexanol **12** as a chiral auxiliary have limited applicability due to the fact that this chiral alcohol is rather expensive.

A few diastereomerically pure sulfinates **16** and **17** were prepared in high chemical yields in the reaction between chiral cyclic sulfite **18** and various organometallic reagents (Scheme 2.1.8).[23]

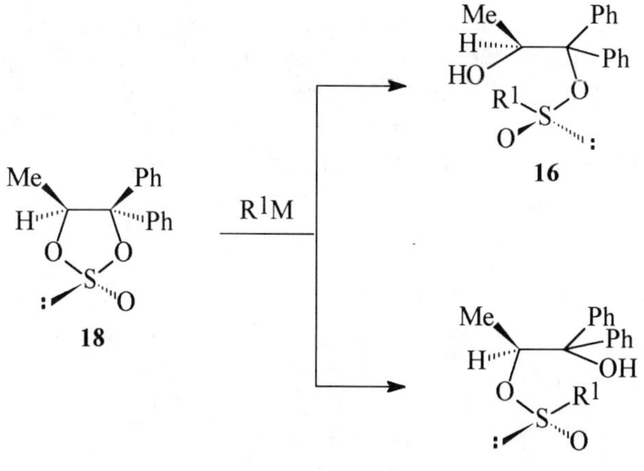

R^1M	**16:17** ratio	Isolated yield [%] of pure sulfinate
MeLi	75:25	55 (**16a**, R^1 = Me)
MeMgI	80:20	70 (**16a**, R^1 = Me)
EtMgBr	92:8	80 (**16b**, R^1 = Et)
n-OctMgBr	95:5	60 (**16c**, R^1 = n-Oct)
t-BuMgBr	10:90	70 (**17d**, R^1 = t-Bu)
t-BuMgCl	5:95	60 (**17d**, R^1 = t-Bu)
$CH_2{=}CHMgCl$	95:5	50 (**17e**, R^1 = $CH_2{=}CH$)
$(CH_3)_3C_6H_2MgBr$	12:88	70 (**17f**, R^1 = $(CH_3)_3C_6H_2$)

Scheme 2.1.8.

2.1.2 Enantiomeric Sulfinic Esters

Considering the use of chiral sulfinic esters as reagents in asymmetric and stereoselective synthesis it should be noted that enantiomeric sulfinates offer in comparison with diastereomeric derivatives at least two advantages. First of all, due to the achiral nature of the alcohol moiety the optically active products from enantiomeric sulfinates are not contaminated by optically active by-products, which as a rule are formed when diastereomeric sulfinates are applied as starting materials. Moreover, enantiomeric sulfinates can be obtained from simple alcohols having a very short carbon chain, like methanol or ethanol. Such sulfinates afford, after conversion, water-soluble by-products, which can be very easily and quantitatively removed from the expected optically active products simply by washing with water. This makes a purification procedure more convenient and less time consuming. For these reasons, the synthesis of enantiomeric sulfinic esters of high optical purity still provides the prime challenge for chemists working in this field. At present, among all known synthetic approaches to this type of sulfinates, only three procedures are synthetically useful and afford products with an acceptable degree of optical purity.

The first is based on the treatment of enantiomerically pure N,N-diethyl p-toluenesulfinamide **19a** with achiral alcohols in the presence of a strong organic acid[24] or $Et_2O{\cdot}BF_3$[25] (Equation 2.1.3).

$$\text{(+)-(S)-}\mathbf{19a}\;(\text{p-Tol}-S(O)-NEt_2) + R^1OH \xrightarrow{H^{\oplus}\text{ or }Et_2OBF_3} \text{(-)-(S)-}\mathbf{20}\;(\text{p-Tol}-S(O)-OR^1) \qquad (2.1.3)$$

This reaction was shown to proceed with inversion of configuration at the sulfinyl sulfur atom and with a high stereoselectivity, which is dependent on the structure of the alcohol used (see Table 2.1.2).

Table 2.1.2 Protonic Acid- and Boron Trifluoride-Catalyzed Esterification of (+)-(S)-N,N-Diethyl p-Toluenesulfinamide **19** into (–)-(S)-p-Toluenesulfinates **20**, p-Tol-S(O)OR^1

Sulfinamide 19a			Sulfinate 20							
$[\alpha]_{589}$ (Me_2CO)	ee [%]	Catalyst	R	No	Yield [%]	$[\alpha]_{589}$ (EtOH)	ee [%]	Stereoselectivity of 19→20	Ref.
+107	88	CF_3COOH	Me	a	94	–192.6	88	100	24
[a]	78	$Et_2O{\cdot}BF_3$	Me	a	99	–170.7	78	100	25
+105	86	$PhSO_3H$	Et	b	76.5	–179.2	86	100	24
[a]	87	$Et_2O{\cdot}BF_3$	Et	b	91.0	–178.8	85.8	98.7	25
+105	86	$PhSO_3H$	n-Pr	c	80.0	–161.2	84	98.2	24
[a]	87	$Et_2O{\cdot}BF_3$	n-Pr	c	92.0	–169.9	85.4	98.2	25
106	87	$PhSO_3H$	Allyl	d	95.0	–106.5	73	84.0	24
[a]	58	$Et_2O{\cdot}BF_3$	Allyl	d	93.0	–80.4	55.1	95.0	25
+96	78.5	CF_3COOH	i-Pr	e	87.0	–109.3	54.0	69.5	24
[a]	87.2	$Et_2O{\cdot}BF_3$	i-Pr	e	89.0	–165.7	82.9	96.3	25
+104.7	86	CF_3COOH	t-Bu	f	55.0	–29.88	23	27.4	24
[a]	58	$Et_2O{\cdot}BF_3$	t-Bu	f	70.0	–69.2	53.4	91.9	25
[a]	60	$Et_2O{\cdot}BF_3$	[b]	g	99.0	–124.6	59.3	98.8	25
[a]	63.4	$Et_2O{\cdot}BF_3$	[c]	h	94.0	–40.7	62.7	98.9	25
+106	96.4	$Et_2O{\cdot}BF_3$	[c]	h	70.0	–34.2	92.3	95.7	26

[a] Not given.
[b] CH=CH(Me).
[c] Cinnamyl.

Interestingly, the acid-catalyzed alcoholysis of optically active N,N-di-isopropyl p-toluenesulfinamide **19b**[27] is accompanied by an unexpectedly high degree of retention of configuration at the sulfinyl sulfur (Equation 2.1.4, Table 2.1.3).

$$(+)\text{-}(S)\text{-}\mathbf{19b}\;[\text{p-Tol-S(O)-N(i-Pr)}_2] + R^1OH \xrightarrow{CF_3COOH} (-)\text{-}(S)\text{-}\mathbf{20}\;[\text{p-Tol-S(O)-OR}^1] + (+)\text{-}(R)\text{-}\mathbf{20}\;[\text{p-Tol-S(O)-OR}^1] \quad (2.1.4)$$

Table 2.1.3 Esterification of N,N-Diisopropyl p-Toluenesulfinamide **19b** with Alcohols Catalyzed by Trifluoroacetic Acid without and in the Presence of Silver Perchlorate[27]

Sulfinamide 19b				Sulfinate		Predominant
$[\alpha]_{589}$	ee [%]	ROH	Salt	$[\alpha]_{589}$	ee [%]	stereochemistry
+94.4	45.3	Me	—	−35.0	17.0	69% Inv.
+103.4	50.4	Me	$AgClO_4$	−112.0	50.4	100% Inv.
+94.4	45.3	Et	—	−7.1	3.4	54% Inv.
+1–3.4	50.4	Et	$AgClO_4$	−86.0	41.3	91% Inv.
+94.4	45.3	n-Pr		−13.9	7.25	58% Inv.
+103.4	50.4	n-Pr	$AgClO_4$	−112.0	50.4	100% Inv.
+94.4	45.3	i-Pr	—	+15.8	7.90	59% Ret.
+103.4	50.4	i-Pr	$AgClO_4$	−62.8	32.3	82% Inv.
+95.0	45.35	Hex[c]	–	+41.0	22.4	74% Ret.
+103.4	50.4	Hex[c]	$AgClO_4$	−27.1	15.7	65% Inv.

Moreover, the stereochemical course of this reaction was found to be strongly influenced by the addition of inorganic salts. For example, the reactions of (+)-(S)-**19b** with isopropanol and cyclohexanol, which in the absence of a salt gave the corresponding sulfinates (–)-**20e** and (–)-**20j** with predominant retention, in the presence of $AgClO_4$ afforded products having inverted configuration at the sulfinyl sulfur atom. The stereochemical course of the reaction under discussion has been explained in terms of the addition-elimination mechanism involving the formation of sulfurane intermediates.[27]

The second approach for the preparation of enantiomeric sulfinates having a relatively high optical purity involves the reaction of symmetrical sulfites **21** with t-butylmagnesium chloride in the presence of optically active aminoalcohols **23**. This asymmetric synthesis affords t-butanesulfinates **22** with the sulfur atom as a sole center of chirality with 40–70% enantiomeric excess values[28] (Equation 2.1.5 and Table 2.1.4).

$$\underset{\mathbf{21}}{RO\text{-}S(=O)\text{-}OR} + t\text{-BuMgCl} \xrightarrow[\mathbf{23}]{\text{Aminoalcohol}} \underset{\mathbf{22}}{t\text{-Bu-}S(=O)\text{-}OR} \quad (2.1.5)$$

For 21,22	For **23**
a, R = Me	**a**, (–)-Quinine
b, R = Et	**b**, (+)-Quinidine
c, R = Pr^n	**c**, (+)-Cinchonine
d, R = Pr^i	**d**, (–)-N-Methylephedrine
e, R = Bu^n	

Table 2.1.4 Asymmetric Synthesis of Optically Active O-Alkyl t-Butanesulfinates, t-BuS(O)OR, 22

Sulfite		Sulfinate 22				
21, R	**Amine 23**	**No**	**Yield [%]**	**$[\alpha]_{589}$**	**ee [%]**	**Abs. conf.**
21a, Me	**a**	**a**	62	+72	53.3	R
21b, Et	**a**	**b**	62	+100	69.0	R
21c, n-Pr	**a**	**c**	79	+99.7	74.2	R
21c, n-Pr	**b**	**c**	69	−25.2	20.0	S
21c, n-Pr	**c**	**c**	63	−37.9	28.2	S
21d, i-Pr	**d**	**d**	70	+54.3	43.0	R
21e, n-Bu	**a**	**e**	84	+83.3	62.4	R

From Drabowicz, J., Legędź, S., and Mikołajczyk, M., *Tetrahedron,* **44,** 5243, 1988. Reprinted with kind permission from Elsevier Science Ltd, The Boulevard, Langford Lane, Kidlington OX5 1GB, UK.

A very general method for the preparation of enantiomeric sulfinates **24** is based on the reaction of sulfinyl chlorides **2** with achiral alcohols in the presence of optically active tertiary amines **25** as chiral reagents (Equation 2.1.6).[29] Some representative examples of the optically active sulfinates prepared by this asymmetric condensation are collected in Table 2.1.5. The usefulness of this approach was demonstrated by the synthesis of optically active O-methyl p-toluenesulfinate **26** labeled with the ^{14}C atom in the methoxy group[30] (Equation 2.1.7).

$$\underset{\mathbf{2}}{R\text{-}\underset{\underset{O}{\|}}{S}\text{-}Cl} + R^1OH \xrightarrow[\mathbf{25}]{Me_2N\overset{*}{R}{}^3} \underset{\mathbf{24}}{R\text{-}\underset{\underset{O}{\|}}{S}\text{-}OR^1} + Me_2N\overset{*}{R}{}^3\cdot HCl \qquad (2.1.6)$$

For **25** **a** R^3 = α-phenylethyl
b R^3 = menthyl
c R^3 = dihydroabietyl

Table 2.1.5 Asymmetric Synthesis of O-Alkyl Sulfinates RS(O)OR^1 **24** in the Reaction of Sulfinyl Chlorides, RS(O)Cl, **2**, and Achiral Alcohols in the Presence of Chiral Tertiary Amines, Me_2NR^3, **25**

Sulfinyl chloride 2		Alcohol,	Me_2NR^3	Sulfinate 24			
No	**R**	**R^1OH**	**25**	**No**	**$[\alpha]_{589}$**	**ee [%]**	**Abs. conf.**
a	p-Tol	Me	(−)**b**	**a**	+45.23	21	R
a	p-Tol	Et	(−)**b**	**b**	+91.80	43.6	R
a	p-Tol	Et	(+)**c**	**b**	+20.66	9.8	R
a	p-Tol	n-Pr	(+)**c**	**c**	+23.10	12.1	R
b	Ph	n-Pr	(−)**b**	**d**	+23.40	13.8	R
h	Me	n-Pr	(−)**a**	**e**	+26.92	19.3	R
k	Et	n-Pr	(−)**a**	**f**	+38.2	23.9	R
n	i-Pr	n-Pr	(−)**a**	**g**	+37.1	29.2	R

From Mikołajczyk, M. and Drabowicz, J., *J. Chem. Soc., Chem. Commun.,* 547, 1974. With permission from the Royal Society of Chemistry.

$$\underset{\mathbf{2a}}{p\text{-Tol-}\underset{\underset{O}{\|}}{S}\text{-Cl}} + {}^{14}CH_3OH \xrightarrow{\mathbf{25b}} :\!\cdots\overset{\overset{O}{\|}}{S}(\text{Tol-p})(O^{14}CH_3) \qquad (2.1.7)$$

26, $[\alpha]_{589}$=+60.6 (e.e.= 27.3%)

2.1.3 Synthetic Application of Optically Active Sulfinate Esters

Diastereomeric arene (alkane) sulfinates are most commonly used as substrates for the preparation of a very rich family of optically active, dialkyl, aryl alkyl, and diaryl sulfoxides. This highly stereoselective synthesis of optically active sulfoxides, first described by Andersen in 1962, involves the reaction of the above mentioned esters (diastereomerically pure or strongly enriched in one diastereomer) with Grignard reagents or other organometallics. In his original report Andersen[31] described the preparation of (+)-(R)-ethyl p-tolyl sulfoxide **27a** from (–)-(S)-O-menthyl p-toluenesulfinate **1a** and ethylmagnesium iodide (Equation 2.1.8).

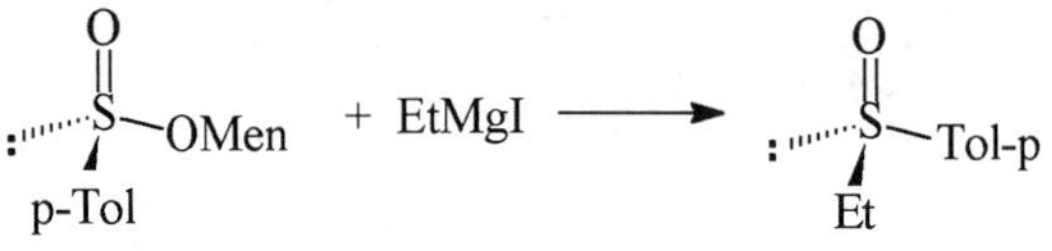

(-)-(S)-**1a** + EtMgI → (+)-(R)-**27a** (2.1.8)

Soon thereafter it became evident that the Andersen approach to chiral sulfoxides is general in scope, and a large number of optically active aryl alkyl and unsymmetrical diaryl sulfoxides were prepared starting from diastereomerically pure O-menthyl arenesulfinates **1a, 1b, 1c**, and **1f** (see Table 2.1.6). The original Andersen[31] assumption that the reaction of Grignard reagents with arene (alkane) sulfinates proceeds with a full inversion of configuration at sulfur was later firmly established by Mislow et al.[12] and other investigators.[32,33]

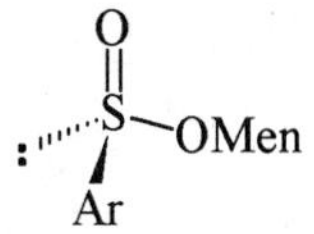

(-)-(S)-**1a-c, 1f**

a, Ar=p-Tol[1]
b, Ar=Ph[9a]
c, Ar=p-An[10]
f, Ar=1-$C_{10}H_7$[10]

Very recently, diastereomerically pure alkane- and arenesulfinates **10a–f** were used as substrates for the synthesis of both enantiomers of a given sulfoxide[20] (Scheme 2.1.9 and Table 2.1.6).

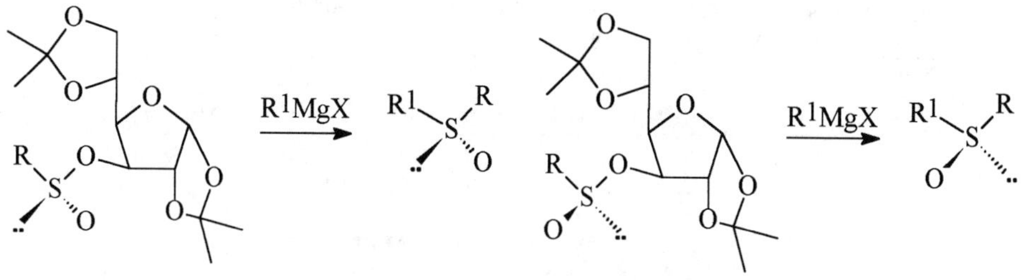

(-)-(S)-**10a-f** $\xrightarrow{R^1MgX}$ (+)-(R)-**10a-f** $\xrightarrow{R^1MgX}$

a, R = p-Tol
b, R = Me
c, R = Et
d, R = n-Pr
e, R = i-Pr
f, R = t-Bu

Scheme 2.1.9

Table 2.1.6 Synthesis of Optically Active Sulfoxides, RS(O)R[1], from Diastereomeric Aryl(Alkyl)Sulfinates, RS(O)OR[2], and Organometallic Reagents, R[1]M

	Sulfinate				Sulfoxide			
R	R[2]	$[\alpha]_{589}$	de [%]	R[1]M	Yield [%]	$[\alpha]_{589}$	ee [%]	Ref.
Ph	Men	–206.1	100	MeMgI	60	–178.3	100	9b
		–205.5	>99	Me_2CuLi	16	+133.9	[a]	32
		[a]		EtMgI	72	+176.2	[a]	67
		–206.1	100	i-PrMgCl	60	+170.0	100	68
		–206.1	100	t-BuMgCl	60	+180.0	100	9b
				$C_5H_{11}MgBr$	[a]	+199.6	[a]	7
				$C_6H_{13}MgBr$	[a]	+184.0	[a]	7
p-Tol	Men	–198.	>95	$MeMgI/Et_2O$	[a]	+145.5	95	12
		–195.0	>94	MeMgI/PhH	82	+150.0	[a]	33
		–210.0	~100	Me_2CuLi	55	+143.2	[a]	32
		–198.0	>95	$EtMgBr/Et_2O$	[a]	+187.5	[a]	12
		–195.0	>94	EtMgBr/PhH	92	+198.0	[a]	33
		[a]	—	n-PrMgBr	[a]	+201.0	[a]	44
		–198.0	>95	$i\text{-}PrMgBr/Et_2O$	22		>95	12
		–195.0	>94	i-PrMgBr/PhH	40	+173.2	[a]	33
		–195.0	>94	n-BuMgBr/PhH	73	+186.0	[a]	33
		–198.0	>95	t-BuMgCl/	[a]	+190.0	>95	12
		–198.0	>95	$t\text{-}BuCH_2MgBr$	[a]	+220.0	>95	69
		–198.0	>95	$C_6H_{11}MgI$	[a]	+176.0	>95	70
		–198.0	>95	$CH_2 = CHCH_2MgBr$	[a]	+212.0	>95	44
		–198.0	>95	PhMgBr	[a]	+27.0	>95	12
		–198.0	>95	o-TolMgBr	[a]	+75.6	>95	12
		–198.0	>95	m-TolMgBr	[a]	+24.4	>95	12
		–198.0	>95	$2,4,6\text{-}Me_2C_6H_2MgBr$	[a]	–259.0	>95	12
		–198.0	>95	9-AnthrylMgBr	[a]	–309.0	>95	12
		–198.0	>95	$4\text{-}CF_3C_6H_4MgBr$	[a]	+57.0	>95	60
		–198.0	>95	$3\text{-}CF_3C_6H_4MgBr$	[a]	+58.0	>95	69
		–198.0	>95	$4\text{-}ClC_6H_4MgBr$	[a]	+250.0	>95	69
		–198.0	>95	$2\text{-}ClC_6H_4MgBr$	[a]	–120	>95	69
		–198.0	>95	$2\text{-}MeOC_6H_4$	[a]	–221.0	>95	69
		–199.2	>95	$4\text{-}MeOC_6H_4$	[a]	–25.1	>95	10
		–199.2	>95	$4\text{-}Me_2NC_6H_4MgBr$	[a]	+85.2	>95	10
		–199.2	>95	1-NaphMgBr	[a]	–414.2	>95	10
		–210.0	~100	$4\text{-}MeCC_6H_{10}CH_2MgBr$	61	+204.0	~100	71
		–210.0	~100	$4\text{-}MeO\text{-}c\text{-}C_6H_{10}CH_2MgBr$	70	+182.0	~100	71
		–210.0	~100	$4Cl\text{-}c\text{-}C_6H_{10}CH_2MgBr$	17	+173.0	~100	71
		–210.0	~100	$4\text{-}C_5H_{11}\text{-}C_6H_{10}MgBr$	72	+169.0	~100	71
		–210.0	~100	$4\text{-}t\text{-}Bu\text{-}c\text{-}C_6H_{10}CH_2MgBr$	41	+155.0		71
p-Tol	DAG	+10.5	72	EtMgBr		–137.0	67	20b
p-Tol		–125.0	100	EtMgBr		+196.0	96	20b
p-Tol	PCH[b]	+82.0	100	$4\text{-}C_6H_5O\text{-}C_6H_4MgBr$		–2.0	100	21
$4\text{-}MeOC_6H_4$	Men	–189.1	~100	$2\text{-}MeO\text{-}C_6H_4MgBr$	[a]	–217.2	~100	10
		–189.1	~100	p-TolMgBr	[a]	+24.2	~100	10
1-Naph	Men	–433.2	~100	p-TolMgBr	[a]	+416.2	~100	10
Mesityl	DCG	+28.8	100	MeMgI	93	–200.1		19
	DCG	+28.8	100	i-PrMgI	71	–176.9		19
Me	DAG	+17.0	100	p-TolMgBr	84	+145.0	100	20b
		+17.0	100	PhMgBr	78	+149.0	100	20b
		+17.0	100	$PhCH_2MgCl$	83	–105.0	100	20b
		+17.0	100	n-Pr	66	–137.0	100	20b
	DAG	–60.0	100	p-TolMgBr	90	–145.0	100	20b
Et	DAG	+12.0	100	p-TolMgBr	96	+195.0	99	20b
n-Pr	DAG	+6.5	100	p-TolMgBr	88	+203.0	100	20b
i-Pr	DAG	+11.0	100	p-TolMgBr	98	+188.0	100	20b
t-Bu	DAG	[a]	100	PhMgBr	85	–155.0	90	20c
		[a]	100	p-TolMgBr	87	–178.0	93	20c
		[a]	100	MeMgI	60	+8.7	100	20c

Table 2.1.6 (continued) Synthesis of Optically Active Sulfoxides, RS(O)R[1], from Diastereomeric Aryl(Alkyl)Sulfinates, RS(O)OR[2], and Organometallic Reagents, R[1]M

Sulfinate					Sulfoxide			
R	R²	$[\alpha]_{589}$	de [%]	R¹M	Yield [%]	$[\alpha]_{589}$	ee [%]	Ref.
Me	Cholesteryl	+77.35	100	n-PrMgBr	32	−139.0	100	17
		+77.35	100	p-TolMgBr	35	+148.0	95	17
		−113.0	100	n-BuMgBr	52	+110.3	100	17
		−113.0	100	i-BuMgBr	50	+138.0	100	17
		−11.85	99	$PhCH_2MgBr$	36	+106.0	99	17
$PhCH_2$	Men	+105.0	~90	MeMgI	a	+96.0	~90	13
		+105.0	~90	EtMgI	a	+47.0	~90	13
		+123.-	~95	n-PrMgI	a	+55.0	~95	14
		+105.0	~90	i-PrMgI	a	+119.0	~90	13
		+105.0	~90	n-BuMgI	a	+16.0	~90	13
		+123.0	~95	i-BuMgI	a	−110.0	~95	14
		+105.0	~90	t-BuMgI	a	+281.0	~90	13
	16a(R = Me)	−49.0	100	n-OctMgBr	a	−63.0	100	23
	16b(R = Et)	−31.0	100	PhLi	a	+176.0	100	23
	16b(R = Et)	−31.0	100	$PhCH_2MgBr$	a	+105.0	100	23
	16c(R = n-Oct)	−46.0	100	MeMgI	a	+62.5	100	23
	17d(R = t-Bu)	−120.0	100	MeLi	a	−10.5	100	23
	17d(R = t-Bu)	−120.0	100	PhLi	a	−175.0	100	23
	17d(R = t-Bu)	−120.0	100	n-BuLi	a	+125.0	100	23
	17d(R = t-Bu)	−120.0	100	$CH_2 = CHMgCl$	a	+283.0	100	23
	17d(R = t-Bu)	−120.0	100	$PhCH_2MgBr$	a	+279.0	100	23
	17d(R = t-Bu)	−120.0	100	$PhCH_2CH_2MgBr$	a	+95.0	100	23
	17f(R = mesityl)	−113.0	100	MeLi		+45.0	100	23
	17f(R = mesityl)	−113.0	100	PhMgBr		+256.0	100	23

[a] Not given.
[b] trans-2-Phenylcyclohexanol.

Similarly, a few enantiomerically pure sulfoxides were obtained by the reaction of diastereomerically pure β-hydroxysulfinates **16** and **17** with organometallics[23] (Scheme 2.1.10).

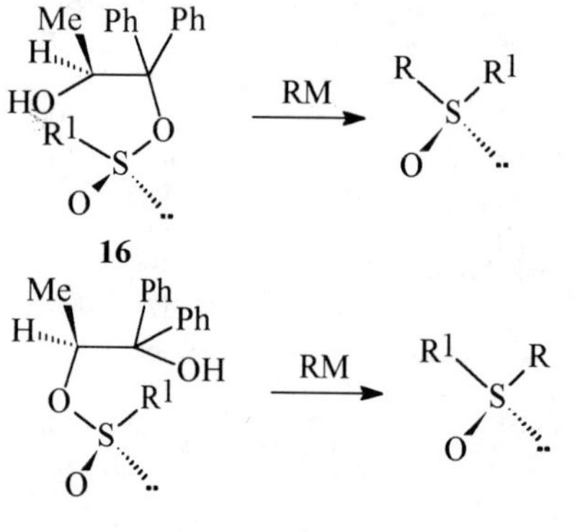

16

a, R^1 = Me
b, R^1 = Et
c, R^1 = n-Oct
d, R^1 = t-Bu
e, R^1 = CH_2=CH
f, R^1 = $(CH_3)_3C_6H_2$

17

Scheme 2.1.10

Alkane and arenesulfinates **13a–d** derived from trans-2-phenylcyclohexanol afforded upon treatment with Grignard reagents the corresponding optically active sulfoxides with e.e. values above 90% (Scheme 2.1.11).[21]

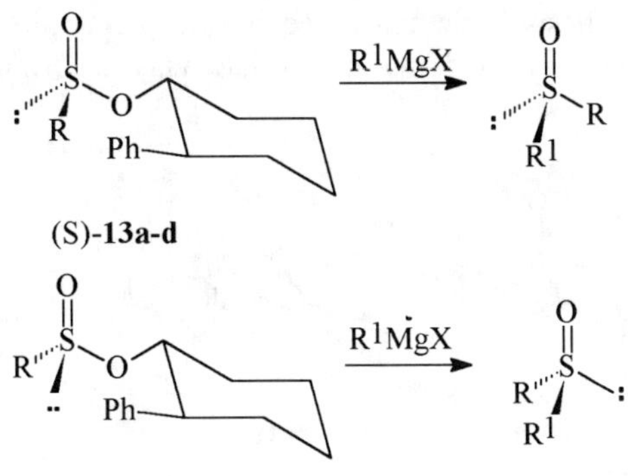

(S)-**13a-d** $\xrightarrow{R^1MgX}$ R(R¹)S=O

(R)-**13a-d** $\xrightarrow{R^1MgX}$ R(R¹)S=O

a, R = p-Tol
b, R = Me
c, R = i-Pr
d, R = 2-$C_{10}H_7$

Scheme 2.1.11

A series of dialkyl sulfoxides of high optical purity was prepared from O-cholesteryl methanesulfinates **6** and Grignard reagents.[17]

$$\underset{\substack{\mathbf{6}\\(+)\text{-}(R)\text{ or }(-)\text{-}(S)}}{Me\text{-}S(=O)\text{-}OChol} \xrightarrow{RMgX} R\text{-}\overset{*}{S}(=O)\text{-}Me \qquad R=n\text{-}Pr,\ n\text{-}Bu,\ PhCH_2,\ p\text{-}Tol \tag{2.1.9}$$

Usually, the reaction of chiral sulfinates with Grignard reagents is carried out in diethyl ether solution. However, in this solvent optically active sulfoxides are formed very often in moderate to low yields depending on the structure of both reaction components. Harpp and co-workers reported[34] that the use of lithium-copper reagents (R_2CuLi) instead of Grignard reagents gives a cleaner conversion of sulfinates to optically active sulfoxides. Even better results are obtained when the reaction of O-menthyl sulfinates with Grignard reagents is carried out in a benzene solution.[35]

A highly stereoselective conversion of O-menthyl arenesulfinate into chiral methyl aryl sulfoxides was also accomplished by means of methyllithium.[36-38] The preparation of optically active sulfoxides **28, 29** and **30**, in which chirality at sulfur is due to isotopic substitution (H→D and $^{12}C \rightarrow {}^{13}C$, respectively), involves the reaction of the appropriate nonlabeled O-menthyl sulfinates **1g, 1h**, and **1i** with fully deuteriated methyl[39] (Equation 2.1.10) and n-butylmagnesium[40] (Equation 2.1.11) iodides and with benzylmagnesium chloride labeled with ^{13}C in the methylene group, respectively (Equation 2.1.12).[41]

$$\underset{\mathbf{1h}}{CH_3S(=O)\text{-}OMen} + CD_3MgI \longrightarrow \underset{\mathbf{28}}{CH_3\text{-}S(=O)\text{-}CD_3} \tag{2.1.10}$$

$$\underset{\mathbf{1i}}{n\text{-}BuS(=O)\text{-}OMen} + n\text{-}C_4D_9MgBr \longrightarrow \underset{\mathbf{29}}{n\text{-}C_4H_9\text{-}S(=O)\text{-}n\text{-}C_4D_9} \tag{2.1.11}$$

$$\underset{\mathbf{1g}}{PhCH_2S(=O)\text{-}OMen} + Ph^{13}CH_2MgCl \longrightarrow \underset{\mathbf{30}}{PhCH_2S(=O)\text{-}^{13}CH_2Ph} \tag{2.1.12}$$

A great value of the Andersen sulfoxides synthesis is demonstrated by the preparation of optically active vinyl sulfoxides **31** from sulfinates **1a,b** and the appropriate vinylic Grignard reagents (Equation 2.1.13).[42,43]

1a,b + $RCH{=}CR^1MgBr$ ⟶ **E-31** + **Z-31** (2.1.13)

a, Ar=p-Tol
b, Ar=Ph

a, Ar = p-Tol, R = R^1 = H	Ref 42
b, Ar = p-Tol, R = H, R^1 = Me	Ref 43a
c, Ar = p-Tol, R = H, R^1 = C_6H_{13}	Ref 43a
d, Ar = p-Tol, R = H, R^1 = Bu_3Sn	Ref 43b
e, Ar = Ph, R = n-C_6H_{13}, R^1 = H	Ref 43c
f, Ar = Ph, R = Me, R^1 = H	Ref 43c
g, Ar = Ph, R = Ph, R^1 = H	Ref 43d
h, Ar = Ph, R = t-Bu, R^1 = H	Ref 43d

Allyl Grignard reagents form with the sulfinate (–)-(S)-**1a** optically active allyl sulfoxides **32**, however, with the rearranged allylic group. It was shown[44] that this rearrangement takes place via the transition state **33** (Equation 2.1.14).

(-)-(S)-**1a** + $RCH{=}CH{-}CH_2MgX$ ⟶ [**33**] ⟶ **32** (2.1.14)

A similar transition state **34** was proposed to explain the formation of allenic sulfoxides **35**, together with the expected alkynylmethyl sulfoxides **36**, when (–)-(S)-**1a** was reacted with the Grignard reagents prepared from α-alkynylmethyl halides[45] (Equation 2.1.15).

$$(-)\text{-}(S)\text{-}\mathbf{1a} + XMgCH_2C{\equiv}CR \longrightarrow p\text{-Tol-}\underset{\underset{O}{\|}}{S}\text{-}CH_2C{\equiv}CR \ (\mathbf{36})$$

$$\downarrow$$

$$[\mathbf{34}] \longrightarrow p\text{-Tol-}\underset{\underset{O}{\|}}{S}\text{-}CR{=}C{=}CH_2 \ (\mathbf{35})$$

(2.1.15)

It should be noted that the reaction of (–)-(S)-**1a** with 1-alkynylmagnesium bromides[45] results in the exclusive formation of 1-alkynyl p-tolyl sulfoxide **37** (Equation 2.1.16).

$$(-)\text{-}(S)\text{-}\mathbf{1a} + BrMgC{\equiv}C\text{-}R \longrightarrow (+)\text{-}\mathbf{37}$$

(2.1.16)

a, R = n-Pr, $[\alpha]_{589} = +88.6$
b, R = n-Bu, $[\alpha]_{589} = +77.6$
c, R = n-C_5H_{11}, $[\alpha]_{589} = +73.8$
d, R = n-C_6H_{13}, $[\alpha]_{589} = +70.0$
e, R = Me_3Si, $[\alpha]_{589} = +154.0$

The Andersen approach is also useful for the preparation of a variety of α-heteroatom-substituted sulfoxides starting from α-heteroatom-stabilized carbanions and the sulfinate (–)-(S)-**1a**. Some selected examples are shown in Scheme 2.1.12. More detailed discussion on the synthesis of a particular group of these sulfoxides is included in the appropriate chapters of this book.

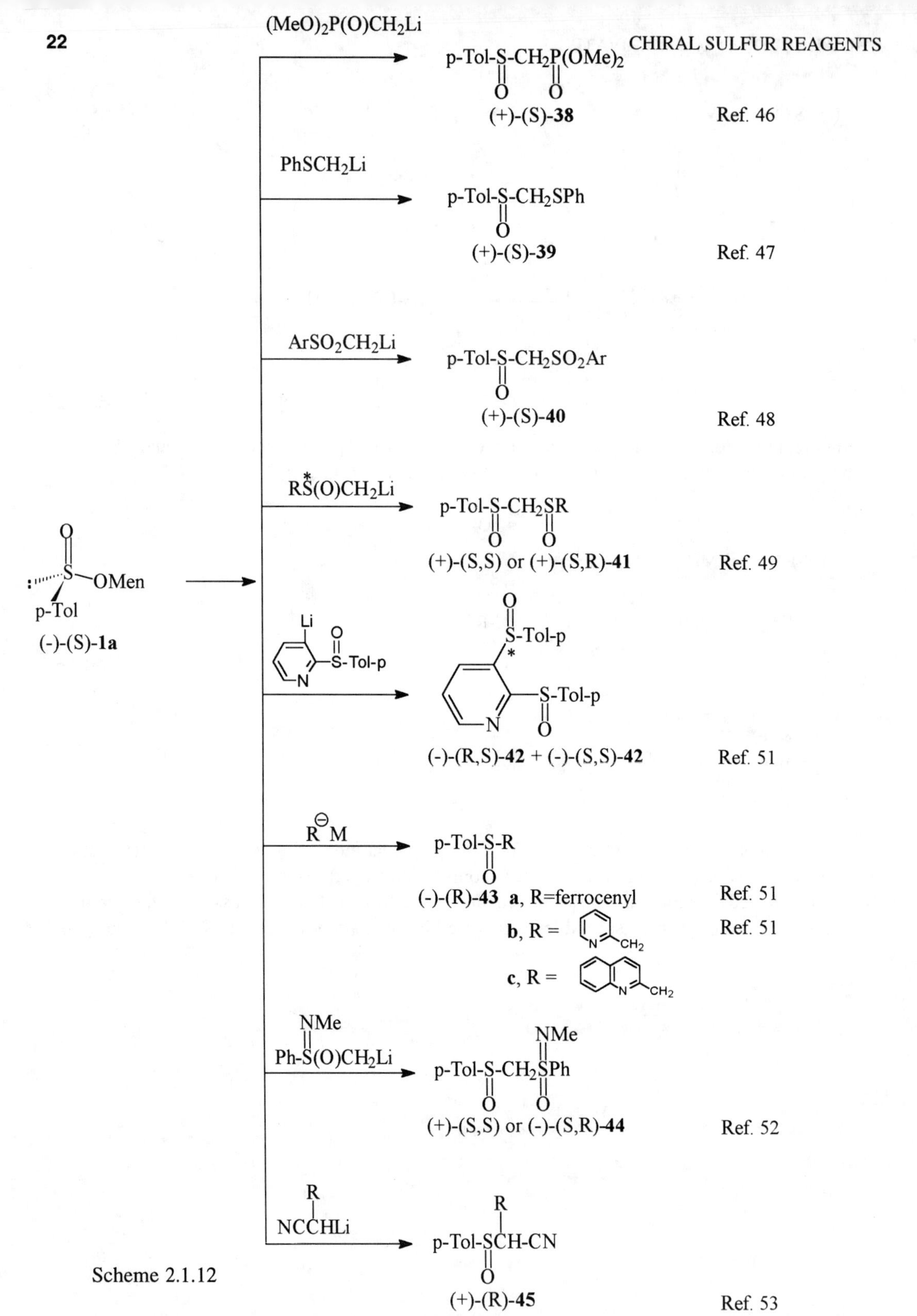

Scheme 2.1.12

The sulfinate (–)-(S)-**1a** has also been used for the preparation of (S)-2-p-toluenesulfinyl-1,4-dimethoxybenzene **46** or the naphthalene derivatives **47b-c** (Equation 2.1.17 and Scheme 2.1.13).[54-57]

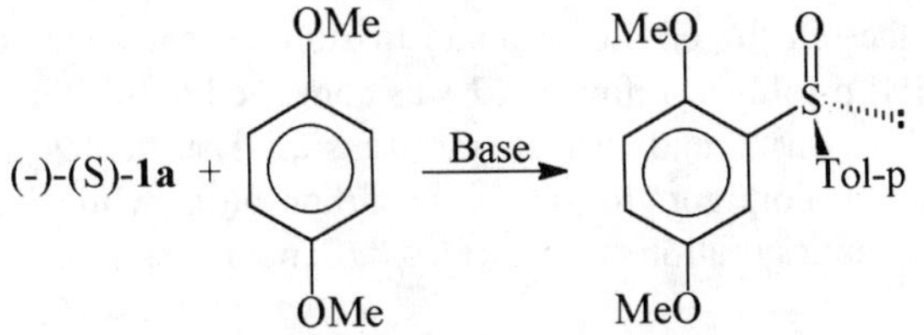

(-)-(S)-**1a** + **48a** —Base→ (-)-(S)-**46** (2.1.17)

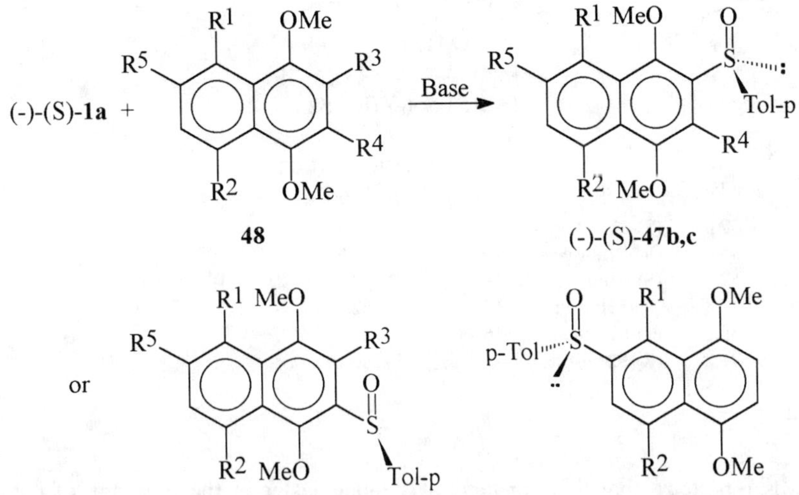

(-)-(S)-**1a** + **48** —Base→ (-)-(S)-**47b,c**

or (-)-(S)-**47d** (-)-(S)-**47e**

48	R^1	R^2	R^3	R^4	47	R^1	R^2	R^3	R^4	R^5
b	H	H	H	H	**b**	H	H		H	H
c	H	OMe	Br	H	**c**	H	OMe		H	H
d	H	OMe	H	H	**d**	H	OMe	H		H
e	OMe	OMe	H	H	**e**	OMe	OMe	H	H	

Scheme 2.1.13

The sulfinylation reactions of 1,4-dimethoxybenzene **48a** or the naphthalene derivatives **48b,d,e** are carried out by their direct metallation with n-butyllithium, followed by treatment with (–)-(S)-**1a** to give the corresponding (S)-**46** or (S)-**47b,d,e**. However, the isomer **47c** is not available by this procedure. The substrate for **47c** is 2-bromo-1,4,5-trimethoxynaphthalene **48c**, which is lithiated with n-butyllithium at –78°C and then reacted with (–)-(S)-**1a**.

A one-pot synthesis of chiral sulfinyl cyclohexanones **50** reported by a Spanish group[58] is based on the reaction of cyclohexanone with (–)-(S)-**1a** in benzene in the presence of i-Pr_2NMgBr (Equation 2.1.18).

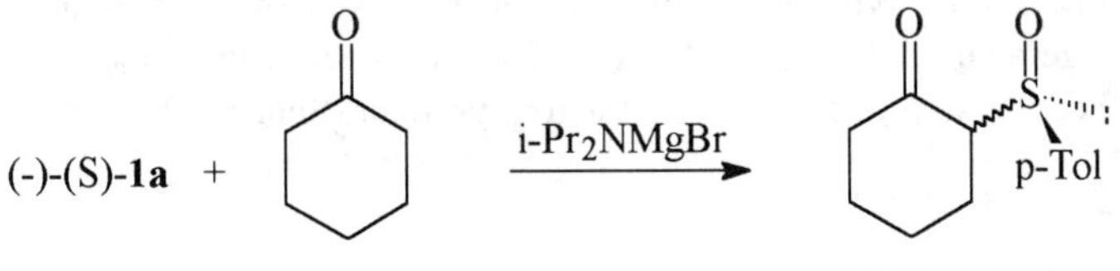

(-)-(S)-**1a** + cyclohexanone —i-Pr_2NMgBr→ **50** (75:25 mixture) (2.1.18)

In the chemical literature one can find only a few procedures which are based on the use of enantiomeric sulfinates as useful chiral sulfur reagents. The first of them is based on the chirality transfer from sulfur to carbon taking place in the allylic sulfinate to sulfone rearrangement.[59] A series of optically active trans and cis-allyl p-toluenesulfinates **51** was converted to the corresponding sulfones **52** upon heating in DMF. It was found that all the trans-allyl sulfinates gave the corresponding sulfones with the S-absolute configuration at the α-carbon atom while from cis-isomers the sulfones with the R-absolute configuration were formed[59] (Scheme 2.1.14).

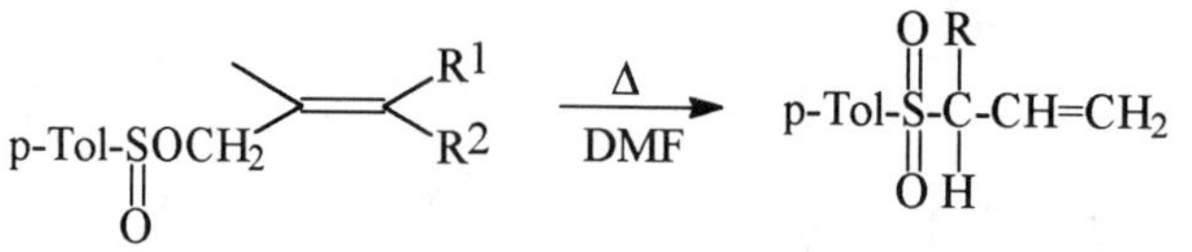

p-Tol-S(O)OCH₂–CH=C(R1)(R2) —Δ, DMF→ p-Tol-SO₂-CH(R)-CH=CH₂

(-)-(S)- **51** (+)-(S) or (-)-(R)- **52**

	R^1	R^2	ee [%]	R	$[\alpha]_{589}$	% Stereoselectivity **51→52**
a	Me	H	57.4	(S)-**52a** Me	+4.4	86.8
b	H	Me	45.9	(R)-**52a** Me	–4.1	88.9
c	n-Pr	H	57.5	(S)-**52b** n-Pr	+18.0	83.6
d	H	n-Pr	47.8	(R)-**52b** n-Pr	–14,7	82.2
e	n-Bu	H	56.8	(S)-**52c** n-Bu	+16.7	80.8
f	H	n-Bu	91.2	(R)-**52c** n-Bu	–28.1	86.1

Scheme 2.1.14

Later on, it was reported[60] that this conversion is much faster in the presence of catalytic amounts of palladium complexes. For instance, heating trans-(–)-(S)-**51a** at 50°C for 10 h in tetrahydrofuran (THF) in the presence of tetrakis (triphenylphosphine) palladium **53** (0.15 molar equivalent) gave (+)-(S)-sulfone **52a** in 74% yield and with a stereoselectivity higher than 90%. The rearrangement of the trans isomer (–)-(S)-**51b** under the same conditions afforded the sulfone (–)-(S)-**52a** in 69% yield and with a lower stereoselectivity equal to 86.4%.

In both conversions the α-rearranged sulfones were also formed. The palladium-catalyzed rearrangement of the sulfinates **51b–e**, which have bulky substituents connected with the α-carbon atom of the allyl moiety, was much less regio- and stereoselective. These differences were explained in terms of the transition state having more ionic character than for the typical [2,3]-sigmatropic rearrangement.

These studies were extended on palladium-catalyzed conversion of chiral allylic sulfinates to the corresponding chiral unsaturated hydrocarbon systems.[61] Thus, the reaction of the sulfinate (–)-(S)-**51a** with sodium dimethyl malonate carried out in the presence of **53** in refluxing tetrahydrofuran for 10 h gave 91:1 mixture of (+)-(S)-dimethyl 2-buten-3-yl-malonate **54a** and dimethyl 1-buten-2-ylmalonate **55** in 75% yield and with a selectivity of the **51a→54a** conversion equal to 83%. When the cis-sulfinate **51b** was reacted with sodium dimethyl malonate under the same conditions, the enantiomeric (–)-(R)-**54a** was formed with 75% stereoselectivity. With other sulfinates **51** this conversion occurred with much lower selectivity. It is obvious that the first step of the reaction under discussion involves a rearrangement of allyl sulfinate **51** to allyl sulfone **52** which undergoes, in turn, the palladium-catalyzed[62] conversion to the final sulfur-free product (Scheme 2.1.15).

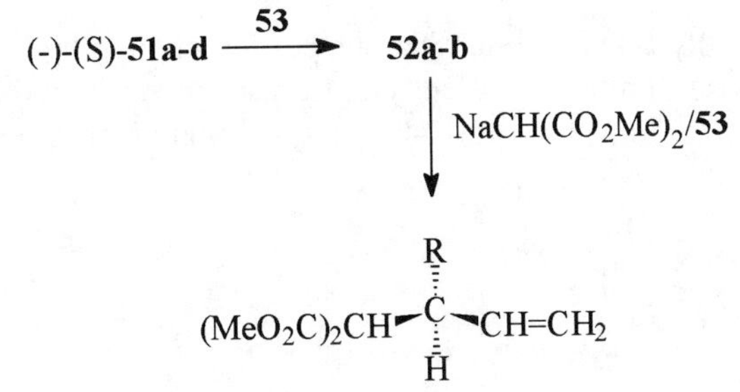

(-)-(S)-**51a-d** →(**53**) **52a-b**

↓ $NaCH(CO_2Me)_2$/**53**

$(MeO_2C)_2CH$–C(R)(H)–$CH=CH_2$

(+)-(S)-**54a-b**

Scheme 2.1.15

Another asymmetric synthesis of optically active unsaturated hydrocarbons is based on the allylic biomolecular nucleophilic substitution at the sulfinyl sulfur atom, which occurs in the reaction of O-alkenyl sulfinates with Grignard reagents carried out in the presence of a mixture of lithium chloride and copper cyanide. When optically active O-cinnamyl p-toluenesulfinate **20h** was allowed to react with a series of Grignard reagents, the optically active α,β-unsaturated hydrocarbons **56** were always formed (Scheme 2.1.16).[26] Unfortunately, so far their optical purities and absolute configurations have not been determined.

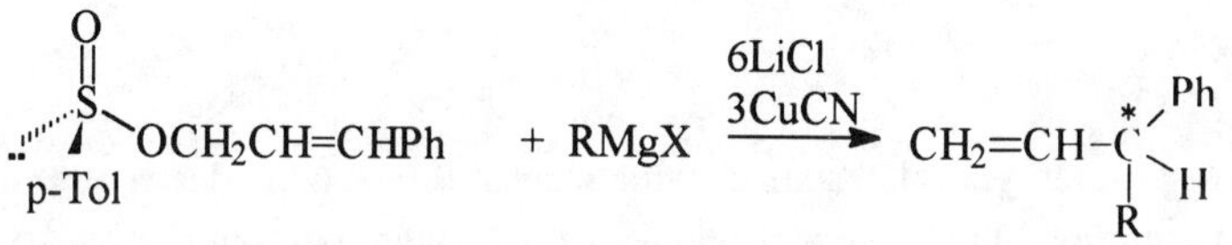

p-Tol–S(O)–$OCH_2CH=CHPh$ + RMgX →(6LiCl, 3CuCN) $CH_2=CH-C^*(Ph)(H)(R)$

(-)-(S)-**20h** **56**

20h	
$[\alpha]_{589}$	ee [%]
−35.5	95.8
−35.5	95.8
−37.6	100
−34.2	92.3

56		
R	Yield [%]	$[\alpha]_{589}$
n-Pr	52.3	+5.20
i-Pr	12.5	+8.65
n-Bu	53.4	+5.45
t-Bu	56.3	+2.59

Scheme 2.1.16

Earlier, this type of nucleophilic substitution was also observed in the reaction of organocuprates with a number of methanesulfinates derived from racemic and optically active propargylic alcohols. For example, treatment of the epimeric methanesulfinates **57** derived from epimestranol with equimolar amounts of the heterocuprate for 1 h at 0°C afforded the corresponding allenes **58** in nearly quantitative yields[63] (Equation 2.1.19).

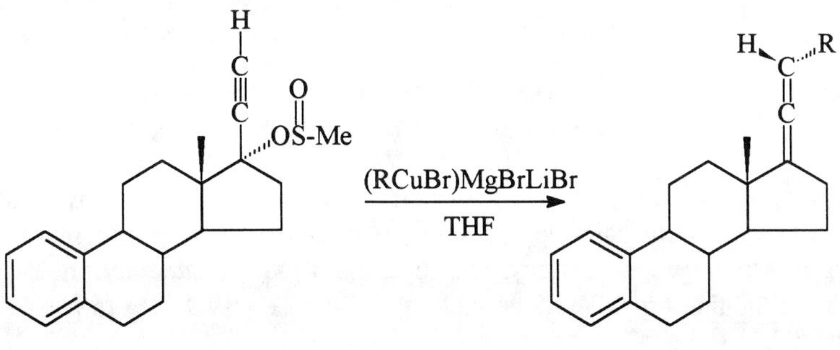

57 (H–C≡C–, OS(O)-Me) →((RCuBr)MgBrLiBr, THF) **58** (H(R)C=C=) (2.1.19)

A highly stereoselective synthesis of optically active nonsteroidal allenes **59** reported by the same group[64] consists in the preferred anti-1,3-substitution by an attack of organocuprates on the diastereomeric methanesulfinates **60** derived from (–)-(R)-3-hydroxy-3-phenyl-propyne (Equation 2.1.20).

H—C≡C—C(Ph)(H)—OS(=O)Me —(RCuBr)MgX,THF→ R(H)C=C=C(Ph)(H)

60 **59**

a, R = Me
b, R = Ph (2.1.20)

In this context, it is noteworthy that (+)-(R)-α-methylpropargyl p-toluenesulfinate **61** rearranged to (–)-γ-methylallyl p-tolyl sulfone **62** on heating at 130°C[65] (Equation 2.1.21).

p-Tol-S(=O)OCH(CH3)-C≡CH —Δ→ p-TolSO2(H)C=C=C(Me)(H)

(+)-(R_C)-**61** (-)-**62** (2.1.21)

Similarly, optically active α,γ-dimethylallyl 2,4-dimethylbenzenesulfinate **63** underwent a clean rearrangement to the corresponding optically active sulfone **64** with practically complete inversion of configuration[66] at carbon (Equation 2.1.22).

63 —Δ→ **64**

Ar = 2,4-dimethylphenyl (2.1.22)

2.1.4 Examples of Experimental Procedures

2.1.4.1 Preparation of (–)-(S)-O-Menthyl p-Toluenesulfinate 1a

2.1.4.1.a From p-Toluenesulfinyl Chloride[5]

A solution of (–)-menthol (15.6 g; 0.1 mol) in pyridine (10 g) was cooled to –10°C and p-toluenesulfinyl chloride (17.2 g; 0.1 mol) added in small quantities. After 12 h, water was added, and the liberated oil extracted, washed, and dried in ether in the usual way. The oil, on cooling after removal of the solvent, deposited crystals (12 g). The residue was dissolved in ethyl ether and a few crystals of tetraethylammonium chloride and some hydrogen chloride gas were added. A second crop of crystals was removed after 1 day, and retreatment with quaternary chloride and hydrogen chloride gave yet another crop. The total yield of (–)-(S)-**1a** was 90.7% $[a]_D = -199.19$ (2, acetone) mp 106–107°C.

2.1.4.1.b *From Sodium p-Toluenesulfinate*[6]

Sodium p-toluenesulfinate (18 g, 0.08 mol, carefully dried by heating overnight at 140°C under vacuum) is added to thionyl chloride (20 mL) at 0°C over 1 h. After 1 h at 0°C, excess thionyl chloride is removed by azeotropic distillation with benzene at 25–30°C under reduced pressure (this operation is performed three times with 100 mL of benzene each time). The residue is diluted with diethyl ether (50 mL) and added at 0°C to a solution of 1-menthol (15.6 g, 0.1 mol) in pyridine (14 mL). After 2 h at room temperature, the mixture is decomposed by addition of water (50 mL), the organic layer is separated, washed with 20% aqueous hydrochloric acid (3 × 20 mL), dried with sodium sulfate, and concentrated under vacuum. The oily residue is diluted with hot acetone (30 mL) and cooled in a refrigerator to give crystals of the product which are washed with cold petroleum ether. The mother liquor is concentrated and the residue dissolved in hot acetone to give a further batch of product as described above. This process is repeated three times. Then, concentrated hydrochloric acid (2 drops) is added before each new recrystallization. The combined crystals are then recrystallized once from hot acetone; yield: 90%; mp 110°C; $[\alpha]_D^{20} = -202$ (c 2.1, acetone).

2.1.4.1.c *From p-Toluenesulfonyl Chloride*[16]

A dry, 2-L, three-necked, round-bottomed flask equipped with a reflux condenser and nitrogen inlet was charged with l-menthol (31.25 g, 0.20 mol), p-toluenesulfonyl chloride (45.76 g, 0.24 mol), triethylamine (33.5 mL, 0.24 mol), and CH_2Cl_2 (1000 ml). Trimethyl phosphite (33.5 mL, 0.30 mol) was added, and the reaction mixture was heated to reflux. After 10 h the reaction mixture was allowed to cool to room temperature, washed with 1 N HCl (2 × 100 ml), saturated $NaHCO_3$ (100 mL), and saturated NaCl (2 × 100 mL), dried ($MgSO_4$), and concentrated. Kugelrohr distillation (45–50°C [0.2 mm]) effected removal of trimethyl phosphate and some remaining menthol to afford a yellow oil (61.3 g), which solidified upon standing. This was dissolved in ether-petroleum ether (ca. 1:2, 300 mL) and filtered. p-Tolyl disulfone (1.32 g) was collected: mp 215°C dec (lit. mp 212°C). The residue after removal of solvent was crystallized from acetone at –20°C to afford 29.5 g of white crystals in four crops, with concentration after each filtration. HCl (g) was then bubbled through the neat mother liquor for 10 min to effect epimerization at sulfur. A white solid separated out and was recrystallized from acetone to give an additional 12.2 g of product. The combined crops recrystallized from acetone to yield 39.11 g (66%) of pure (S)-(–)-menthyl p-toluenesulfinate in two crops: mp 103–105°C; $[\alpha]_D^{25} = -200.2$ (c 1.23, acetone).

2.1.4.2 General Procedure for the Preparation of Diastereomeric Alkane- and Arenesulfinates Derived from Diacetone-d-Glucose (DAG)[20]

Method A — To a solution of the chiral alcohol (1 mmol) and pyridine (1,2 equiv) in THF (5 mL) at –78°C, a solution of the corresponding sulfinyl chloride (1.2 equiv) was added dropwise with vigorous stirring. After being stirred for 1–3 h at –78°C, the reaction mixture was quenched with water and diluted with CH_2Cl_2. The organic phase was washed with 5% HCl solution, 2% $NaHCO_3$ solution, and saturated NaCl solution and dried over Na_2SO_4. After the solvent was removed *in vacuo,* the sulfinate obtained was purified by recrystallization or by flash chromatography.

Method B — The methodology is similar to that described for method A, but i-Pr_2NEt is used as the base and toluene as the solvent.

2.1.4.3 The Trifluoroacetic Acid-Catalyzed Alcoholysis of (+)-(S)-N,N-Diethyl p-Toluenesulfinamide 19a: General Procedure[24]

A dry, 10-mL, two-necked flask is charged with a solution of (+)-(S)-N,N-diethyl p-toluenesulfinamide (**19a**, 1 mmol) in 5 mL of the appropriate alcohol. To this solution trifluoroacetic acid (2 mmol) is added and progress of the reaction is followed by polarimetry. After the reaction is complete the reaction mixture is diluted with water (25 mL) and extracted with hexane (2 × 20 mL) and chloroform (20 mL). The combined organic extracts are dried over magnesium sulfate. The solvents are removed under reduced pressure to leave an oil. The product is purified by column chromatography on silica gel using 10:1 hexane/ethyl ether as eluent.

2.1.4.4 Synthesis of Chiral Sulfoxides

2.1.4.4.a Synthesis of Chiral Sulfoxides from O-Menthyl p-Toluenesulfinate (–)-(S)-1a: General Procedure with Benzene as a Solvent[35]

A solution of O-menthyl p-toluenesulfinate **1a** (1 equiv) in benzene (30 mL/10 mmol of **1**) was added dropwise to a benzene solution of Grignard reagent (2 equiv) at room temperature. After completion of the addition, the reaction mixture was stirred for 0.5 h at room temperature. Then, the reaction mixture was treated with aqueous ammonium chloride solution, saturated with sodium chloride, and extracted with chloroform. The combined organic layers were washed with 5% sodium carbonate solution and water and dried over magnesium sulfate. Evaporation of the solvents yielded an oil which was purified by chromatography on silica gel with hexane, hexane-ether, and chloroform as eluents.

2.1.4.4.b Synthesis of Optically Active Alkyl p-Tolyl Sulfoxides from (–)-(S)-1a with Ethyl Ether as a Solvent: (+)-(R)-Methyl p-Tolyl Sulfoxide[10,12]

To a solution of (–)-(S)-**1a** (10 g, 3 mmol) in diethyl ether (50 mL) at 0°C is added dropwise a solution of methylmagnesium iodide prepared from magnesium (2.16 g, 90 mmol) and methyl iodide (12.8 g, 90 mmol) in diethyl ether (70 mL). After the addition is completed, the mixture is stirred at room temperature for 2 h and then decomposed by addition of saturated ammonium chloride solution (50 mL). The organic layer is separated, dried over magnesium sulfate, and evaporated under reduced pressure. The crude product is purified by chromatography on silica gel or recrystallization: $[\alpha]_{589} = +145.5$ (acetone) mp 71–72°C.

2.2 CHIRAL SULFINAMIDES AND N-ALKYLIDENESULFINAMIDES

Like sulfinic esters, sulfinamides of the general structure $RS(O)NR^1R^2$ as well as N-alkylidene-sulfinamides having the general formula $R\text{-}S(O)N{=}CR^1R^2$ can exist depending on the nature of $R^{1'}$ and/or R^2 substituents, as enantiomers (if R^1 and R^2 are achiral) or as diastereomers (if R^1 and R^2 are chiral). Therefore, it is reasonable to divide the presentation of the synthetic routes to these two groups of sulfinyl derivatives and their application into two separate subchapters as in the case of sulfinates.

2.2.1 Diastereomeric Sulfinamides

The first diastereomeric sulfinamides **65** were obtained in the reaction of racemic benzenesulfinyl chloride **2b** with chiral N-methyl-2-phenylpropylamine **66** (Equation 2.2.1).[72] The mixture of diastereomers of **65** formed was separated into pure components by fractional crystallization.

$$\underset{\mathbf{2b}}{Ph\text{-}S(=O)\text{-}Cl} + \underset{\mathbf{66}}{MeNH\text{-}CH_2\text{-}\overset{*}{C}H(Me)\text{-}Ph} \longrightarrow \underset{\mathbf{65}}{Ph\text{-}\overset{*}{S}(=O)\text{-}N(Me)\text{-}CH_2\text{-}\overset{*}{C}H(Me)Ph} \quad (2.2.1)$$

It is interesting to note that the extent of asymmetric induction in this reaction is strongly temperature dependent. For example, the condensation carried out at 0°C yielded the mixture of diastereomeric sulfinamides **65** in a 3:1 ratio, whereas this ratio was 1:1 when the reaction was performed at –70°C.

In a similar way, diastereomeric N-α-methylbenzyl p-toluenesulfinamide **67a** was formed upon treatment of p-toluenesulfinyl chloride **2a** with (–)-α-methylbenzylamine **68a** (Equation 2.2.2).[73]

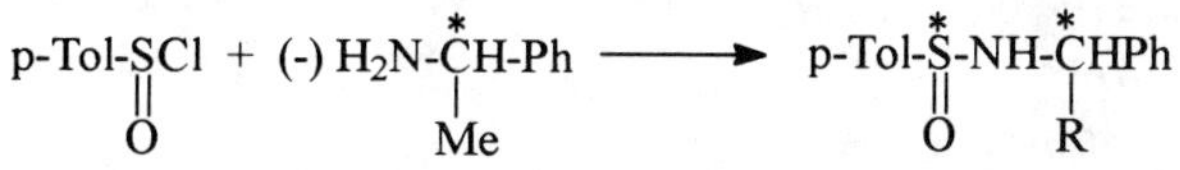

p-Tol-S(O)Cl + (-) H_2N-C*H(Me)-Ph ⟶ p-Tol-S*(O)-NH-C*H(R)Ph

2a **68a** **67a** R=Me (2.2.2)

An interesting asymmetric synthesis of diastereomeric sulfinamides **67b–e** involving the reduction of enantiomeric N-alkylidenesulfinamides is presented in Section 2.2.4. A few diastereomeric N-sulfinyloxazolidinones **69** and **70** were synthesized by sulfinylation of the metallated oxazolidinones **71** and **72** with the appropriate sulfinyl chlorides **2a–c** (Schemes 2.2.1 and 2.2.2).[74,75a]

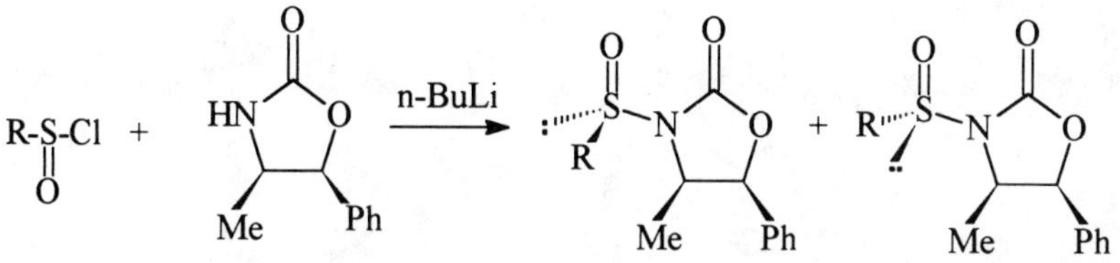

R-S(O)-Cl + **71** —n-BuLi→ **(R)-69a-c** + **(S)-69a-c**

2a-c

No	R	R/S ratio	Isolated yield [%]	
			R_S isomer	S_S isomer
a	Ph	77/23	61	4
b	p-Tol	66/34	69	1
c	t-Bu		41	28

Scheme 2.2.1

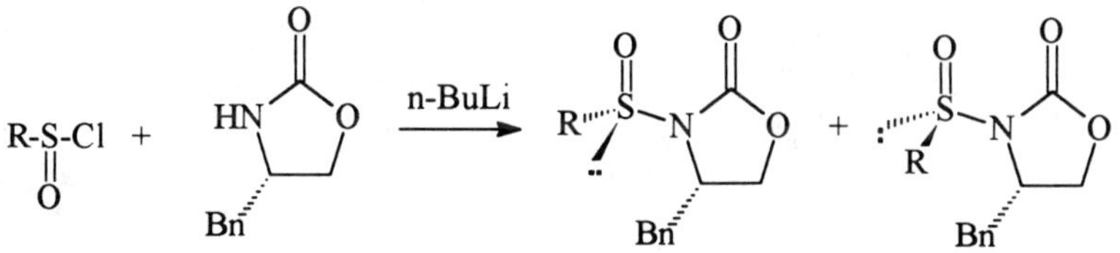

R-S(O)-Cl + **72** —n-BuLi→ **(S)-70a,b** + **(R)-70a,b**

2a,b

No	R	R/S ratio	Isolated yield [%]	
			S_S isomer	R_S isomer
a	Ph	67:23	50	20
b	p-Tol	68:22	61	9

Scheme 2.2.2

The condensation products were obtained as mixtures of diastereomers from which the major diastereomers (>99% ee) could be isolated after a single recrystallization or by flash or medium-pressure chromatography. Alternatively, the N-sulfinyloxazolidinones **70a,c,d** were prepared by oxidation of the N-(arylthio) and N-(alkylthio)oxazolidinones **73a–c** with m-chloroperbenzoic acid[74] (Scheme 2.2.3).

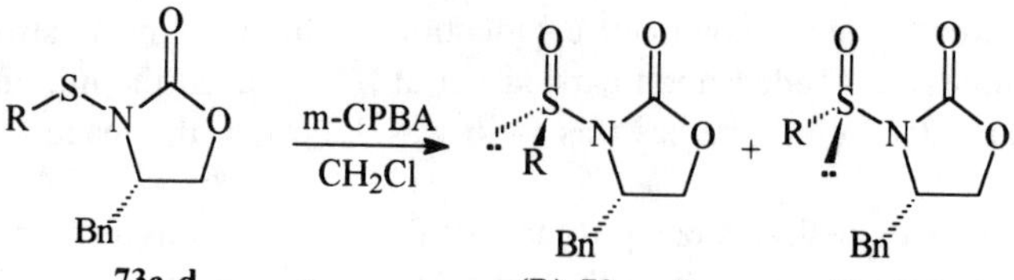

73a-d → (m-CPBA, CH_2Cl) → (R)-70a,c,d + (S)-70a,c,d

No	R	S/R ratio	Isolated yield [%] S_S isomer	R_S isomer
a	Ph	29:71	28	68
b	t-Bu	42:58	49	35
c	Me	42:58	33	

Scheme 2.2.3

Recently, diastereomerically pure, heterocyclic sulfinamide **74** was prepared by stereoselective cyclization of (–)-(R)-sulfinic acid **75** (Equation 2.2.3).[75b]

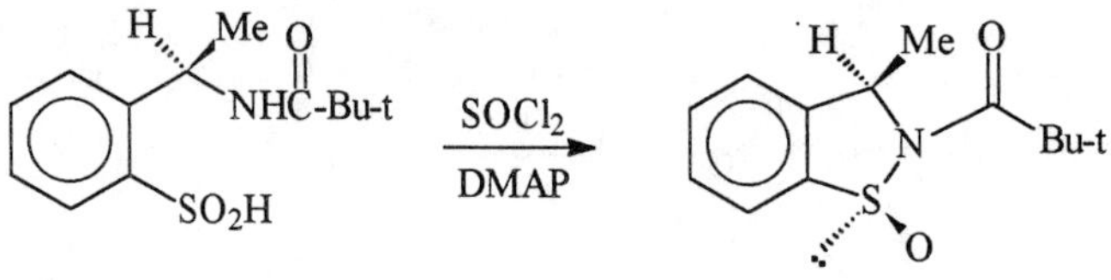

(-)-(R_C)-75 → ($SOCl_2$, DMAP) → (S_SR_C)-trans-74 (2.2.3)

2.2.2 Enantiomeric Sulfinamides

Most of the known enantiomeric sulfinamides were prepared by the reaction of diastereomerically pure O-menthyl arenesulfinates with organometallics containing the nitrogen-metal bond. Montanari and co-workers[76] showed that the reaction of (–)-(S)-menthyl p-toluenesulfinate **1a** with dialkylaminomagnesium halides carried out at 0°C in ethyl ether gives the corresponding optically active sulfinamides (+)-(S)-**19a–c** in yields around 60% (Equation 2.2.4 and Table 2.2.1). Very recently, Colonna et al.[77] used in this reaction primary and secondary lithium amides (Equation 2.2.3 and Table 2.2.1). They found that stereoselectivity of this conversion is strongly influenced by the nature of substituents connected with the nitrogen atom. For example, whereas the reaction of N-methyllithium anilide with (–)-(S)-**1** gave the corresponding sulfinamide **19d** with 5% ee only, N,N-dialkylaminolithium afforded the corresponding sulfinamide **19h** having optical purity as high as 93%.

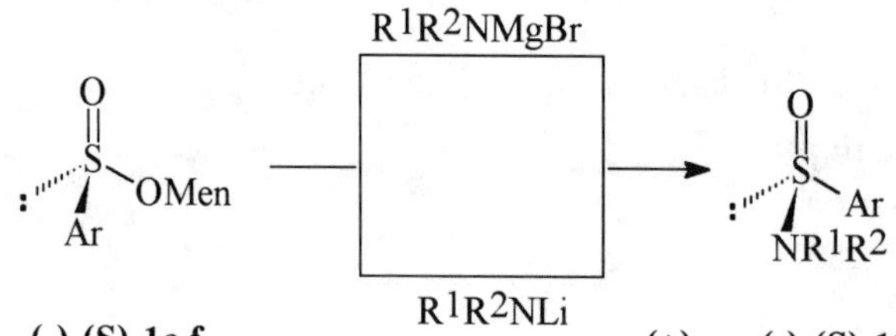

(-)-(S)-1a,f → (R^1R^2NMgBr or R^1R^2NLi) → **(+)- or (-)-(S)-19 or 76** (2.2.4)

a, Ar = p-Tol
f, Ar = 1-Naph

A different approach to highly stereoselective synthesis of the enantiomeric sulfinamides **77** is based on the reduction of optically pure methyl phenyl sulfoximine **78** by means of aluminum amalgam[81] (Equation 2.2.5).

Table 2.2.1 Synthesis of Optically Active Aryl Sulfinamides, ArS(O)NR^{1}R^2 **19** and **76** from Diastereomerically Pure O-Menthyl Arenesulfinates **1a,f** and Nitrogen-Containing Organometallic Reagents, R^{1}R^{2}NX

Sulfinate			Sulfinamide 19 and 76						
Ar	$[\alpha]_{589}$	X	No	R^1	R^2	Yield (%)	$[\alpha]_{589}$	ee (%)	Ref.
p-Tol	-196.0	MgBr	**19a**	Me	Me	~60	+157.0	~100	76
	–196.0	MgBr	**19b**	Et	Et	~60	+110.0	~100	76
	–196.0	MgBr	**19c**	i-Pr	i-Pr	~60	+205.0	~100	76
	–202.0	Li	**19d**	Ph	Me	27	–5.64	4	77
	–202.0	Li	**19e**	i-Pr	Me	26	+85.9	31	77
	–202.0	Li	**19f**	Allyl	Me	62	+57.3	45	77
	–202.0	Li	**19g**	Allyl	Ph	91	–164.0	a	77
	–202.0	Li	**19h**	Allyl	Allyl	89	+49.2	93	77
	–200.6	MgBr	**19d**	Ph	Me	70	–109.3	98	73
	–202.0	Li	**19i**	Ph	H	60	+199.0	89	77
	–197.3	Li	**19i**	Ph	H	41	+216.9	100	73
	–202.0	Li	**19j**	i-Pr	H	66	+167.4	100	73
	–202.0	Li	**19k**	Allyl	H	55	+145.5	100	77
		MgBr		$(CH_2)_5$					78
		Li		c-C_6H_{11}	H				79
1-Naph	–426.5	Li	**76**	1-Naph	H	34	+561.0	a	80

[a] Not given.

Ph-S*(=O)(=NR)-Me **78** —(Al-Hg)→ Ph-S*(=O)NHR **77**

78
a, R=Me
b, R=H
c, R=Tos

(2.2.5)

Optically active N,N-dimethyl p-toluenesulfinamide **19a** was obtained in a stereoselective way in the reaction of dimethyl fumarate with the ylide **79** prepared from dimethylamino-methyl-p-tolueneoxosulfonium tetrafluoroborate[82] (Equation 2.2.6).

p-Tol-S*(=O)(NMe_2)$^{\oplus}$-$CH_2^{\ominus}$ **79** —(CO_2Me-CH=CH-CO_2Me)→ p-Tol-S*(=O)-NMe_2 (+)-(S)-**19a**

(2.2.6)

2.2.3 Enantiomeric N-Alkylidenesulfinamides

Optically active N-alkylidenesulfinamides **80a–e** were obtained for the first time by treatment of the sulfinate (–)-(S)-**1a** with the imino-Grignard reagents generated from alkyl or aryl Grignard reagents and benzonitrile[83] (Equation 2.2.7 and Table 2.2.2).

They were formed in moderate to good yields and with very high stereoselectivity. An improved procedure consisting in treating benzonitrile with alkyllithium in ether at 0°C followed by addition of (–)-(S)-**1a** at 0°C allowed to obtain chiral **80a** and **80f** in higher yields.[84a] The absolute configuration at the sulfinyl sulfur in **80** was assigned by assuming that the reaction described by the Equation 2.2.7 proceeds with inversion of configuration as it had been established for the reaction of Grignard reagent with sulfinate esters.

PhCN —RMgX(RLi)→ R(Ph)C=NMgX(Li) + (−)-(S)-**1a** → (S)-**80a-f**

(2.2.7)

Table 2.2.2 Synthesis of Optically Active N-Alkylidene Sulfinamides, p-TolS(O)N=CPhR **80a–f** from (–)-(S)-**1a** and Iminoorganometallic Reagents Ph(R)C=NX

No	R	X	Yield	$[\alpha]_{589}$	Abs. conf.	Ref.
a	Me	MgX	20	–98.0	S	83
a	Me	Li	50	[a]	S	84
b	Et	MgBr	52	+26.0	S	83
c	i-Pr	MgBr	27	–288.0	S	83
d	n-Bu	Li	75	[a]	S	84a
e	Ph	MgBr	70	–56.2	S	83
f	1-Naph	MgBr	70	+71	S	83

[a] Not given.

Two other procedures for the preparation of enantiomerically pure alkylidenesulfinamides **81** from the sulfinate (–)-**1a** as a substrate were described by Davis and co-workers.[85] The first of them (A) is based on the reaction between the sulfinate (–)-(S)-**1a** and the ate complex **82**, prepared *in situ* by treatment of aryl(alkyl)cyanide with diisobutylaluminumhydride (DIBAL) and methyllithium (Equation 2.2.8). The second method (B) involves a "one-pot" reaction of lithium bis(trimethylsilyl)amide (LiHMDS) with (–)-(S)-**1a** to give an intermediate sulfinamide **83** which, upon treatment with an excess of aldehyde and cesium fluoride, affords **81a–f** in good to excellent yields (Equation 2.2.8). The alkylidenesulfinamides **81** prepared according to the above discussed procedures are collected in Table 2.2.3.

(A): (-)-(S)-**1a** + $[RCH{=}N{-}Al(Me){-}i{-}Bu_2]^-Li^+$ (**82**) → (+)-(S)-**81a-f**

(B): (-)-(S)-**1a** —$[(Me_3)_2Si]_2NLi$→ p-TolS(O)$N(SiMe_3)_2$ (**83**) —RCHO, CsF→ (+)-(S)-**81a-f**

a, R=Ph
b, R=p-$MeOC_6H_4$
c, R=o-$MeOC_6H_4$
d, R=E-PhCH=CH
e, R=E-MeCH=CH
f, R=n-Bu

(2.2.8)

Table 2.2.3 Synthesis of N-Alkylidenesulfinamides, p-Tol-S(O)N = CHR-**81**, from (–)-(S)-1a and Nitriles and Aldehydes

			N-Alkylidenesulfinamide 81[a]		
Nitrile/aldehyde	**Method**	**Symbol**	**Yield [%]**	**$[\alpha]_{589}$ ($CHCl_3$)**	**ee [%]**
PhCN	A	**81a**	36	+119.8	>95
PhCHO	B	**81a**	82	+117.0	>95
p-$MeOC_6H_4CN$	A	**81b**	37	+41.1	>95
p-$MeOC_6H_4CHO$	B	**81b**	90	+37.9	>95
o-$MeOC_6H_4$	B	**81c**	81	+362.9	>95
E-PhCH = CH-CN	A	**81d**	42	+336.0	>95
E-MeCH = CHCN	A	**81e**	33	+610.7	>95
n-Bu-CHO	B	**81f**	30	+308.8	>95

[a] All sulfinamides **81** have the (S) absolute configuration.

From Davis, F. A., Reddy, R. E., Szewczyk, J. M., and Portonovo, P. S., *Tetrahedron Lett.*, **34,** 6229, 1993. Reprinted with kind permission from Elsevier Science Ltd, The Boulevard, Langford Lane, Kidlington OX5 1GB, UK.

The N-alkylidenesulfinamides **84** were also prepared by asymmetric oxidation of the corresponding N-alkylidenesulfenamides **85** with (+) or (–) N-(phenylsulfonyl)(3,3-dichlorocamphoryl)oxaziridine **86** (Scheme 2.2.4 and Table 2.2.4).[86]

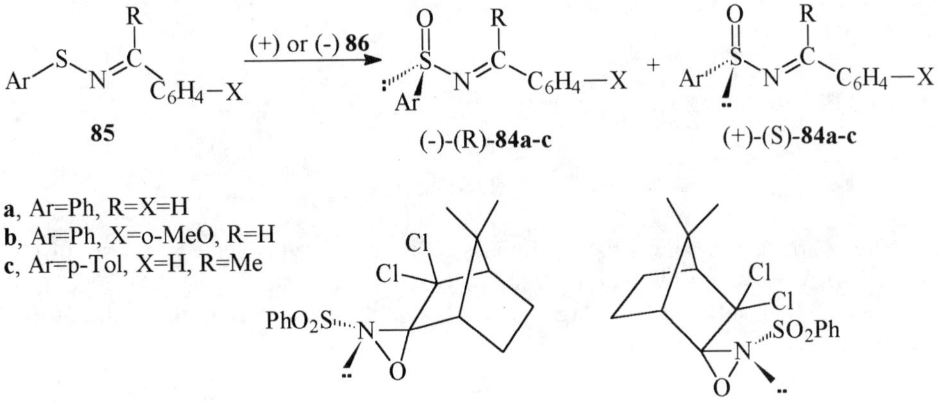

a, Ar=Ph, R=X=H
b, Ar=Ph, X=o-MeO, R=H
c, Ar=p-Tol, X=H, R=Me

Scheme 2.2.4

Table 2.2.4 Asymmetric Oxidation of Sulfenimides **85** to Sulfinamides **84** by Chiral Oxaziridine **86**

		Sulfinamide 84				
Sulfenimide	**Oxaziridine**	**No**	**Yield [%]**	**$[\alpha]_{589}$ ($CHCl_3$)**	**ee [%]**	**Abs. conf.**
85a	(–)-**86**	**84a**	55	–112.0	88	R
85a	(+)-**86**	**84a**	72	+112.0	88	S
85b	(+)-**86**	**84b**	89	–300.5	85	R
85c	(–)-**86**	**84b**	90	–85.9	87	R

From Davis, F. A., Reddy, R. T., and Reddy, R. E., *J. Org. Chem.*, 57, 6387, 1992. Copyright 1992 American Chemical Society. With permission.

2.2.4 Synthetic Application of Chiral Sulfinamides and N-Alkylidenesulfinamides

The use of optically active enantiomeric N,N-diethyl p-toluenesulfinamide **19b** and N,N-diisopropyl analog **19c** for the preparation of optically active enantiomeric sulfinic esters has already been discussed in Section 2.1.2. However, the chiral N-sulfinyloxazolidinones **69** and **70** are better substrates for nucleophilic substitution at sulfur than the sulfinamide **19b** because the cyclic oxazolidinone moiety is an excellent leaving group. For instance, methyl and isopropyl p-toluenesulfinates **20a,e** were obtained by reacting the corresponding lithium alkoxides with (S)-**70b** in the presence of an excess of alcohol (Equation 2.2.9) in 80 and 92% yields, respectively, and with enantiomeric purities above 95%.[74] The N-sulfinyloxazolidinones **70a,b** are also efficient reagents for the synthesis of other classes of chiral sulfinyl derivatives[74] (Equations 2.2.10 and 2.2.11).

ROLi, ROH

(-)-(S)-20, **a**, R=Me; **e**, R=i-Pr (2.2.9)

(S)-70a,b
a, Ar=Ph
b, Ar=p-Tol

Et_2NMgBr

(-)-(R)-19b
Yield 91%, ee>98% (2.2.10)

Zn(O)/40°C

87
Yield 81%, ee>98% (2.2.11)

Especially useful are N-sulfinyloxazolidinones **69b,c** and **70a–d** in the synthesis of chiral sulfoxides.[74,75] They react rapidly with a variety of Grignard reagents to give the corresponding aryl alkyl and dialkyl sulfoxides in high yields (78–92%) and enantioselectivities (>90%) (Equations 2.2.12 and 2.2.13 and Table 2.2.5).

R^1MgX, -78°C/ether

70a-d (2.2.12)

R^1MgX, rt, ether

69b,c (2.2.13)

The first example of the use of chiral N-alkylidenesulfinamide **80** in asymmetric synthesis involves their highly efficient asymmetric reduction which occurs when these compounds are treated with $LiAlH_4$.[83] The diastereomeric sulfinamides **67a** and **88b,c** produced were easily cleaved to

Table 2.2.5 Synthesis of Chiral Sulfoxides R, S(O)R[1], from Diastereomeric N-Sulfinyloxazolidinones **69b,c** and **70b–d** and Grignard reagents R^1MgX

N-Sulfinyloxazolidinone	R^1MgX	Sulfoxides				
		Yield (%)	$[\alpha]_{589}$	ee [%]	Abs. conf.	Ref.
(S)-**70b** (R = p-Tol)	Me	90	–132	99	S	74
	Et	90	–204	98	S	74
	i-Pr	91	–181	97	S	74
	t-Bu	88	–185	97	S	74
	$PhCH_2$	86	–213	99	S	74
(S)-**70c** (R = Me)	Ph	87	+120	90	R	74
	t-Bu	78	–7.3	93	R	74
	$PhCH_2$	82	+50	91	R	74
	C_8H_{17}	78	–79.7	100	R	74
(S)-**70d** (R = t-Bu)	Me	92	+7.8		S	74
	n-Bu	91	–129	100	S	74
(R)-**69b** (R = p-Tol)	E-PhCH = CH	E-66	+120	[a]	R	75
		Z-20	[a]	[a]	R	
(R)-**69c** (R = t-Bu)	E-PhCH = CH	E-61	[a]	[a]	R	75
		Z-8	[a]	[a]	R	

[a] Not given.

chiral α-alkylbenzylamines **68a–c** by treatment with methanol in the presence of trifluoroacetic acid (Scheme 2.2.5).

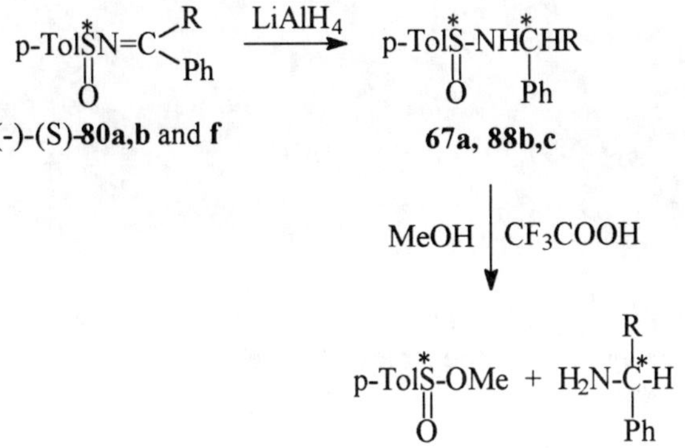

p-TolS*(=O)N=C(R)Ph —$LiAlH_4$→ p-TolS*(=O)-NHC*HR(Ph)

(-)-(S)-**80a,b** and **f** → **67a, 88b,c**

↓ MeOH | CF_3COOH

p-TolS*(=O)-OMe + H_2N-C*H(R)(Ph)

68a-c

Scheme 2.2.5

Recently, N-alkylidenesulfinamides **80a** and **80d** were also stereoselectively reduced with DIBAL in THF at –30°C.[84] Thus, **80a** gave upon reduction **67a** in 92% yield and with a diastereomeric ratio of 96:4. From **80d, 88d** was obtained in 96% yield with a diastereomeric ratio of 94:6. The diastereomerically pure sulfinamides **67a** and **88c** were separated on a silica gel column using hexane-ether mixtures as eluent and then converted into optically active (S)-amines **68a** and **68c**.[84]

In addition to asymmetric reduction, N-alkylidenesulfinamides undergo addition reactions of nucleophilic reagents. For example, allylmagnesium bromide and **80a** and **80d** produce diastereomeric sulfinamides **89a,b** with the S-absolute configuration at the sulfinyl sulfur atom. A single diastereomer (+)-(S_SR_C)-**89a** was isolated in a 98% yield from **80a** while diastereomers of **89b** formed from **80d** were readily separated by silica gel chromatography (Scheme 2.2.6), giving (+)-(S_S,R_C)-**89b** in 84% yield.

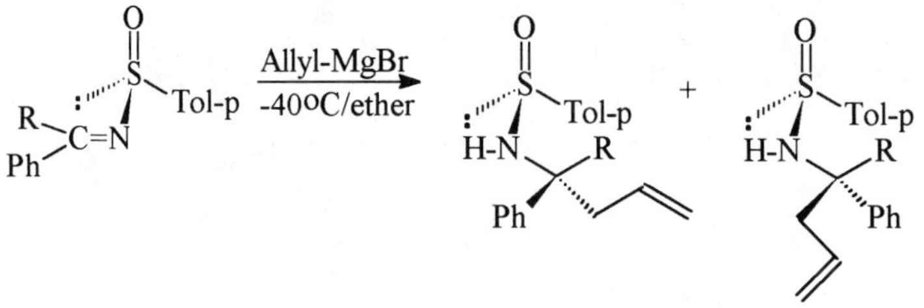

O ‖ S Tol-p; R, Ph C=N → Allyl-MgBr, -40°C/ether → O ‖ S Tol-p, H-N, R, Ph (allyl) + O ‖ S Tol-p, H-N, R, Ph (allyl)

(+)-(S_S, R_C)-**89** (+)-(S_S, S_C)-**89**

a, R=Me, $[\alpha]_{589}$=+43.2 -

b, R=n-Bu $[\alpha]_{589}$=+59.2 $[\alpha]_{589}$=+1.1

Scheme 2.2.6

The diastereomerically pure **89a** was converted into enantiomerically pure allylamine **90a** from which the known (+)-(R)-2-phenylbutylamine **91** was obtained as shown in Scheme 2.2.7.

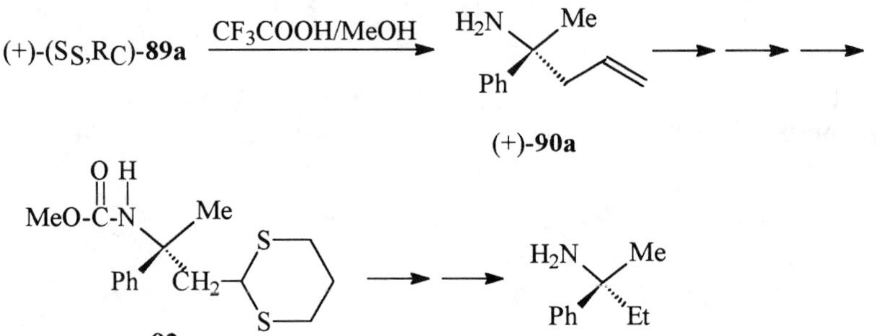

(+)-(S_S,R_C)-**89a** → $CF_3COOH/MeOH$ → H_2N, Me, Ph (allyl) → → →

(+)-**90a**

O H, MeO-C-N, Me, Ph, CH_2-(1,3-dithiane, S, S) → → H_2N, Me, Ph, Et

92

(+)-(R)-**91,** $[\alpha]_{589}$=+18 ($CHCl_3$)

Scheme 2.2.7

The enantiomerically pure allyl amines (+)-(R)-**90a,b** prepared as presented above were readily converted into β- and γ-amino acids (–)-**93** and (+)-**94** (Schemes 2.2.8 and 2.2.9).[84a]

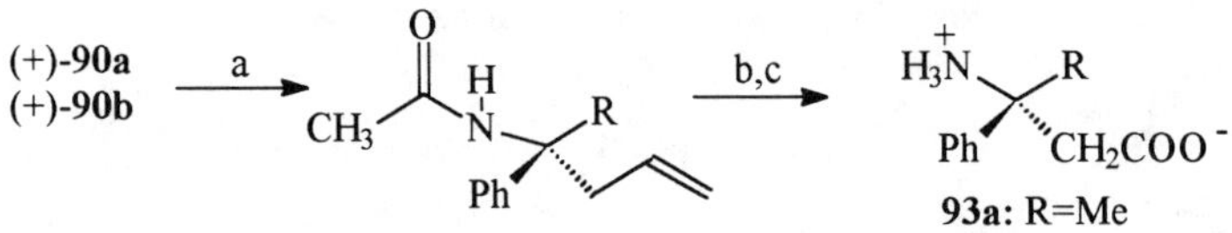

(+)-**90a**
(+)-**90b**
→ a → O, CH_3, H, N, R, Ph (allyl) → b,c → $H_3\overset{+}{N}$, R, Ph, CH_2COO^-

95a: R=Me
95b: R=n-Bu

93a: R=Me
93b: R=n-Bu

[a] (a) Ac_2O, Et_3N, ether, 0°C, 1 h (R = CH_3, 95% yield; R = n-Bu, 93% yield); (b) (i) O_3, CH_2Cl_2 –78°C; (ii) $AgNO_3$, KOH, EtOH, 0°C (R = CH_3, 70% yield; R = n-Bu, 69% yield); (c) (i) 1 N HCl, H_2O, 100°C, 12 h; (ii) NH_4OH, Rexyn-101 (H^+) (R = CH_3, 80% yield; R = n-Bu, 87% yield).

Scheme 2.2.8

95a, 95b —(a-c)→ $H_3\overset{+}{N}$(R)(Ph)C$CH_2CH_2COO^-$

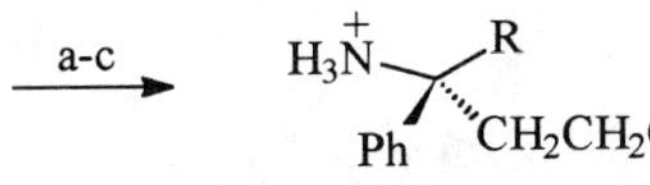

94a: R=Me
94b: R=n-Bu

[a] (a) (i) BH_3·THF, 0°C, 3h; (ii) NaOH, 30% H_2O_2 (R = CH_3, 55% yield; R = n-Bu, 60% yield); (b) (i) pyridinium chlorochromate, CH_2Cl_2, 25°C, (ii) $AgNO_3$, KOH, EtOH, 0°C (R = CH_3, 65% yield; R = n-Bu, 80% yield); (c) (i) 1 N HCl, H_2O, 100°C, 12 h; (ii) NH_4OH; Rexyn-101 (H^+) (R = CH_3 81% yield; R = n-Bu, 85% yield).

Scheme 2.2.9

Another interesting asymmetric synthesis of β-amino acids **94a,b** involves addition of enolates to chiral N-alkylidenesulfinamides **84a** and **84b**.[86] Thus, addition of lithium methyl acetate **96** (1.5 equivalent) to the carbon-nitrogen double bond in these amides gave sulfinamides **97a** and **b**. The first of them was obtained after flash chromatography as a 9:1 mixture of diastereomers from which on crystallization from n-hexane the diastereomerically pure (R_S,3S)-**97a** was isolated in 74% yield. The sulfinamide (R_S,3S)-**97b** was obtained diastereomerically pure in 90% yield. These sulfinamides were then converted, without epimerization, to β-amino acids (3S)-**98a** and (3S)-**98b** with 4 equivalents of CF_3COOH/MeOH (Scheme 2.2.10).[86,87]

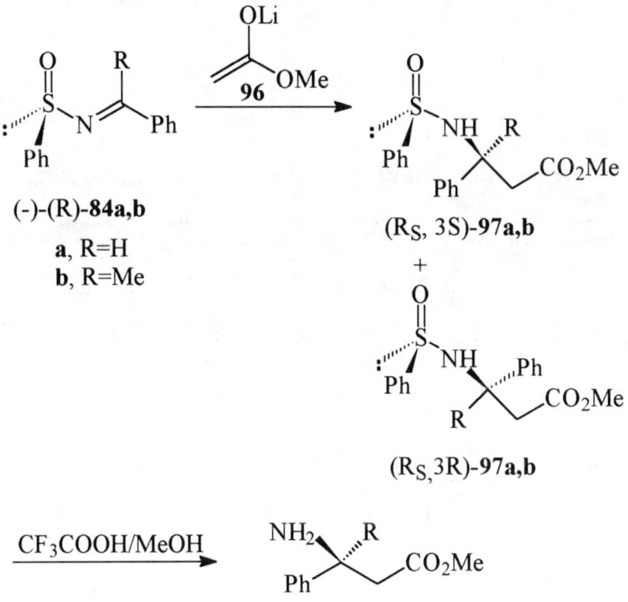

(-)-(R)-**84a,b**
a, R=H
b, R=Me

—(**96**: OLi, OMe)→

(R_S, 3S)-**97a,b**
+
(R_S,3R)-**97a,b**

—(CF_3COOH/MeOH)→

(-)-(3S)-**98a,b** (ee>95%)
98a, 73% yield
98b, 85% yield

Scheme 2.2.10

A convenient synthesis of optically active α-amino acids **102** is based on a highly stereoselective addition of diethylaluminum cyanide to optically pure N-alkylidene-sulfinamides **81** and **99** (Scheme 2.2.11).[88] The major diastereomers of α-aminonitriles **100** and **101** were isolated in high yields by chromatography on silica gel. They were gently refluxed in 6 N HCl for 6–8 h followed by washing with ether and passing the aqueous solution through a Dowex® 50X-8–100 ion exchange resin to give (+)-(S)-phenylglycine **102a**, (+)-(S)-norvaline **102b**, and (+)-(S)-leucine **102c** in >95% ee and 71–81% isolated yield (Scheme 2.2.11 and Table 2.2.6).

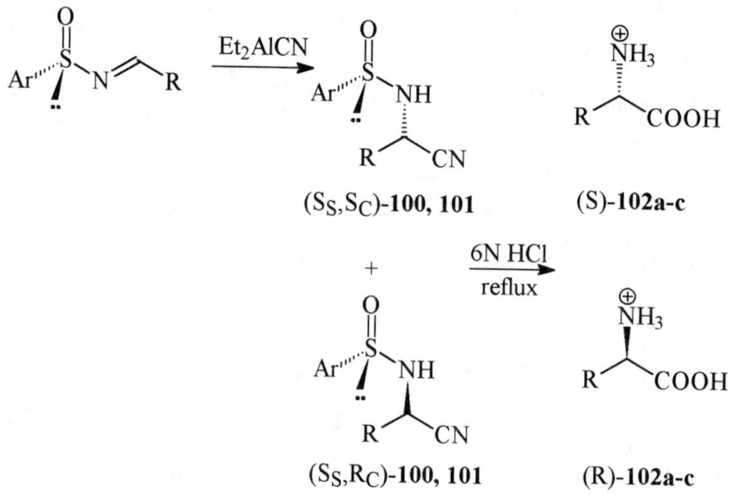

Et_2AlCN

(S_S,S_C)-**100, 101**

(S)-**102a-c**

+

6N HCl

reflux

(S_S,R_C)-**100, 101**

(R)-**102a-c**

(S)-**81**, Ar = p-Tol
(S)-**99**, Ar = 2-Methoxy-1-naphthyl
For **81 a**, R = Ph, **f**, R = n-Pr, **g**, R = i-Bu
For **100**, Ar = p-Tol
For **101**, Ar = 2-Methoxy-1-naphthyl
For **100, 101, 102, a**, R = Ph, **b**, R = n-Pr, **c**, R = i-Bu

Scheme 2.2.11

Table 2.2.6 Stereoselective Addition of Diethylaluminum Cyanide to N-Alkylidenesulfinamides **81** and **99**

	α-Amino nitrile 100, 101		α-Amino acid 102			
N-Alkylidenesulfinamide	**Yield [%]**	**$[S_S,S]$ $[S_S,R]$**	**R**	**No**	**Yield [%]**	**ee (cont.)**
(+)-(S)-**81a**	72	70:30	Ph	**a**	71	>95 (S)
(+)-(S)-**81f**	67	69:31	n-Pr	**b**	79	>95 (S)
(+)-(S)-**81g**	62	71:29	i-Bu	**c**	67	>95 (S)
(+)-(S)-**99a**	78	80:20	Ph	**a**	81	>95 (S)
(+)-(S)-**99b**	75	83:17	n-Pr	**b**	73	>95 (S)
(+)-(S)-**99c**	72	83:17	i-Bu	**c**	71	>95 (S)

From Davis, F. A., Reddy, R. E., and Portonovo, P. S., *Tetrahedron Lett.*, **35,** 9351, 1994. Reprinted with kind permission from Elsevier Science Ltd, The Boulevard, Langford Lane, Kidlington OX5 1GB, UK.

A similar approach to the asymmetric synthesis of optically active α-amino acids via chiral N-alkylidenesulfinamides utilizes (–)-(R)-N[1-(triethoxymethyl)ethylidene]-p-toluenesulfinamide **103** as a substrate.[84b] Its reduction with various reducing agents gave the corresponding sulfinamide **104** with a very high stereoselectivity. A six-membered ring transition state **105** was proposed to explain the exclusive formation of the (R_C,R_S)-**104** diastereomer with 9-borabicyclo[3.3.1]nonane (9-BBN). The absolute configuration of the new chiral center in **104** was determined through ethanolysis of (R_C,R_S)-**106** (prepared by the silica gel-catalyzed deblocking of (R_C,R_S)-**104**) which gave (R)-alanine ethyl ester-**107** (Scheme 2.2.12).

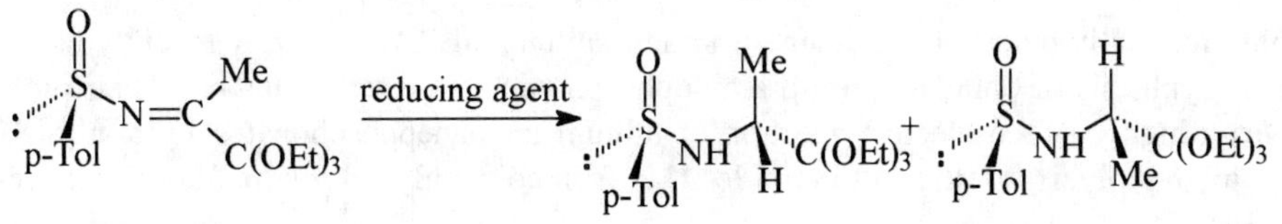

(R)-**103** —reducing agent→ (R_C,R_S)-**104** + (S_C,R_S)-**104**

	(R_C,R_S)-**104**	(S_C,R_S)-**104**
$LiAlH_4$	64.5% yield	21.5% yield
DIBAL-H	64.5% yield	18.5% yield
9-BBN	95% yield	none

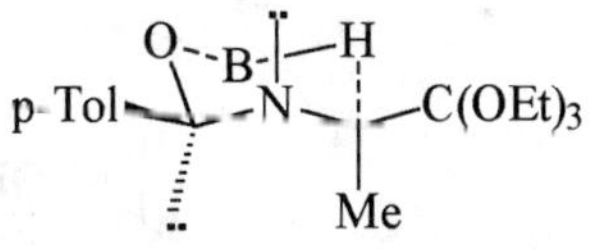

105

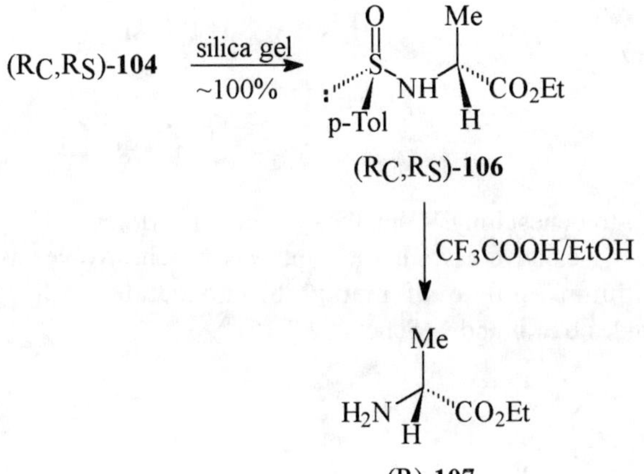

(R_C,R_S)-**104** —silica gel, ~100%→ (R_C,R_S)-**106** —CF_3COOH/EtOH→ (R)-**107**

Scheme 2.2.12

N-Alkylidenesulfinamide **103** also undergoes a stereoselective addition with allylmagnesium bromide to give the sulfinamide **108** in 95% yield (based on 48% recovery of **103**).[84b] The chirality of the newly formed carbon center in **108** was determined by converting it into (S)-2-amino-2-methyl-4-butenoic acid **109**, by a two-step sequence shown in Scheme 2.2.13.

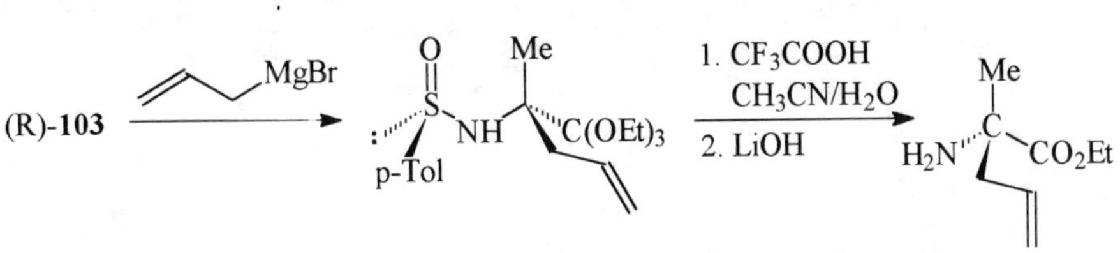

(R)-**103** —allyl-MgBr→ **108** —1. CF_3COOH, CH_3CN/H_2O; 2. LiOH→ (S)-**109**

Scheme 2.2.13

The enantiomerically pure N-benzylidene-p-toluenesulfinamide **81a** was very recently used in the asymmetric synthesis of optically pure β-aminophosphonic acid esters **110a–c**.[89] Their preparation involves a highly stereoselective addition of lithium methanephosphonates **111a–c** to **81a** followed by methanolysis of the sulfinamide **112** formed catalyzed by trifluoracetic acid (Scheme 2.2.14).

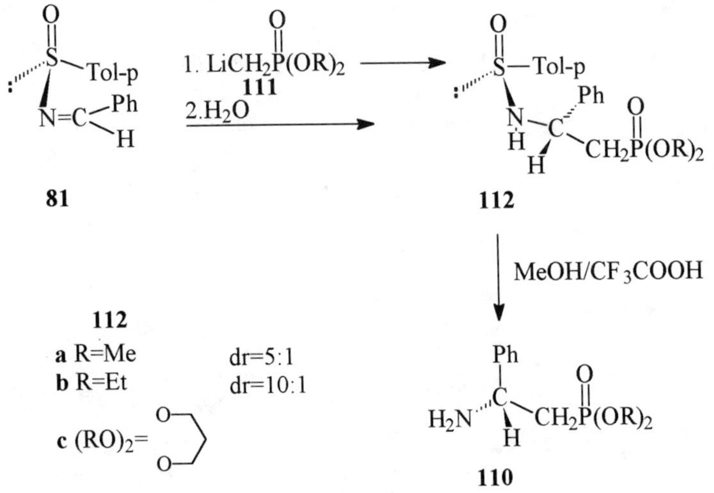

1. $LiCH_2P(O)(OR)_2$ **111**
2. H_2O

81 → **112** → ($MeOH/CF_3COOH$) → **110**

112
a R=Me dr=5:1
b R=Et dr=10:1
c $(RO)_2$=

Scheme 2.2.14

A few of cis-N-(p-toluenesulfinyl)-2-methoxycarbonylaziridines **113** were prepared by simple one-pot, highly diastereoselective asymmetric synthesis which involves as a key step a Darzens-type reaction of lithium enolate of methyl bromoacetate with enantiomerically pure N-alkylidenesulfinamides **81a,b** and **f** (Scheme 2.2.15).[90]

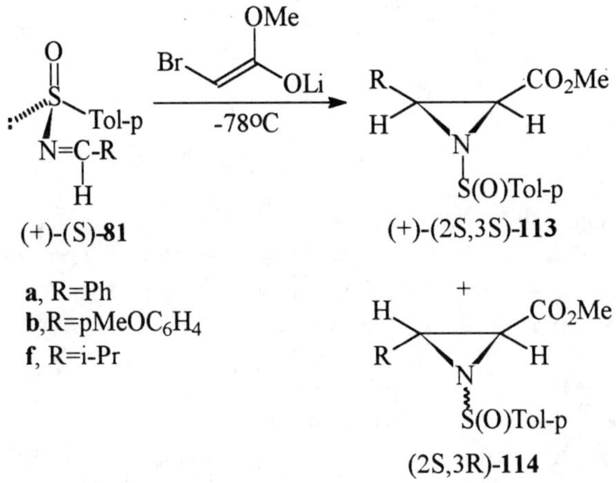

(+)-(S)-**81** → (-78°C) → (+)-(2S,3S)-**113** + (2S,3R)-**114**

a, R=Ph
b,R=pMeOC_6H_4
f, R=i-Pr

No	R	**113:114** ratio	**113** Yield [%]	**113** $[\alpha]_D$ ($CHCl_3$)
a	Ph	97:3	65	+51.4
b	p-MeO-C_6H_4	99:1	74	+26.4
c	i-Pr	99:1	64	+110.7

Scheme 2.2.15

The aziridine-2-carboxylates **113a,c** were easily oxidized to the corresponding N-sulfonyl derivatives **115a,c** with 1,5 equivalents of MCPBA. Transfer hydrogenation of **115a** for 3 h afforded the (S)-phenylalanine derivative **116a** and the acid-catalyzed hydrolysis at 100°C gave the phenylserine derivative **117a** as an 84:16 mixture of diastereomers. On the other hand, heating **113a** at 45°C in 50% aqueous TFA afforded a 71% yield of syn-β-phenylserine derivative **118a** as a 93:7 mixture of diastereomers. Under similar conditions the aziridine **113c** gave only the 1H-aziridine **119c** (Scheme 2.2.16).

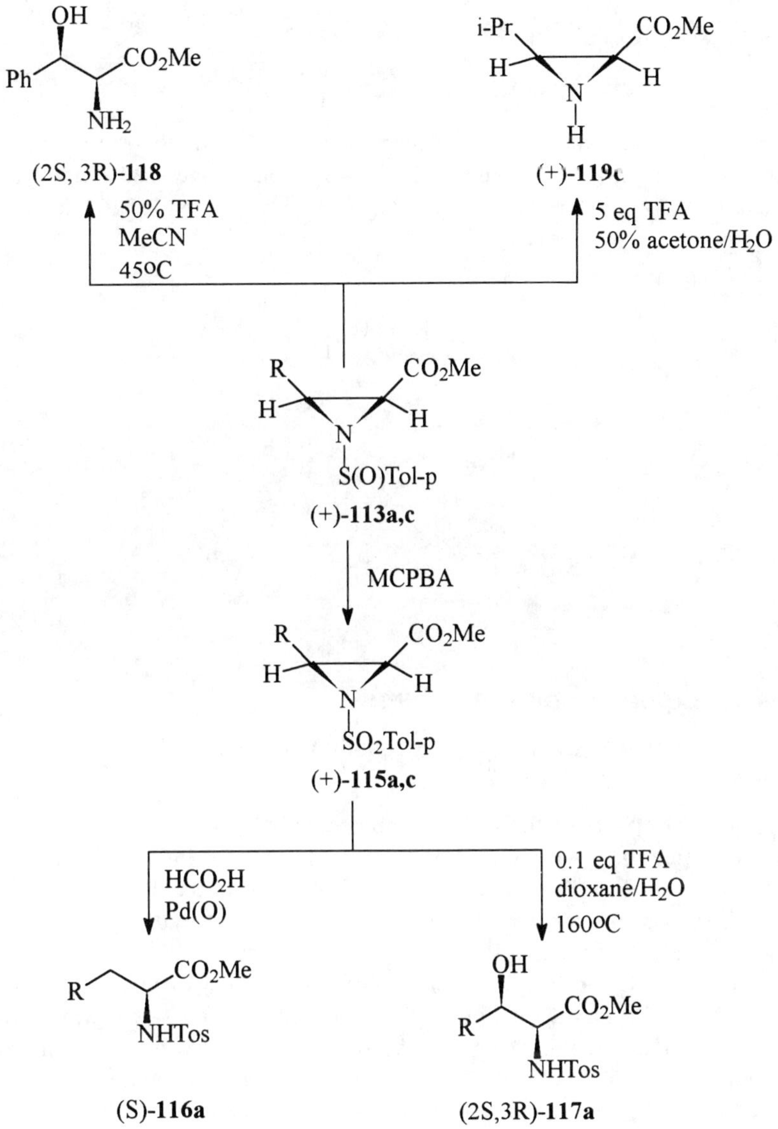

(2S, 3R)-**118**

50% TFA
MeCN
45°C

(+)-**119c**

5 eq TFA
50% acetone/H_2O

S(O)Tol-p
(+)-**113a,c**

MCPBA

SO_2Tol-p
(+)-**115a,c**

HCO_2H
Pd(O)

0.1 eq TFA
dioxane/H_2O
160°C

(S)-**116a**

(2S,3R)-**117a**

Scheme 2.2.16

A new asymmetric synthesis of 2H-azirines **120** and **121** is based on the reaction of the cis-(N-p-toluenesulfinyl)aziridines **113** and **122** with lithium diisopropylamide (LDA) followed by the addition of methyl iodide and hydrolysis (Equations 2.2.14 and 2.2.15).[91]

(2S,3S)-**113** → (1. LDA, 2. MeI, 3. H_2O, THF) (+)-(S)-**120** (~50%) + p-Tol-S(O)-Me (~80%)

(2.2.14)

122 → (1. LDA, 2. MeI, 3. H_2O, THF) **121**

(2.2.15)

The reaction of optically active N-alkylidenesulfinamide **81a** with dimethyloxosulfonium methylide or dimethylsulfonium methylide yields a mixture of N-sulfinylaziridines **123** epimers at C-2 with a moderate stereoselectivity (highest **123a**:**123b** ratio 67:33)[92] (Equation 2.2.16).

81a → ($Me_3\overset{\oplus}{S}(O)_nI^{\ominus}$/Base) **123a** + **123b**

n=0,1

(2.2.16)

2.2.5 Examples of Experimental Procedures

2.2.5.1 p-Toluenesulfinylation of the (4R,5S)-Norephedrine-Derived Oxazolidin-2-one 71; Synthesis of 69b[74]

To a solution of 6.75 g (38.1 mmol) of the (4R, 5S)-norephedrine-derived oxazolidin-2-one in 50 mL of THF under nitrogen at 0°C was added dropwise 18.1 mL (2.0 *M* in hexane, 36.2 mmol) of n-butyllithium over a 15-min period. The resultant suspension was stirred at 0°C for 10 min and then cooled to –78°C. A suspension of 53.4 mmol (1.4 equiv) of p-toluenesulfinyl chloride was added over a 2-min period (as a slurry with NaCl) to the enolate suspension by a dropping funnel. The reaction was stirred at –78°C for 10 min, quenched by the addition of 80 mL of saturated sodium bicarbonate, and diluted with 120 mL of ethyl acetate. The aqueous layer was extracted with 50 mL of ethyl acetate, and the combined organic layers were washed sequentially with 100 mL of saturated aqueous ammonium chloride and 100 mL of saturated aqueous sodium chloride. The organic layer was dried ($MgSO_4$) and the solvent removed *in vacuo* to leave 12.74 g of a white solid. Analysis of HPLC (4.6 mm × 15 cm Du Pont Zorbax®, 5 mm silica gel, isooctane/isopropyl alcohol (97:3), 2.0 mL/min) afforded a 4.6:1 ratio of **69a** (R) (R_t = 4.4 min) to a **69b** (S) (R_t = 4.3 min). The reaction mixture was triturated with 300 mL of diethyl ether, and the solid material collected proved to be 98% diastereomerically pure **69a** (R). It was recrystallized from hexane/ethyl acetate to give pure **69a** (7.44 g, 65%). The residues derived from filtration and recrystallization were collected and purified by flash chromatography (7 × 15 cm, hexane/isopropyl alcohol [95:5], 125 mL fraction): a 83:17 mixture of **69a** (R) and **69b** (S) was first eluted (2.06 g). This mixture was chromatographed (gradient MPLC, from hexane [100%] to hexane/isopropyl alcohol [9:1], linear gradient, 4.7- × 45-cm column, flash silica gel, 50-mL fractions) to give only one pure fraction of **69b** (S) as a colorless oil which solidified upon standing (162 mg). A mixture enriched in the diastereomer **69b** (R) eluted second, affording after recrystallization from hexane/ethyl acetate 767 mg of pure crystalline **69b** (R). The overall yield of **69b** (R) was 69%.

2.2.5.2 Synthesis of Optically Active N,N-Dialkyl p-Toluenesulfinamides 19a–d: General Procedure[76,78]

Secondary amine (30 mmol) in ether (20 mL) was added to a stirred solution of Grignard reagent prepared from n-PrBr (30 mmol) and magnesium (30 mmol) in ether (50 mL) at room temperature. The mixture was kept at room temperature for 30 min and then was cooled to 0°C. At this temperature a solution of (–)-(S)-**1a** (10 mmol) in ether/THF was added dropwise. After an appropriate time the mixture was hydrolyzed with saturated ammonium chloride. The organic phase was separated and the aqueous phase extracted with diethyl ether (3 × 50 mL). The combined organic extracts were washed with 5% aqueous sodium bicarbonate solution, dried over $MgSO_4$, and evaporated under reduced pressure. The crude product was purified by column chromatography or silica gel with diethyl ether-hexane (1:1 v/v).

2.2.5.3 Synthesis of Optically Active N-Alkyl and N-Phenyl Sulfinamides 76: General Procedure[77]

LDA (1 mmol) in THF was added to a stirred solution of amine (1 mmol) in THF at –30°C, under nitrogen. The mixture was kept at –30°C for 15 min and then added dropwise to a solution of (–)-(S)-menthyl p-toluenesulfinate (1 mmol) in THF at –70°C, under nitrogen. In the case of N-isopropyl p-toluenesulfinamide, butyllithium was added instead of LDA. After an appropriate time the mixture was quenched with saturated aqueous ammonium chloride, the aqueous phase was extracted with dichloromethane, and the combined organic phases were dried ($MgSO_4$) and evaporated under reduced pressure. The crude product was purified by flash chromatography (in diethyl ether-light petroleum, 8:2 v/v). In the case of N-phenyl p-toluenesulfinamide the crude product was purified by column chromatography (in diethyl ether-light petroleum 4:6 v/v). Details are in Table 2.2.1.

2.2.5.4 Synthesis of Optically Active N-Alkylidenesulfinamides 80a–f[83]

Nitrile (20 mmol) in diethyl ether was added dropwise to a saturated solution of Grignard reagent (20 mmol) at 0°C. The mixture was kept at room temperature for 2–6 h and then cooled to –40°C. Sulfinate ester **1** (10 mmol) was added in one portion and the mixture was stirred overnight at room temperature. A usual workup afforded a syrupy oil which was chromatographed (silica gel, diethyl ether/light petroleum) to give the product listed in Table 2.2.2.

2.2.5.5 Reduction of N-Alkylidenesulfinamides 80: General Procedures[83]

(A) Sodium borohydride (1 mmol) was added to a stirred solution or suspension of the N-alkylidenesulfinamide (1 mmol in absolute EtOH [10 mL]). The reaction mixture was kept at room temperature for 10–15 h. Methanol (5 mL) was then added and solvents evaporated under reduced pressure. Water and dichloromethane were added to the residue. The organic phase was dried and the solvent evaporated to give the crude product which was chromatographed (silica gel, ethyl ether/light petroleum) to give 90–100% yield of diastereomeric mixtures of the sulfinamides.

(B) To a stirred solution of lithium aluminum hydride (1 mmol) in diethyl ether (5 mL) was added as drops, the N-alkylidenesulfinamide (1 mmol) in diethyl ether (10 mL) at room temperature. The mixture was stirred for 10–15 h under nitrogen and ethyl acetate (1 mL) was added. Workup gave the crude product which was chromatographed (silica gel, diethyl ether/light petroleum) to give 80–90% yield of diastereomeric mixtures of the sulfinamides.

REFERENCES

1. Mikołajczyk, M. and Drabowicz, J., *Top. Stereochem.*, 13, 333, 1982.
2. Drabowicz, J. and Oae, S., *Tetrahedron*, 34, 63, 1978.
3. a) Drabowicz, J., Kiełbasiński, P. and Mikołajczyk, M., in *The Chemistry of Sulfinic Acids, Esters and their Derivatives*, S. Patai, Ed., John Wiley & Sons, Chichester, 1990, pp. 351–429.
   b) Nudelman, A., in *The Chemistry of Sulfinic Acids, Esters and their Derivatives*, S. Patai, Ed., John Wiley & Sons, Chichester, 1990, pp. 35–85.
   c) Zoller, U., in *The Chemistry of Sulfinic Acids, Esters and their Derivatives*, S. Patai, Ed., John Wiley & Sons, Chichester, 1990, pp. 217–237.
   d) Nudelman, A., *The Chemistry of Optically Active Sulfur Compounds*, Gordon and Breach, New York, 1984.
   e) Colonna, S., Annunziata, R. and Cinquini, M., *Phosphorus Sulfur*, 10, 197, 1981.
   f) Kice, J.L., *Adv. Phys. Org. Chem.*, 17, 66, 1980.
   g) Wenschuh, E., Dölling, K., Mikołajczyk, M. and Drabowicz, J., *Z. Chem.*, 20, 122, 1980.
4. Phillips, H., *J. Chem. Soc.*, 127, 2552, 1925.
5. a) Herbrandson, H.F. and Dickerson, R.T., Jr., *J. Am. Chem. Soc.*, 81, 4102, 1959.
   b) Herbrandson, H.F., Dickerson, R.T., Jr. and Weinstein, J., *J. Am. Chem. Soc.*, 78, 2576, 1956.
6. a) Mioskowski, C. and Solladie, G., *Tetrahedron*, 36, 227, 1980.
   b) Solladie, G., *Synthesis*, 185, 1981.
7. a) Drabowicz, J., Dudziński, B., Łyźwa, P. and Mikołajczyk, M., unpublished results.
   b) Drabowicz, J., Dudziński, B., Bujnicki, B., Mikołajczyk, M. and Biscarini, P., unpublished results.
8. Estep, R.E. and Tavares, D.F., *Int. J. Sulfur Chem.*, 8, 279, 1973.
9. a) Herbrandson, H.F. and Cusano, C.M., *J. Am. Chem. Soc.*, 81, 2124, 1961.
   b) Folli, U., Iarrosi, D., Montanari, F. and Torre, U., *J. Chem. Soc. (C).*, 1317, 1968.
10. a) Andersen, K.K., Gaffield, W., Papanikolaou, N.E., Foley, J.W. and Perkins, R.I., *J. Am. Chem. Soc.*, 86, 5637, 1964.
    b) Andersen, K.K., *J. Org. Chem.*, 29, 1953, 1964.
11. Burgess, K. and Henderson, I., *Tetrahedron Lett.*, 30, 4235, 1989.
12. Mislow, K., Green, M.M., Laur, P., Mellilo, J.T., Simmons, T. and Ternay, A.L., Jr., *J. Am. Chem. Soc.*, 87, 1958, 1965.
13. Mislow, K., Green, M.M. and Raban, M., *J. Am. Chem. Soc.*, 87, 2761, 1965.
14. Mikołajczyk, M. and Drabowicz, J., *J. Am. Chem. Soc.*, 100, 2518, 1978.
15. Harpp, D.H., Friedlander, B.T., Larsen, C., Steliou, K. and Stockton, A., *J. Org. Chem.*, 43, 3481, 1978.
16. Klunder, J.M. and Sharpless, K.B., *J. Org. Chem.*, 52, 2598, 1987.
17. Andersen, K.K., Bujnicki, B., Drabowicz, J., Mikołajczyk, M. and O'Brien, J.B., *J. Org. Chem.*, 49, 4070, 1984.
18. Pyne, S.G., Hajipour, A.R. and Prabakaran, K., *Tetrahedron Lett.*, 35, 645, 1994.
19. a) Ridley, D.D. and Small, M.A., *J. Chem. Soc., Chem. Commun.*, 505, 1981.
    b) Ridley, D.D. and Small, M.A., *Aust. J. Chem.*, 35, 496, 1982.
20. a) Llera, J.M., Fernandez, I. and Alcudia, F., *Tetrahedron Lett.*, 32, 7299, 1991.
    b) Fernandez, I., Khiar, N., Llera, J.M. and Alcudia, F., *J. Org. Chem.*, 57, 6789, 1992.
    c) Khiar, N., Fernandez, I. and Alcudia, F., *Tetrahedron Lett.*, 35, 5719, 1994.
21. Whitesell, J.K. and Wong, M.S., *J. Org. Chem.*, *56*, 4552, 1991.
22. Whitesell, J.K. and Wong, M.S., *J. Org. Chem.*, 59, 597, 1994.
23. Rebiere, F., Samuel, O., Ricard, L. and Kagan, H.B., *J. Org. Chem.*, 56, 5991, 1991.
24. Mikołajczyk, M., Drabowicz, J. and Bujnicki, B., *J. Chem. Soc., Chem. Commun.*, 568, 1976.
25. Hioi, K., Kitayama, R. and Sato, T., *Synthesis*, 1983, 1040.
26. Drabowicz, J., and Łyźwa, P., XVI ISOCS, Merseburg, Germany, July 1994, Abstract Book, p.242.
27. a) Mikołajczyk, M., Drabowicz, J. and Bujnicki, B., *Tetrahedron Lett.*, 26, 5699, 1985.
    b) Mikołajczyk, M., in *Perspectives in the Organic Chemistry of Sulfur*, Zwanenburg, B. and Klunder, A.J.H., Eds., Elsevier, Amsterdam, 1987, pp.23–40.
28. Drabowicz, J., Legędź, S. and Mikołajczyk, M., *Tetrahedron*, 44, 5243, 1988.
29. Mikołajczyk, M. and Drabowicz, J., *J. Chem. Soc., Chem. Commun.*, 547, 1974.
30. Mikołajczyk, M., Drabowicz, J. and H. Ślebocka-Tilk, *J. Am. Chem. Soc.*, *101*, 1302, 1979.
31. Andersen, K.K., *Tetrahedron Lett.*, 93, 1962.
32. Hope, H., de la Camp, H., Homer, G., Messing, A.W. and Sommer, L.H., *Angew. Chem.*, 81, 619, 1969; *Angew. Chem., Int. Ed. Engl.*, 8, 6121, 1969.
33. de la Camp, H. and Hope, H., *Acta Crystallogr., Sect.B.*, 26, 846, 1970.
34. Harpp, D.N., Vines, S.M., Montillier, J.P. and Chan, T.H., *J. Org. Chem.*, 41, 3987, 1976.
35. Drabowicz, J., Bujnicki, B. and Mikołajczyk, M., *J. Org. Chem.*, 47, 3325, 1982.
36. Jacobus, J. and Mislow, K., *J. Am. Chem. Soc.*, 89, 5228, 1967.
37. Colonna, S., Giovini, R. and Montanari, F., *J. Chem. Soc., Chem. Commun.*, 865, 1968.
38. Jacobus, J. and Mislow, K., *J. Chem. Soc., Chem. Commun.*, 253, 1968.
39. Pirkle, W.H. and Beare, S.D., *J. Am. Chem. Soc.*, 90, 6250, 1968.
40. Buist, P.H. and Marecak, D.M., *J. Am. Chem. Soc.*, 114, 5073, 1992.
41. Andersen, K.K., Colonna, S. and Stirling, C.J.M., *J. Chem. Soc., Chem. Commun.*, 645, 1973.

42. Abbot, D.J., Colonna, S. and Stirling, C. J. M., *J. Chem. Soc., Perkin Trans. 1.*, 492, 1972.
43. a) Posner, G.H. and Tang, P.W., *J. Org. Chem.*, 43, 4131, 1978.
b) Marino, J.P., Laborde, E., Deering, C.F., Paley, R.S. and Ventura, M.P., *J. Org. Chem.*, 59, 3193, 1994.
c) Posner, G.H., Mallamo, J.P. and Miura, K., *J. Am. Chem. Soc.*, 103, 2886, 1981.
d) Fawcett, J., House, S., Jenkins, P.R., Lawrance, N.J. and Russell, D.R., *J. Chem. Soc., Perkin Trans. 1*, 67, 1993.
44. Bickart, P., Carson, F.W., Jacobus, J., Miller, E.G. and Mislow, K., *J. Am. Chem. Soc.*, 90, 4869, 1968.
45. a) Cinquini, M., Colonna, S., Cozzi, F. and Stirling, C.J.M., *J. Chem. Soc., Perkin Trans. 1*, 2061, 1976.
b) Kosugi, H., Kitaoka, M., Tagami, K., Takahashi, A. and Uda, H., *J. Org. Chem.*, 52, 1078, 1987.
46. Mikołajczyk, M., Midura, W. Grzejszczak, S., Zatorski, A. and Chefczyńska, A., *J. Org. Chem.*, 43, 478, 1978.
47. Colombo, G.L., Gennari, C. and Narrisano, E., *Tetrahedron Lett.*, 3861, 1978.
48. Annunziata, R., Cinquini, M. and Cozzi, F., *Synthesis*, 535, 1979.
49. Kunieda, N., Nokami, J. and Kinoshita, M., *Bull. Chem. Soc., Jpn.*, 49, 256, 1976.
50. Shibutani, T., Fujihara, H. and Furukawa, N., *Tetrahedron Lett.*, 32, 2947, 1991.
51. Rebiere, F., Samuel, O. and Kagan, H.B., *Tetrahedron Lett.*, 31, 312, 1990.
52. Annunziata, R., Cinquini, M. and Cozzi, F., *Synthesis*, 767, 1982.
53. Annunziata, R., Cinquini, M., Colonna, S. and Cozzi, F., *J. Chem. Soc., Perkin Trans. 1*, 614, 1981.
54. Carreno, M.C., Garcia Ruano, J.L. and Urbano, A., *Synthesis*, 651, 1992.
55. Carreno, M.C., Garcia Ruano, J.L. and Urbano, N., *Tetrahedron Lett.*, 30, 4003, 1989.
56. Carreno, M.C., Garcia Ruano, J.L., Mata, M. and Urbano, A., *Tetrahedron*, 47, 605, 1991.
57. Carreno, M.C., Garcia Ruano, J.L. and Urbano, A., *J. Org. Chem.*, 57, 6870, 1992.
58. Carreno, M.C., Garcia Ruano J.L. and Rubio, A., *Tetrahedron Lett.*, 28, 4861, 1987.
59. Hiroi, K., Kitayama, R. and Sato, S., *J. Chem. Soc., Chem. Commun.*, 1470, 1983.
60. Hiroi, K., Kitayama, R. and Sato, S., *Chem. Pharm. Bull.*, 32, 2628, 1984.
61. Hiroi, K., Kitayama, R. and Sato, S., *Chemistry Lett.*, 929, 1984.
62. Trost, B., Schmull, N.R. and Miller, M.J., *J. Am. Chem. Soc.*, 102, 5979, 1980.
63. Westmijze, H. and Vermeer, P., *Tetrahedron Lett.*, 410, 1979.
64. Tadema, G., Everhardus, R.H., Westmijze, H. and Vermeer, P., *Tetrahedron Lett.*, 3935, 1978.
65. a) Stirling, C.J.M., *J. Chem. Soc., Chem. Commun.*, 131, 1967.
b) Smith, G. and Stirling, C.J.M., *J. Chem. Soc. (C)*, 1530, 1971.
66. Braverman, S., *Int. J. Sulfur Chem., (C)*, 6, 149, 1971.
67. Komiyama, K, Minato, H. and Kobayashi, M., *Bull. Chem. Soc., Jpn.*, 46, 3895, 1993.
68. Bonvicini, P., Levi, A. and Scorrano, G., *Gazz. Chim. Ital.*, 102, 621, 1972.
69. Rayner, D.R., Gordon, A.J. and Mislow, K., *J. Am. Chem. Soc.*, 90, 4854, 1968.
70. Bravo, P., Resnati, G., Viani, F. and Arone, A., *Tetrahedron*, 43, 4635, 1987.
71. Solladie, G., Zimmerman, R., Bartsch, R. and Walborsky, H.M., *Synthesis*, 662, 1985.
72. Jacobus, J. and Mislow, K., *J. Chem. Soc., Chem. Commun.*, 253, 1968.
73. Nudelman, A. and Cram, D.J., *J. Am. Chem. Soc.*, 90, 3869, 1968.
74. Evans, D.A., Faul, M.M., Colombo, L., Bisaha, J.J., Clardy, J. and Cherry, D., *J. Am. Chem. Soc.*, 114, 5977, 1992.
75. a) Fawcett, J., House, S., Jenkins, P.R., Lawrance, N.J. and Russell, D.R., *J. Chem. Soc., Perkin Trans. 1*, 67, 1993.
b) Buttlin, R.J., Linney, I.D., Critcher, D.J., Mahon, M.F., Molloy, K.C. and Willis, M., *J. Chem. Soc., Perkin Trans. 1*, 1581, 1993.
76. Colonna, S., Giovini, R. and Montanari, F., *J. Chem. Soc., Chem. Commun.*, 865, 1968.
77. Colonna, S., Germinario, G., Manfredi, A. and Stirling, C.J.M., *J. Chem. Soc., Perkin Trans. 1*, 1695, 1988.
78. Mikołajczyk, M., Drabowicz, J. and Bujnicki, B., *Bull. Acad. Pol. Sci., Ser. Chem.*, 25, 267, 1977.
79. Wenschuh, E., Winter, H., Mendel, G. and Kolbe, A., *Phosphorus Sulfur*, 7, 321, 1977.
80. Booms, R.E. and Cram, D.J., *J. Am. Chem. Soc.*, 94, 5438, 1979.
81. Schroeck, C.W. and Johnson, C.R., *J. Am. Chem. Soc.*, 93, 5305 1971.
82. Johnson, C.R. and Schroeck, C.W., *J. Am. Chem. Soc.*, 95, 7418, 1973.
83. a) Cinquini, M. and Cozzi, F., *J. Chem. Soc., Chem. Commun.*, 502, 1977.
b) Cinquini, M. and Cozzi, F., *J. Chem. Soc., Chem. Commun.*, 723, 1977.
c) Annunziata, R., Cinquini, M. and Cozzi, F., *J. Chem. Soc., Perkin Trans. 1*, 339, 1982.
84. a) Hua, D.H., Miao, S.W., Chen, J.S. and Iguchi, S., *J. Org. Chem.*, 56, 4, 1991.
b) Hua, D.H., Lagneau, N., Wang, H. and Chen.J., *Tetrahedron: Asymmetry*, 6, 349, 1995.
85. Davis, F.A., Reddy, R.E., Szewczyk, J.M. and Portonovo, P.S., *Tetrahedron Lett.*, 34, 6229, 1993.
86. Davis, F.A., Reddy, R.T. and Reddy, R.E., *J. Org. Chem.*, 57, 6387, 1992.
87. Jiang, J., Schumacher, K.K., Joullie, M.M., Davis, F.A. and Reddy, R.E., *Tetrahedron Lett.*, 35, 2121, 1994.
88. Davis, F.A., Reddy, R.E. and Portonovo, P.S., *Tetrahedron Lett.*, 35, 9351, 1994.
89. Mikołajczyk, M., Łyżwa, P., Drabowicz, J., Wieczorek, M.W., and Błaszczyk, J., *J. Chem. Soc., Chem. Commun.*, 1503, 1996.
90. Davis, F.A., Zhou, P. and Reddy, G.V., *J. Org. Chem.*, 59, 3243, 1994.
91. Davis, F.A., Reddy, V. and Lin, H., *J. Am. Chem. Soc.*, 117, 3651, 1995.
92. Garcia Ruano, J.L., Fernandez, I. and Hamdouchi, C., *Tetrahedron Lett.*, 36, 295, 1995.

Chapter 3

Chiral Sulfoxides

Chiral sulfoxides belong to the class of chiral organosulfur compounds which are most widely used in asymmetric synthesis. Their application as chiral synthons has now become a well-established and reliable strategy. This is mainly due to their easy availability and high asymmetric induction exerted by the chiral sulfinyl group.

There are three main directions of application of chiral sulfoxides: (a) reactions of α-sulfinyl carbanions with a broad variety of electrophiles; (b) reactions of α,β-unsaturated sulfoxides; (c) introduction of heteroatomic groups to sulfoxides and their transformation.

Since the application of a large variety of sulfoxides (e.g., those discussed in Sections 3.2, 3.3, 3.4, 3.5, 3.6, 3.7, and 3.8) is based on the use of their α-carbanions, it seems necessary to give a general background concerning generation and properties of the latter.

3.1 α-SULFINYL CARBANIONS*

There is a large number of papers devoted to α-sulfinyl carbanions. Of particular importance are publications dealing with the structure and stereochemical properties of these anions. Careful inspection of all the contributions to this subject reveals that an evolution of ideas concerning this matter has been taking place and some models and former explanations have been found to be inconsistent with new findings. Therefore, in the present section a short historical overview and recent achievements in this field will be presented.

α-Sulfinyl carbanions are intrinsically diastereomeric (when $R^2 \neq H$) due to the attachment of the carbanion center to the chiral sulfinyl moiety. For the same reason the α-methylene hydrogens are diastereotopic.

O
HA HB
: R1
R2

Base

O
: HB
: R1
R2

and/or

O
HA :
: R1
R2

This feature was confirmed as early as 1965 by Wolfe et al.,[1] who found out that the rates of H/D exchange of the diastereotopic methylene protons in benzyl methyl sulfoxide were different. Moreover, subsequent investigations of the reactions of α-sulfinyl carbanions with electrophiles

* References follow each of the 10 sections in this chapter.

revealed that the absolute configuration around the α-carbon atom in the major diastereomers produced depends on the kind of electrophile used. Thus, the reaction of α-lithiobenzyl sulfoxides with H_2O, D_2O, CO_2, R_2CO, and $(MeO)_3PO$ was found to proceed with retention while with MeI, with inversion of configuration at the anionic α-carbon atom (Scheme 3.1.1.).[2-7] Additionally, it was found that the reactive hydrogen in benzyl methyl sulfoxide changes with a change in the reaction medium — in polar, protic solvents this is the *pro*-R hydrogen which undergoes exchange by deuterium.

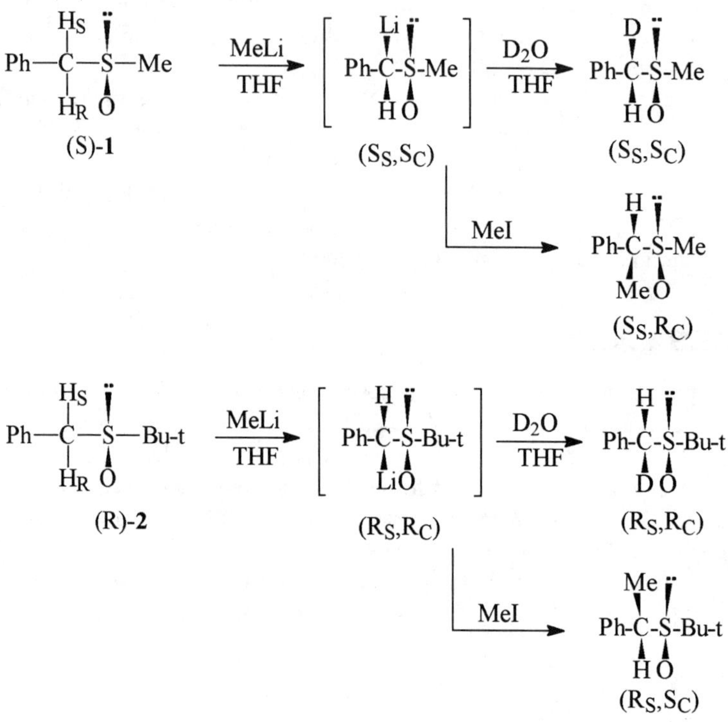

Scheme 3.1.1

To explain the aforementioned features of α-sulfinyl carbanions, a four-center chelate model **3** has been proposed in which the lithium cation is bound both to the sulfoxide oxygen and to the α-carbon atom, and the carbanionic center is nearly planar.[8-10]

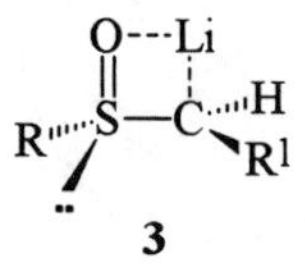

3

Such a structure has been deduced on the basis of various spectroscopic studies. For example, the coupling constant values $J_{^{13}C-^{1}H}$, which are higher for the carbanionic center than for the parent starting compound, have been taken as a proof for the higher sp^2 character of the anionic α-carbon atom. The ^{13}C NMR spectra of an α-lithiosulfide recorded for comparison show that the coupling constant values J_{13C-1H} in this case are substantially lower and therefore the hybridization state of the anionic α-carbon atom is assumed to be sp^3. Similarly, the carbanionic centers in an α-lithiosulfone and in an α-lithiosulfoximine have been deduced to be hybridized between sp^2 and sp^3 (Table 3.1.1).[11]

Table 3.1.1 $J_{^{13}C-^{1}H}$ of the Anionic α-Carbon Atoms and Their Differences with Respect to the Parent Compounds[11]

Anionic species	Solvent	J_{13C-1H} [Hz]	Δ J [Hz]
$PhSCH_2Li$	Toluene	118	–20
$PhSCH_2Li$	THF	120	–18
$PhS(O)CH_2Li$	Toluene	153	+14
$PhS(O)CH_2Li$	THF	155.5	+16.5
MeS(O)CH(Ph)Li	THF	160	+20
$PhSO_2CH_2Li$	Tolene	132	–6
$PhSO_2CH_2Li$	THF	139	0
$PhSO(NMe)CH_2Li$	THF	137	–2

In support of the chelate structure[3] are also the vibrational spectra in which the C-Li and O-Li stretching vibrations have been observed.[12] On the other hand, however, no coupling constants $J_{^{13}C-^{7}Li}$ were observed for α-lithiosulfoxides, although they are clearly visible in the ^{7}Li and ^{13}C NMR spectra of methyllithium and butyllithium.[11] Moreover, based on the ^{13}C NMR spectra of fluorophenyl lithiomethyl sulfoxide, Seebach et al.[13] have drawn a conclusion that the lithium cation is coordinated only by the sulfinyl oxygen atom.

Nevertheless, the chelate model has been for quite a long time successfully applied to explain different stereochemical outcomes of the reactions of α-sulfinyl carbanions with various electrophiles. Thus, according to this model, the electrophiles which can chelate the lithium cation approach the anion from the lithium site, giving products with retention of configuration. On the contrary, the nonchelating electrophiles react with inversion of configuration.[8] It should be emphasized that planarity of the anionic center is a prerequisite of the divergent course of this reaction. If the carbanion were sp^3 hybridized, it would react exclusively with retention of configuration, as does, for example, the sp^3 anion derived from 1,3-dithiane.

retention: H_2O, D_2O, RCHO, R_2CO,CO_2, $(MeO)_3PO$

O---Li
R—S—C—H
R1

3

inversion: MeI

This commonly accepted model has been, however, disturbed by the finding of Japanese workers[14] that the absolute configuration of the deuterated derivative of benzyl t-butyl sulfoxide **2** (obtained according to Scheme 3.1.1) was erroneously established by Durst et al.[2] Thus, it has now turned out that the benzyl t-butyl sulfoxide anion reacts in tetrahydrofuran (THF) with both chelating (D_2O) and nonchelating (MeI) electrophiles with the same stereochemistry.

Precise and detailed investigations of Ohno et al.[15] revealed that the stereochemistry of the reaction of α-sulfinyl carbanions with electrophiles is a function of several factors, the main being the structure of a parent sulfoxide, the kind of an electrophile, and the reaction medium. The results obtained with (S)-benzyl methyl sulfoxide **1** and (R)-benzyl t-butyl sulfoxide **2** are summarized in Scheme 3.1.2.[15]

BuLi/THF/R_2CO
BuLi/THF/D_2O
O
H_S H_R
: Me
Ph
(S)-**1**
$^-OD/D_2O$, CD_3OD
BuLi/THF/MeI

BuLi/THF/D_2O
t-BuOK/THF/D_2O
BuLi/THF/MeI
Bu^t
H_S H_R
O :
Ph
(R)-**2**
$^-OD/CD_3OD$

Scheme 3.1.2[15]

The carbanion generated from (S)-**1** in a nonpolar solvent reacts with D_2O or carbonyl compounds ("chelating electrophiles") to give the corresponding products arising from the substitution of the *pro*-(S) hydrogen, while with MeI (a "nonchelating electrophile"), a product of the substitution of the *pro*-(R) hydrogen; the *pro*-(R) hydrogen is also substituted when the reaction is carried out in polar protic solvents. On the other hand, however, the carbanion generated from (R)-**2** in a nonpolar solvent gives the products resulting from the substitution of the *pro*-(S) hydrogen, irrespective of the electrophile used (compare with the incorrect data in Scheme 3.1.1), while in polar protic solvents the products of the substitution of the *pro*-(R) hydrogen (Scheme 3.1.2).

The authors put forward a coherent explanation of these results which is based on the assumption that the kinetic acidity controlling the stereochemistry of the initially formed carbanions can be neglected for the reaction performed in THF (but not in protic polar solvents), since the carbanion has enough time to reorganize into its most stable form before it reacts with an electrophile. This assumption was later confirmed by the existence of an equilibrium between two diastereomeric carbanions derived from α-methylbenzyl phenyl sulfoxide (vide infra)[17,18] and isopropyl α-methylbenzyl sulfoxide.[16] The detailed explanation is shown in Scheme 3.1.3 and discussed below.[15]

O
H_S H_R
: Me
Ph
(S)-**1**

O Li^+ H_R : Me Ph
Li-(S_C, S_S)-**1**
Deuteration
(in THF)

O Li^+ H_S : Me Ph
Li-(R_C, S_S)-**1**
Methylation (in THF)
Deuteration
(in polar media)

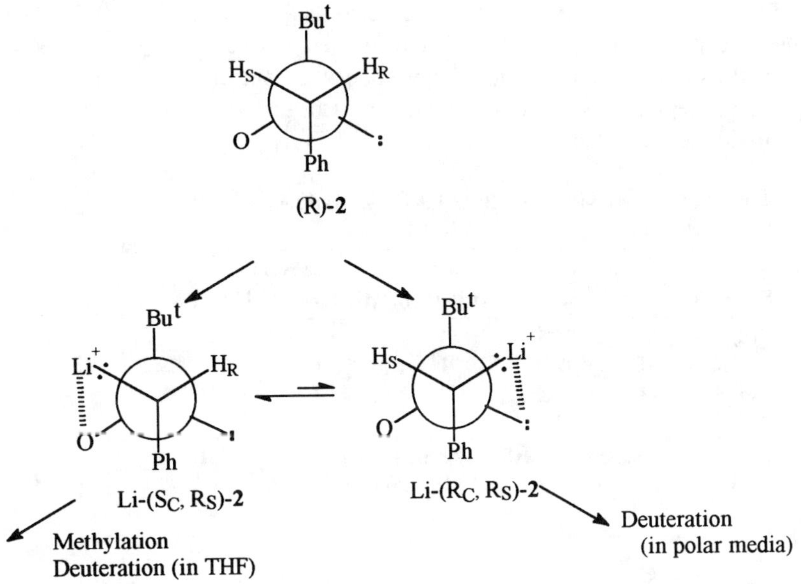

Scheme 3.1.3[15]

First of all, the most stable conformations adopted by the sulfoxides (S)-**1** and (R)-**2** have been predicted on the basis of some spectral data. For **1** it is the conformation in which the methyl group is *gauche* to the phenyl group, while for **2** it is the conformation in which the *tert*-butyl group and the sulfinyl oxygen are *anti* and *gauche* to the phenyl ring, respectively. When a base used to abstract a proton approaches the sulfoxide, its counterion is initially trapped (coordinated) by the sulfinyl oxygen. Hence, in (R)-**2** the H_S is more reactive than the H_R, and the carbanion formed in this way is more stable than the other one. The situation in (S)-**1** is somewhat different since both hydrogen atoms are similar with respect to distance from the sulfinyl oxygen. However, the carbanion generated by abstraction of the H_S seems to be thermodynamically more stable due to the possible coordination of the lithium cation by both the sulfur and oxygen lone electron pairs, although the H_R should be more susceptible to abstraction because of the relatively lower electrostatic repulsion between the developing negative charge and the sulfur lone electron pair.

As a result of the aforementioned considerations, it may be deduced that the predominant configurations (and conformations) of the carbanions derived from (S)-**1** and (R)-**2** are Li-(S_CS_S)-**1** and Li-(S_CR_S)-**2**, respectively. "Chelating" electrophiles (H_2O, D_2O, R_2CO) approach the carbanion from its lithiated side since they can initially interact with the lithium cation. On the contrary, methyl iodide (a "nonchelating" electrophile) prefers to react on the more nucleophilic side. Due to the so called α-effect exerted by the sulfur lone electron pair, this is the *anti* pair of the carbanion which becomes more polarizable and hence more nucleophilic. Hence, in the case of (S)-**1** the carbanion generated on the H_R side is more nucleophilic, which ultimately results in the opposite stereochemical courses for deuteration and methylation. In the case of (R)-**2**, however, the more nucleophilic carbanion is the one generated on the H_S side, thus the same which is approached by the "chelating" electrophiles. Therefore, the stereochemistry of the reaction is the same for both types of electrophiles.

The situation in polar solvents is quite different. In this case, a base employed is usually dissociated and therefore not attracted by the sulfinyl oxygen via coordination of a counterion. The anionic base tends rather to keep away from the negatively charged face of a sulfoxide. For this reason, both in (S)-**1** and (R)-**2** the *pro*-(R) hydrogen undergoes abstraction. Since the carbanion formed in a polar solvent can react with an electrophile without any previous reorganization into another form, this result can be considered as a reflection of a kinetic acidity.[15]

Further investigations by the same group[16] of other benzyl sulfoxides (Table 3.1.2) have led to formulation of the concept which explains the aforementioned and new results in terms of the HSAB principle, as well as elimination of some doubt about the conformations adopted by the starting sulfoxides. This concept is illustrated in Scheme 3.1.4 (for the sake of simplicity, only one enantiomer of the sulfoxide is presented).

Table 3.1.2 Reaction of Alkyl Benzyl Sulfoxide Carbanions with Electrophiles

		Deuteration		Methylation	
Sulfoxide	**R**	**H_S:H_R**	**Yield [%]**	**H_S:H_R**	**Yield [%]**
(S)-**1**	Me	95:5	55	5:95	67
(R)-**2**	t-Bu	>99:<1	91	>99:<1	99
(S)-**4**	Et	92:8	74	18:82	89
(R)-**5**	i-Pr	95:5	83	42:58	99

From Higaki, M., Goto, M., Ohno, A., *Heteroatom. Chem.*, 1, 181, 1990. With permission from VCH Verlagsgesellschaft mbH.

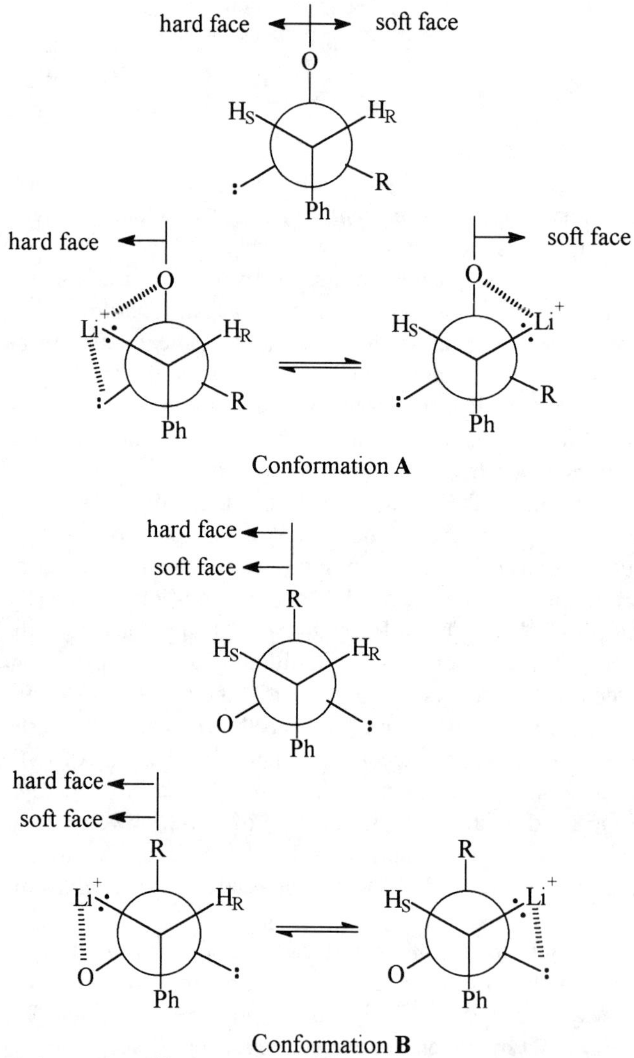

Conformation **A**

Conformation **B**

Scheme 3.1.4[16]

According to the definition introduced by Ohno et al.,[15,16] the "hard face" of an α-sulfinyl carbanion is the one in which the carbanion lone electron pair is located between the sulfinyl oxygen atom and sulfur lone electron pair, while the "soft face" is that in which the carbanion lone electron pair is located *anti* to the sulfur lone electron pair. As shown in Scheme 3.1.4, the "hard face" of the carbanion is always the H_S side, irrespective of the conformation of the parent sulfoxide. For this reason the stereoselectivity of deuteriation (deuterium cation is a "hard electrophile") is very high for all the sulfoxides (Table 3.1.2).

On the contrary, the "soft face" of the carbanion is different in each conformation: it is the H_R side for the conformation **A** and the H_S for the conformation **B**. This difference is reflected by the reverse stereochemistry of the methylation (reaction with a "soft electrophile") of the carbanions derived from (S)-**1** and (R)-**2** (Table 3.1.2), as it was discussed earlier (cf. Scheme 3.1.3). However, the difference is not so pronounced for (S)-**4** (R=Et) in which, due to a larger steric repulsion between the phenyl and ethyl groups than that between the phenyl and methyl groups in (S)-**1**, the contribution of the conformation **B** becomes more important than in (S)-**1**. Consequently, the isopropyl group present in (R)-**5** exerts such a steric hindrance (yet much lower than the tert-butyl group in (R)-**2**), that both conformations are populated almost equally, which ultimately results in a nearly complete loss of stereoselectivity of the methylation reaction.

It is important to note that although the above discussion is based on the assumption that the carbanionic center is sp^3 hybridized, the same arguments are valid for the sp^2-hybridized anion, provided the coordination of the counterion is asymmetric or distorted by neighboring dipoles.[15]

It should be stressed, however, that on the basis of the theoretical 3-21G* calculations Wolfe et al.[19] predicted quite a different structure of the α-sulfinyl carbanion. The proposed structure has been fully confirmed by X-ray analyses performed independently by two groups: Boche et al.[17] and Boche,[18] using the dimeric tetramethylethylenediamine (TMEDA) complex of α-lithio-α-methylbenzyl phenyl sulfoxide **6**,* and Veya et al.,[20] using a "naked" carbanion of methyl phenyl sulfoxide **7**, obtained by treatment of potassiomethyl phenyl sulfoxide with 18-crown-6.

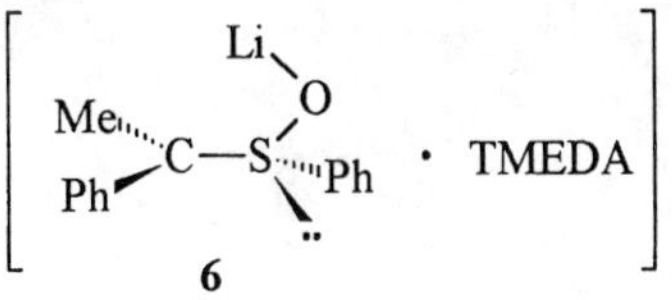

[Me(Ph)C–S(O–Li)(Ph) · TMEDA]$_2$ **6** $H_2\ddot{C}$–S(O····KL)Ph **7**

TMEDA = tetramethylethylenediamine L = 18-Crown-6

* The pure diastereomer **6** (as shown in Figure 3.1.1) was obtained from a 1:1 mixture of diastereomeric of α-methylbenzyl phenyl sulfoxides with n-BuLi in ether/TMEDA. Apparently, in solution from two diastereomeric carbanions which are initially generated, the more thermodynamically stable one is formed and finally isolated as a complex with TMEDA. Note that protonation of the carbanion leads exclusively to one diastereomer of the starting sulfoxide!

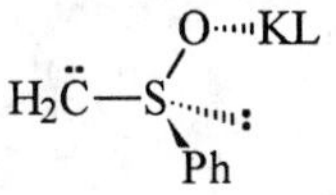

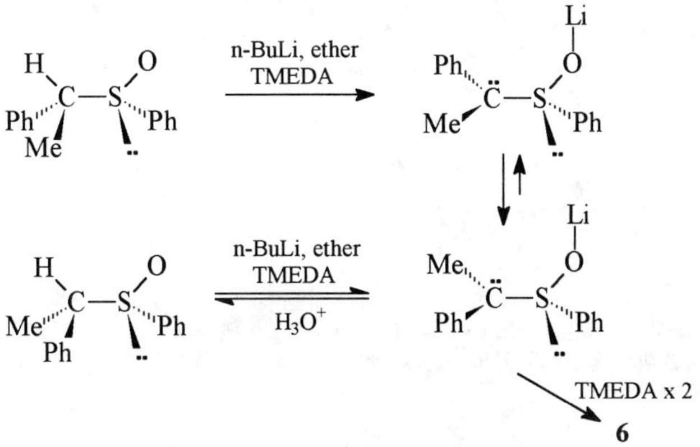

n-BuLi, ether TMEDA

n-BuLi, ether TMEDA H_3O^+

TMEDA x 2

6

Their crystal and molecular structures are shown in Figures 3.1.1 and 3.1.2, respectively.

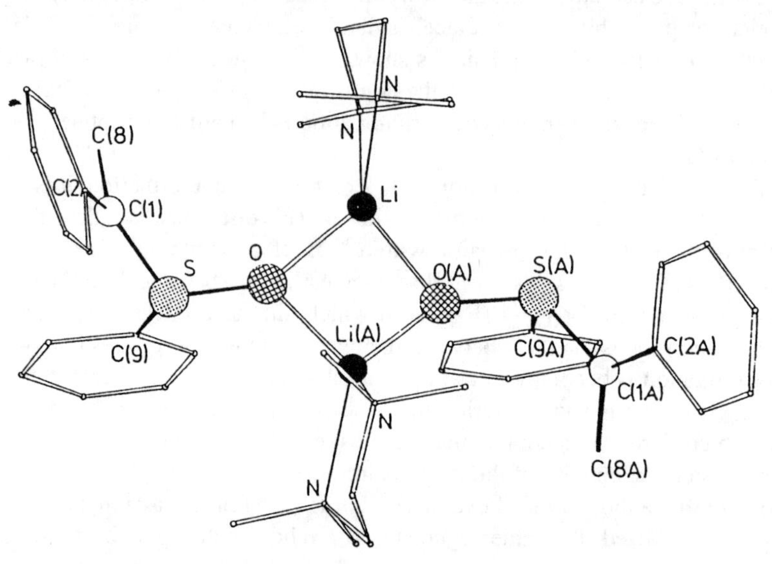

Figure 3.1.1 Crystal structure of the dimer **6**. (From Marsch, M., Massa, W., Harms, K., Baum, G., and Boche, G., *Angew. Chem.*, 98, 1004, 1986; *Angew. Chem., Int. Ed. Engl.*, 25, 1011, 1986. With permission from VCH Verlagsgesellschaft mbH.)

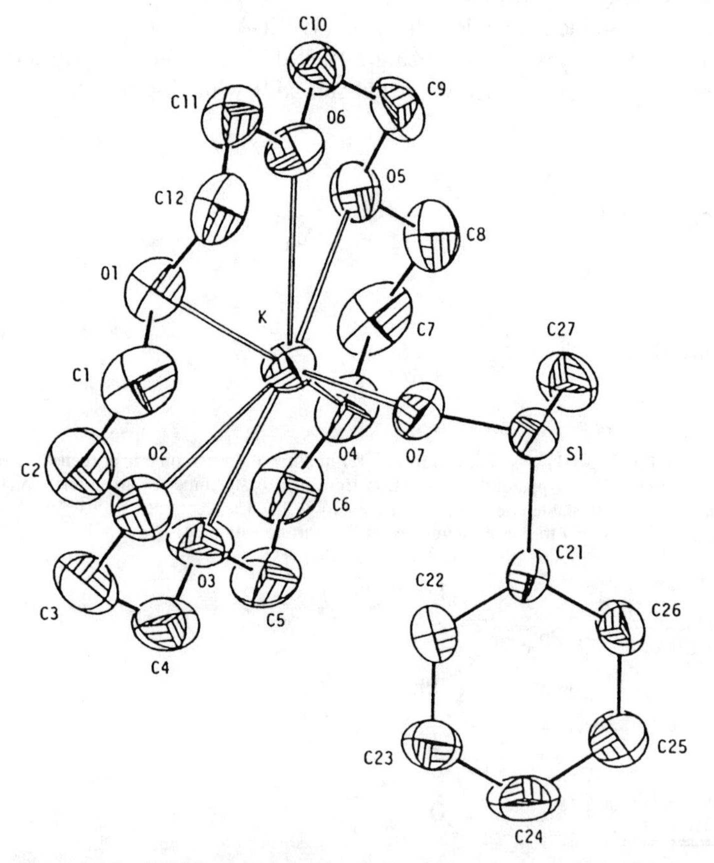

Figure 3.1.2 ORTEP view of complex **7**. (From Veya, P., Floriani, C., Chiesi-Villa, A., and Guastini, C., *Organometallics,* 12, 253, 1993. Copyright 1993 American Chemical Society. With permission.)

The most important finding was that in both investigated cases the metal cations are linked exclusively to the sulfinyl oxygen atom, the distance between the anionic carbon atom and the metal being very large (e.g., C1-Li in **7** is equal to 400 pm, while the normal C-Li bonds are shorter than 250 pm).[18] The bonds between the anionic carbon atoms and sulfur are distinctly shorter and the sulfur oxygen bonds somewhat longer than in DMSO, taken as a reference (Table 3.1.3).

Table 3.1.3 Bond lengths in **6**, **7**, and DMSO[17,18,20]

Compound	Bond length [pm]		
	M–O	S–O	S–C°
DMSO	—	147	180[a]
6	192	158	163
7	266	152	166

[a] S–CH_3.

The lithium atom does not lie exactly in the C1, S, O plane, since the torsional angle C1, S, O, Li is 12°. Moreover, the carbanionic center is not planar and the carbon atom C1 projects out of the S, C2, C8 plane by 12 pm, the substituents being bent toward the oxygen and lithium atoms. This feature is particularly surprising because benzylic carbanionic centers are known to be planar in such carbanions like α-sulfonylbenzyl, α-cyanobenzyl, and α-nitrobenzyl.[17] The origin of this effect is not known yet.

As it was already mentioned, the structure of the carbanion of **6** determined by an X-ray analysis is in agreement with that deduced from theoretical calculations. The theoretical model[19] of the carbanion of DMSO (Figure 3.1.3) predicted nonplanarity of the anionic center and the most probable conformation of the molecule, which agrees surprisingly well with that found in the solid state for **6** (Figure 3.1.4).

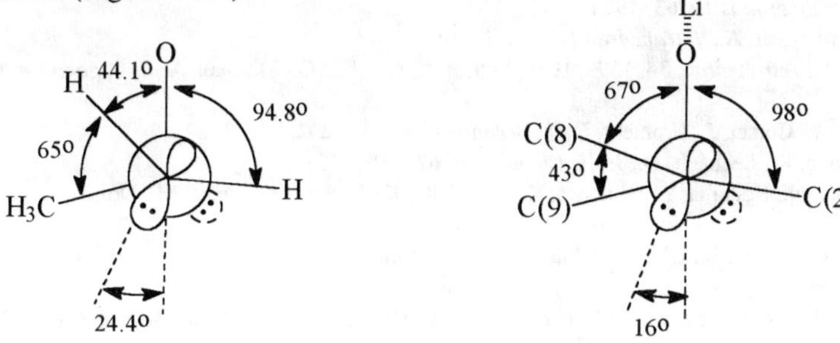

Figure 3.1.3 Figure 3.1.4

(Figures 3.1.3 and 3.1.4 from Marsch, M., Massa, W., Harms, K., Baum, G., and Boche, G., *Angew. Chem.*, 98, 1004, 1986; *Angew. Chem., Int. Ed. Engl.*, 25, 1011, 1986. With permission from VCH Verlagsgesellschaft mbH.

To sum up, according to the results discussed above one should avoid using the term "α-lithio(metalo)sulfoxide" since there is no C-Li (C-metal) bond. This turned out to be true also in solutions. Ohno et al.[16] proved on the basis of ^7Li and ^{17}O-NMR spectra that the anions derived from benzyl sulfoxides behave as oxylate anions and not as carbanions.

In view of all the recent findings, a new attempt has been made to explain stereoselectivity of the reactions of α-sulfinyl carbanions with electrophiles[17,18] (Scheme 3.1.5).

$R^2 > R^1$

A B

Scheme 3.1.5

A "chelating" electrophile forms a complex with lithium which leads to an attack from the lithium side. This is true for all kinds of benzyl sulfoxides (cf. Table 3.1.2). On the other hand, methyl iodide approaches the methyl compound **A** from the face away from the oxygen and lithium, as is normal. However, in the case of **B** the steric hindrance exerted by the *tert*-butyl group is so great that methyl iodide approaches from the otherwise unusual face with the O–Li bond. When the steric hindrance of the substituents at sulfur is between those of the methyl and *tert*-butyl groups, the stereoselectivity of methylation is lower, since the electrophile can approach the anion from both sides with comparable probability.

In spite of the ambiguity concerning explanation of the stereochemical outcomes of the reactions of α-sulfinyl carbanions with electrophiles, there are numerous examples of their successful application in the asymmetric formation of the carbon-carbon bond.[21] In the next several sections the use of α-sulfinyl carbanions derived from various types of sulfoxides will be presented. In each section the following types of reactions will be consecutively discussed: alkylation, Michael addition, hydroxyalkylation, aminoalkylation, and acylation.

REFERENCES (3.1)

1. Rank, A., Buncel, E., Moir, R.Y., Wolfe, S., *J. Am. Chem. Soc.*, **87**, 5498, 1965; for a general summary see: Wolfe, S., Stolow, A., La John, L.A., *Can. J. Chem.*, **62**, 1470, 1984.
2. Durst, T., Viau, R., McClory, M.R., *J. Am. Chem. Soc.*, **93**, 3077, 1971.
3. Nishihata, K., Nishio, M., *Chem. Commun.*, 958, 1971.
4. Durst, T., Viau, R., Van den Elzen, R., Nguyen, C.H., *Chem. Commun.*, 1334, 1971.
5. Nishihata, K., Nishio, M., *Tetrahedron Lett.*, 1965, 1976.
6. Nishihata, K., Nishio, M., *Tetrahedron Lett.*, 4839, 1972.
7. Viau, R., Durst, T., *J. Am. Chem. Soc.*, **95**, 1346, 1973.
8. Biellmann, J.F., Vicens, J.J., *Tetrahedron Lett.*, 2915, 1974; Biellmann, J.F., Vicens, J.J., *Tetrahedron Lett.*, 467, 1978.
9. Durst, T., Molin, M., *Tetrahedron Lett.*, 63, 1975.
10. Chassaing, G., Lett, R., Marquet, A., *Tetrahedron Lett.*, 471, 1978.
11. Chassaing, G., Marquet, A., *Tetrahedron*, **34**, 1399, 1978; Lett, R., Chassaing, G., Marquet, A., *J. Organomet. Chem.*, **111**, C17, 1976.
12. Chassaing, G., Marquet, A., Corset, J., Froment, F., *J. Organomet. Chem.*, **232**, 293, 1982.
13. Najera, C., Yus, M., Hässig, R., Seebach, D., *Helv. Chim. Acta*, **67**, 1100, 1984.
14. Itaka, Y., Itai, A., Tomioka, N., Kodama, Y., Ichikawa, K., Nishihata, K., Nishio, M., Izumi, M., Doi, K., *Bull. Chem. Soc. Jpn.*, **59**, 2801, 1986.
15. Nakamura, K., Higaki, M., Adachi, S., Oka, S., Ohno, A., *J. Org. Chem.*, **52**, 1414, 1987.
16. Higaki, M., Goto, M., Ohno, A., *Heteroatom Chem.*, **1**, 181, 1990.
17. Marsch, M., Massa, W., Harms, K., Baum, G., Boche, G., *Angew. Chem.*, **98**, 1004, 1986; *Angew. Chem., Int. Ed. Engl.*, **25**, 1011, 1986.
18. Boche, G., *Angew. Chem.*, **101**, 286, 1989; *Angew. Chem., Int. Ed. Engl.*, **28**, 277, 1989.
19. Wolfe, S., Stolow, A., La John, L.A., *Tetrahedron Lett.*, **24**, 4071, 1983; Wolfe, S., in *Organic Sulfur Chemistry*, Bernardi, F., Csizmadia, I.G., Mangini, A., (Eds.), Elsevier, Amsterdam, 1985, 133.
20. Veya, P., Floriani, C., Chiesi-Villa, A., Guastini, C., *Organometallics*, **12**, 253, 1993.
21. For recent reviews see:
    Drabowicz, J., Kiełbasiński, P., Łyżwa, P., *Sulfur Reports*, **12**, 213, 1992; Walker, A.J., *Tetrahedron: Asymmetry*, **3**, 961, 1992; Drabowicz, J., Kiełbasiński, P., Mikołajczyk, M., in *Syntheses of Sulphones, Sulphoxides and Cyclic Sulphides*, Patai, S., Rappoport, Z., (Eds.). John Wiley & Sons, Chichester, 1994, pp. 109, 255; Solladie, G., in *Methods of Organic Chemistry (Houben-Weyl)*, Georg Thieme Verlag, Stuttgart, 1995, vol. E21a, p. 1056.

3.2 DIALKYL AND ALKYL ARYL SULFOXIDES

Carbanions of dialkyl and alkyl aryl sulfoxides are usually generated using methyllithium, lithium diisopropylamide (LDA) and n-butyllithium. The latter, as well as t-butyllithium, must be, however, used with caution since they can cause cleavage of the carbon-sulfur bond, resulting in an exchange of the organic substituent in the sulfoxide.

3.2.1 Alkylation

The first successful and very impressive application of stereoselective alkylation of an α-sulfinyl carbanion was presented by Marquet et al.[1] in the total synthesis of biotin **3**. Although they used achiral starting sulfoxide **1** having an internal symmetry plane and the product obtained was racemic, the methodology applied allowed them to introduce the desired alkyl chain *trans* to the sulfoxide oxygen with complete stereoselectivity.

MeLi
HMPA/THF
or HMPA/diglyme

$I{-}(CH_2)_4{-}CO_2Bu^t$
80%

$(CH_2)_4CO_2Bu^t$

1 **2** **3**

(3.2.1)

The same *trans* selectivity has recently been observed in the alkylation and silylation of the anion derived from the thiane oxide **4**. Moreover, the use of chiral lithium amide bases (HCLA) allowed one to obtain the corresponding reaction products **5** in optically active forms with ee up to 69%.[2]

$OSiBu^tPh_2$

1) HCLA
2) R-X
70 ÷ 90%

4 **5**

(3.2.2)

a RX = MeI
b RX = Me_3SiCl

HCLA =

A high diastereoselection was achieved in the alkylation reaction of α-sulfinyl carbanions with lithium α-bromoacrylate. The use of a sterically hindered base, namely lithium tetramethylpiperidide (LTMP), turned out to be crucial in this case and led to the corresponding sulfinyl acids **6** with dr up to 95:5. The latter were transformed into chiral α-methylene butyrolactones **8** via three consecutive reactions: reduction of the sulfoxide moiety, methylation of the sulfide formed to give a sulfonium salt **7**, and finally an intramolecular, fully stereoselective displacement of the sulfonium group.[3]

1) Base
2) $BrCH_2C(=CH_2)CO_2Li$

1) NaI, TFAA
2) $Me_3O^+BF_4^-$, -45 °C

6

t-BuOK, 0 °C
-MeSTol-p

7 **8** (3.2.3)

R	Base	dr [(R,S):(R,R)]**6**	Yield [%]
Me	LDA	82:18	88
	LTMP	95:5	81
n-Pr	LDA	79:12	82
	LTMP	84:16	72

A particularly interesting example is the alkylation of the dianions derived from β-hydroxy sulfoxides, which proceeds with a high 1,2-asymmetric induction. Thus, treatment of 2-hydroxyethyl phenyl sulfoxide **9** with 2.2 equiv. of LDA followed by the addition of an alkylating reagent leads to the formation of both possible diastereomers **10** in a ratio close to 7:1, which are separable by chromatography.[4]

1) LDA (2.2 equiv.); -78 - -20 °C
2) RX; -78 °C → r.t.

9 *syn*-**10** *anti*-**10** (3.2.4)

RX	Yield [%]	Ratio *syn:anti*
MeI	85	7:1
$MeOSO_2CF_3$	95	6.9:1
$CH_2{=}CHCH_2Br$	78	6.9:1
$PhCH_2Br$	85	6.6:1
$C_8H_{17}I$	64	1:1.7

However, when β-substituted-β-hydroxy sulfoxides **11** are used as substrates, the high 1,2-asymmetric induction and stereochemistry of the alkylation are controlled by the hydroxyl group.[5]

1) MeLi, 2 equiv.
2) R^2X

(S_S*, R_C*) - **11** → *threo* - **12** + *erythro* - **12** (3.2.5)

(S_S*, S_C*) - **13** → *threo* - **14** + *erythro* - **14** (3.2.6)

β-Hydroxy Sulfoxide	R^2X	Product: Yield [%]	Product: Ratio *threo:erythro*
11a, R^1 = Me	MeI	**12a** (87)	16:1
11b, R^1 = i-C_7H_{15}	n-$C_{10}H_{21}I$	**12b** (58)	4:1
13a, R_1 = Me	MeI	**14a** (76)	11:1
13a, R_1 = Me	$(MeO)_3PO$	**14a** (50)	10:1
13b, R^1 = i-C_7H_{15}	n-$C_{10}H_{21}I$	**14b** (75)	20:1

3.2.1.1 Synthetic Applications

Selected *anti*-diastereomers of **10** have been isolated and transformed into enantiopure epoxides **15** via a set of reactions depicted in Equation 3.2.7.[4] In a similar way, starting from *threo*-**12b**, (+) disparlure was obtained with ee = 55% (Equation 3.2.8).[5]

anti - **10** — $TiCl_4/CH_2Cl_2/NaBH_4$, DME, 0 °C → PhS — $Me_3O^+BF_4^-$, CH_2Cl_2, r.t. → NaOH → (S) - **15**

a : **R** = $PhCH_2$; $[\alpha]_D^{27}$ – 16.3
b : **R** = C_8H_{17}; $[\alpha]_D^{23}$ – 14.5

(3.2.7)

threo - **12b** — 1) Zn/Me_3SiCl; 2) $Me_3O^+BF_4^-$, NaOH; 66% →

(3.2.8)

3.2.2 Michael Addition to α,β-Unsaturated Carbonyl Compounds

Conjugate additions of the anions of dialkyl and aryl alkyl sulfoxides have been so far the subject of surprisingly narrow interest. The only reports dealing with the stereochemical course of this type of reaction concern the addition of a variety of α-sulfinyl carbanions to α,β-unsaturated carboxylic esters.

The reaction has been found to proceed smoothly and to give in most cases the conjugate addition products **16** without traces of **17** expected from a competing 1,2-addition (acylation).[6,7] A large number of t-butyl and p-tolyl alkyl, aralkyl, and benzyl sulfoxides have been tested. Stereoselection of the reaction was usually high and dependent on the spatial requirements of the unreacting group R^1 connected with sulfur. For instance, when R^1 = t-Bu the diastereomeric ratio of the products **16** was substantially higher (10:1) than that with R^1 = p-Tol, particularly for alkyl and aralkyl sulfoxides (R^1 = alkyl, CH_2Ph). In the case of benzyl sulfoxides (R^2 = Ph) the influence of the group R^1 was very little (Table 3.2.1).[6,7] The major diastereomers were isolated and their configuration determined by X-ray; the exact configuration of minor diastereomers is unknown.

16 **17** (3.2.9)

Table 3.2.1 Michael Addition of α-Sulfinyl Carbanions to α,β-Unsaturated Esters

Sulfoxide		α,β-Unsaturated esters		Pure diastereomer 16	
R^1	R^2	R^3	R^4	yield [%]	Remarks
t-Bu	CH_2Ph	H	Me	64	
t-Bu	CH_2Ph	Me	Me	68	
p-Tol	CH_2Ph	Me	Me	89	1.5:1 mixture of diastereomers
t-Bu	CH_2Ph	Ph	Me	63	
t-Bu	Et	Me	Me	64	
p-Tol	Et	Me	Me	68	1.5:1 mixture of diastereomers
t-Bu	Me	Me	t-Bu	53	
t-Bu	Ph	H	Me	78	
t-Bu	Ph	Me	Me	80	
t-Bu	Ph	Ph	Me	86	
p-Tol	Ph	Me	Me	62	
Me	Ph	Me	Me	79	84:7:5:4 mixture of diastereomers

From Casey, M., Manage, A. C., and Nezhat, L., *Tetrahedron Lett.*, **29**, 5821, 1988. Reprinted with kind permission from Elsevier Science Ltd, The Boulevard, Langford Lane, Kidlington OX5 1GB, UK.

Diastereomerically pure adducts **16** (X = OMe) and their derivatives (X = OH, NMe_2) were transformed in a highly stereoselective way into *trans*-β,γ-disubstituted butyrolactones **18a** by treatment with soft electrophiles, e.g., I_2, N-iodosuccinimide (NIS), or phenyliodonium bis-trifluoroacetate (PIFA). The oxosulfonium salt moiety, which is initially formed in this reaction, undergoes stereoselective intramolecular displacement by the carbonyl group, the mechanism of which is not known yet.[8]

16

18a 18b

(3.2.10)

R	Ar	X	[Y+]	Ratio 18a:18b	Yield [%] 18a
t-Bu	Ph	OH	1.3 equivalent NIS	97:3	69
t-Bu	Ph	OH	1.2 equivalent PIFA	100:0	71
t-Bu	Ph	NMe_2	1 equivalent PIFA	95:5	72
p-Tol	Ph	OMe	1.2 equivalent NIS	98:2	98
t-Bu	3,4,5-$(MeO)_3C_6H_2$	OH	1.2 equivalent NIS	91:9	89

Although all the experiments described above have been performed using racemic substrates, the high diastereoselectivity of the reactions should allow development of a convenient method for the asymmetric synthesis of chiral compounds.

3.2.3 Hydroxyalkylation

The reaction of α-sulfinyl carbanions with aldehydes and ketones, leading to β-hydroxyalkyl sulfoxides, is generally known to be highly diastereoselective with respect to the α-sulfinyl carbon, but poorly diastereofacially selective with respect to attack on the carbonyl component.

Durst and co-workers[9] were the first to report the condensation of chiral α-sulfinyl carbanions with carbonyl compounds. Thus, the anion derived from (+)-(S)-benzyl methyl sulfoxide was quenched with acetone to give a mixture of diastereomeric β-hydroxy sulfoxides in a 15:1 ratio (note that in this case a new chiral center is created only on the α-carbon atom and its stereochemistry is governed by the general rules described in Section 3.1).

1) MeLi 2) Me_2CO

(+)-(S) → (S_S, R_C) - major diastereomer

(3.2.11)

When the anions derived from methyl sulfoxides are reacted with aldehydes or unsymmetrical ketones, the new chiral center is created only on the β-carbon atom. The diastereoselectivity of this process is rather low, but can be substantially increased by the use of properly substituted sulfoxides and by the addition of zinc or magnesium salts (Table 3.2.2). Interestingly, a trend of decreasing diastereoselectivity on increasing the size and branching of the alkyl substituents in ketones is observed in the reaction (Entries 10–12).

(R) → (R, R) prevailing diastereomer (3.2.12)

Table 3.2.2 Hydroxyalkylation of Carbanions Derived from Aryl Methyl Sulfoxides

Entry	Ar	R[1]	R[2]	Base	Yield [%] (R,R+R,S)	Diastereomeric excess [%]	Ref.
1	p-Tol	Ph	H	LDA	95	0	10
2	p-Tol	Ph	H	LDA + $ZnBr_2$	35	60	10
3	p-Tol	Ph	H	LDA + $ZnCl_2$		80	11
4	2-Pyridyl	Ph	H	LDA + $MgBr_2$	88	60	10
5	p-Tol	CF_3	Ph	LDA	~100	50	12
6	1-Naph[a]	Ph	Me	Et_2NLi	96–98	100	13
7	1-Naph[a]	Ph	Et	Et_2NLi	96–98	100	13
8	1-Naph[a]	Ph	Pr	Et_2NLi	96–98	100	13
9	1-Naph[a]	Ph	Bu	Et_2NLi	96–98	100	13
10	1-Naph[a]	Ph	*i*-Pr	Et_2NLi	96–98	44	13
11	1-Naph[a]	Ph	*t*-Bu	Et_2NLi	96–98	50	13
12	1-Naph[a]	Ph	*n*-Hex	Et_2NLi	96–98	60	13
13	1-Naph[a]	Et	Me	Et_2NLi	96–98	6	13

[a] Starting sulfoxide has (S) configuration; hence prevailing diastereomers have (S,S) configuration.

In the case of the reaction of sulfoxides possessing α-methylene group with aldehydes or unsymmetrical ketones, two new chiral centers at the α- and β-carbon atoms are simultaneously created. However, it has been found that when *tert*-butyl sulfoxides are used (instead of the commonly applied p-tolyl sulfoxides), the reaction leads to the formation of only two of four possible diastereomers. This is due to the complete steric control over the α-center and a moderate control at the β-center.[14-16] The results collected in Table 3.2.3 reveal some consistent trends: (a) with unhindered aldehydes diastereoselection is poor, (b) hindered aldehydes or ketones and sulfoxides bearing bulky substituents lead to the products with better diastereoselection, and (c) other factors, like solvent, temperature, transmetallation, and the kind of base applied have only moderate influence on stereoselectivity.

1) Base, THF
2) $R^2C(O)R^3$

19 anti syn (3.2.13)

Table 3.2.3 Hydroxyalkylation of *tert*-Butyl Sulfoxides **19**

Entry	R^1	R^2	R^3	Base	Total yield [%]	Ratio *anti/syn*	Ref.
1	Ph	Et	H	LDA	85	2.0	15
2	Ph	Et	H	LDA (Zn^{2+})	85	4.0	15
3	Ph	Ph	H	LDA	81	1.7	15
4	Ph	Ph	H	LDA (Zn^{2+})	98	5.3	15
5	Ph	i-Pr	H	LDA	90	4.0	15,16
6	Ph	i-Pr	H	LDA (Zn^{2+})	72	10.0	15
7	Ph	Ph	Me	LDA	90	3.0	16
8	n-Pr	Ph	H	LDA	85	1.2	16
9	n-Pr	i-Pr	H	LDA	81	5.0	16
10	i-Pr	i-Pr	H	LDA	76	8.0	16
11	i-Pr	Ph	Me	LDA	98	5.0	16
12	Ph	Ph	Me	LHMDS	83	5.0	16
13	$(CH_2)_4CHMe_2$	$C_{10}H_{21}$	H	BuLi	75	1.5	14

Among aryl alkyl sulfoxides bearing additional functional groups, those containing α-hydroxyl group deserve special attention. Williams et al.[17] have found that the stereoselectivity of the reaction of the dianions of γ-hydroxysulfoxides with aldehydes depends on the relative configuration of the carbinol and the sulfinyl centers. Thus, the diastereomeric ratio observed upon deprotonation of the sulfoxide **20** with LDA at –78°C, quenching with benzaldehyde and workup at –78°C was 91:9, while in the case of the diastereomer with the opposite configuration at sulfur **20a** the reaction led to a mixture of four possible diastereomers in a 67:17:13:3 ratio.

1) LDA
2) PhCHO
85%

20 **21a** 91 : 9 **21b**

1) LDA
2) PhCHO
94%
mixture of four diastereomers

20a (3.2.14)

The results were rationalized in terms of internal coordination of the lithium cation. The role of coordination was indirectly confirmed by performing the same reaction with a desoxy analog of **20**, which resulted in a poor stereoselectivity. Moreover, the use of N-methyl-2-imidazolyl sulfoxides **22** and **23**, in which the imidazolyl group exhibits a high ability to coordinate Li^+ (Equations 3.2.15 and 3.2.16) gave the corresponding adducts with dr = 5:1.[17]

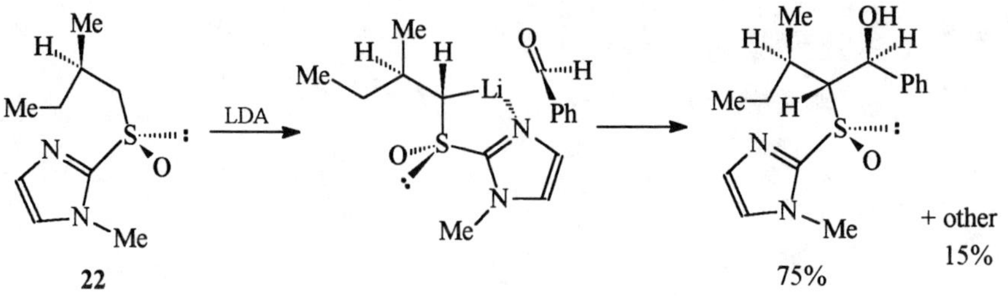

(3.2.15)

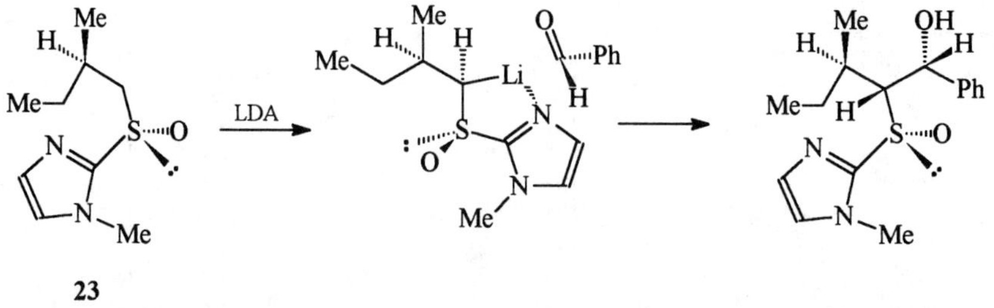

(3.2.16)

3.2.3.1 Synthetic Applications

3.2.3.1.1 Synthesis of Juvabiols[18]

In the title synthesis two diastereomeric, enantiopure sulfoxides **24** and **25** were applied as substrates. Each of them was separately deprotonated and quenched with 3-methylbutanal to give a mixture of diastereomeric adducts in a 3:2 ratio. The sequence of reactions for sulfoxide **24** is shown in Equation 3.2.17.

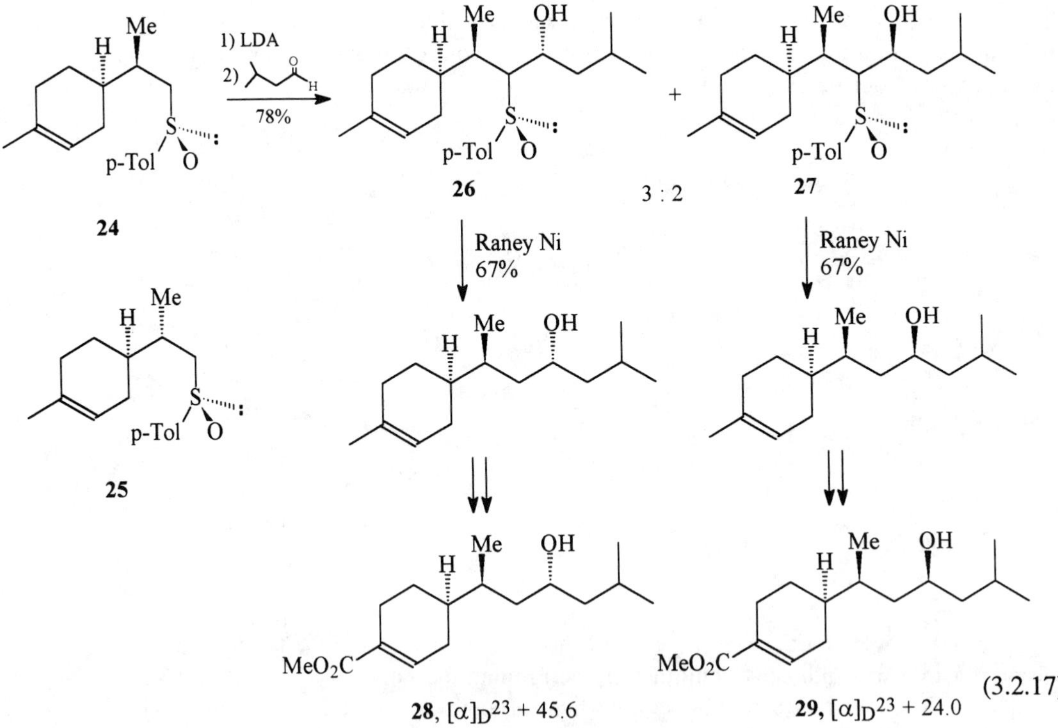

(3.2.17)

The diastereomers **26** and **27** were separated by chromatography and each was individually reduced with Raney nickel. The vinylic methyl group (C-7) was then transformed into a methyl ester via a seven-step procedure to give enantiopure isoepijuvabiol **28** and epijuvabiol **29**.[18]

3.2.3.1.2 General Stereoselective Synthesis of Substituted Tetrahydrofurans[19]

When the adduct **21a** is treated with an excess of acetyl bromide, a dehydration and cyclization occur, giving the tetrahydrofuran **30** as a single stereoisomer.

CH_3COBr, 3 eq., 0 °C→ r.t.
77%

21a **30** (3.2.18)

Cyclizations can be performed more efficiently and in higher yields when the corresponding sulfides, e.g., **31**, are used as substrates (Table 3.2.4) and dimethyl sulfate as a methylating agent. The reaction is assumed to proceed via the episulfonium salt **32** as an intermediate.

$(MeO)_2SO_2$

31 **32**

(3.2.19)

3.2.3.1.3 Synthesis of the Chiral Sequences of Macrolides

Reaction of the anion of the properly substituted racemic sulfoxide **33** with the ketone **34** has been used to create a new asymmetric center at the carbon atom C-6 in the acyclic intermediate for Erythronolide A (**35**).[20] As a result, a 5:1 diastereomeric mixture has been obtained in 80% yield at 50% conversion and separated into a major and minor isomer at C-6. Ozonolysis and desulfurization of the major isomer gave the ketol **35** in 84% yield as a single isomer.

1) LDA, 2 equiv., glyme
2) $O_3/CH_2Cl_2/MeOH/Me_2S$
3) Raney Ni

33 **34**

Table 3.2.4 Synthesis of Selected Tetrahydrofuranes **30** from Bis(Hydroxyalkyl)Sulfides **31**

Substrate 31	**Product 30**	**Yield [%]**
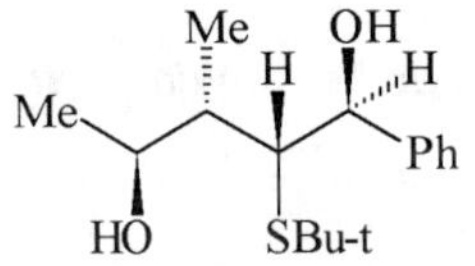	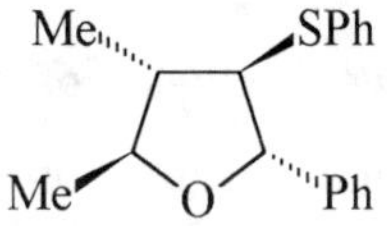	96
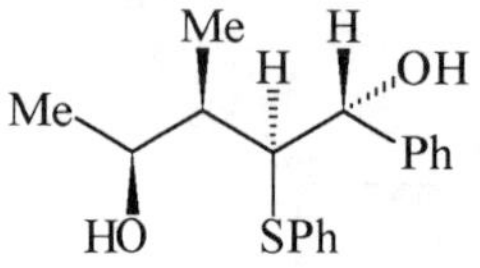	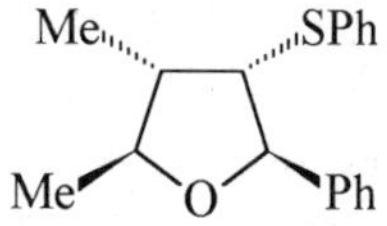	90
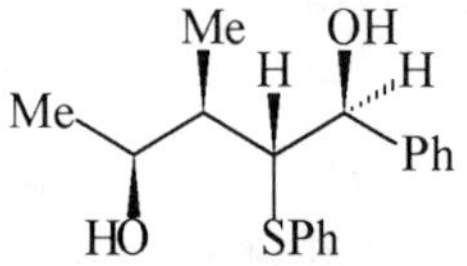	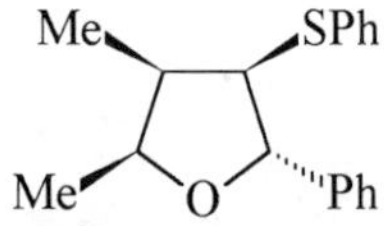	85
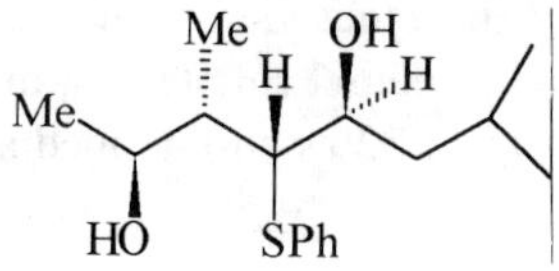	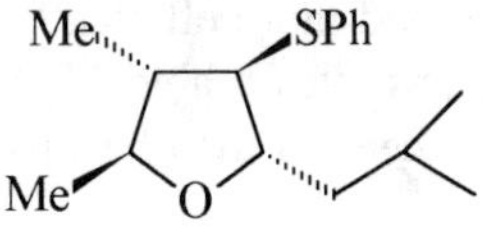	100
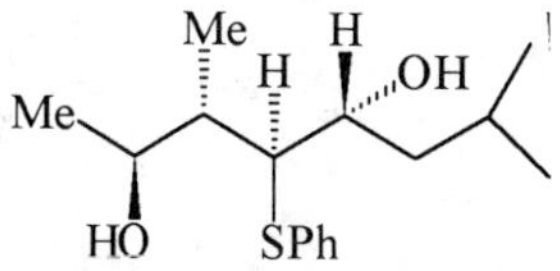	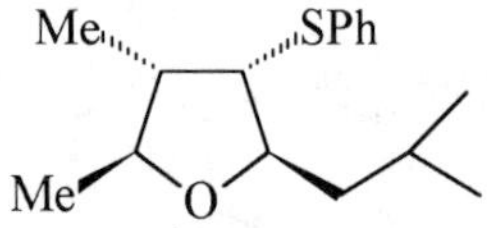	98

From Williams, D. R., Phillips, J. G., *Tetrahedron,* **42**, 3013, 1986. Reprinted with kind permission from Elsevier Science Ltd, The Boulevard, Langford Lane, Kidlington OX5 1GB, UK.

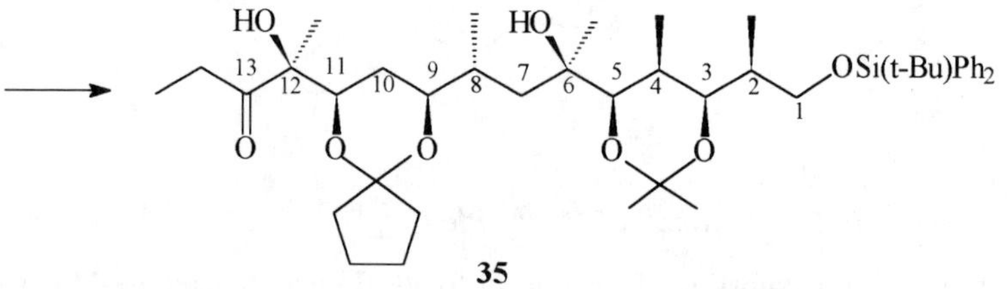

35 (3.2.20)

The above example clearly demonstrates that the stereochemistry of the sulfoxide group exerts no influence on the ratio of diastereomers formed; most probably this is the lithium coordination with carbonyl oxygen atoms which is responsible for the diastereoselection observed. Similar results have been obtained by Masamune et al.[21] in their synthesis of the A fragment of the aglycon of Amphotericin B. In this case, regardless of which enantiomer (or a racemate) of the sulfoxide **36** was applied, the diastereomeric ratio of the product **38** was always 15:1.

BnO O O S—Tol-p + O OSiMe$_2$t-Bu 1) BuLi 2) Raney nickel 90%

36 **37**

OH OSiMe$_2$t-Bu BnO

38 (major diastereomer)

dr 15 : 1 (3.2.21)

However, in a more recent work Kochetkov et al.[22] showed that the absolute configuration of the sulfoxide moiety is crucial in the alternative synthesis of Erythronolides A and B (Equation 3.2.22; cf. Equation 3.2.20). In this case, sulfoxides **39a** (instead of **33**) and **39b** were used as substrates. When the sulfoxide (R_S)-**39b** was lithiated and condensed with the ketone **40**, a mixture of two diastereomeric products **41b** in a 7:1 ratio was formed. On the other hand, no reaction was observed when the diastereomeric sulfoxide (S_S)-**39b** was used. In turn, the sulfoxide **39a** gave under the same conditions the product **41a** in a low yield.[22]

(R_S) - **39** **40**

a : R = OH, R' = H

b : R = H, R' = TBDS

LDA/THF, -78°C

41 (3.2.22)

3.2.3.1.4 Syntheses of Chiral Sulfinylbutadienes

β-Hydroxyalkyl sulfoxides can eliminate water to give the corresponding vinyl sulfoxides. This is usually achieved by introducing proper substituents in the alkyl chain of the sulfoxide which serve as good leaving groups. In this way, Maignan et al.[23] succeeded in synthesizing enantiopure 1-α-hydroxyvinyl p-tolyl sulfoxides **43**, starting from 2-ethoxyethyl p-tolyl sulfoxide **42**.

p-Tol S(O) CH2CH2OEt —[1) LDA, 2 equiv., THF; 2) RC(O)R, THF]→ p-Tol S(O) C(=CH2) C(OH)R R

(R) - **42**

$[\alpha]_D^{20}$ +169 (EtOH)

(S) - **43**

(3.2.23)

R_2	Yield of **43**	$[\alpha]_D^{20}$ (acetone)
Me_2	68	+277
$-(CH_2)_2-$	58	+125

The same substrate was used in the synthesis of 2-sulfinylbutadiene **44**.[24]

(R)-**42** —[1) BuLi; 2) MeCHO; 48%]→ p-Tol S(O) C(=CH2) CH(OH)Me —[MsCl; 93%]→ p-Tol S(O) C(=CH2) CH(OMs)Me —[DABCO; 62%]→ (R)-**44**, $[\alpha]_D^{20}$ +174

(3.2.24)

In turn, 1-sulfinylbutadienes **46** were synthesized in the reaction of the anion of enantiopure methyl p-tolyl sulfoxide with α,β-unsaturated aldehydes, which, unlike that with α,β-unsaturated esters (cf. Section 3.2.2), proceeds exclusively in a 1,2-manner. Dehydration of the β-hydroxy adduct **45** was easily achieved with an excess of sodium hydride and methyl iodide.[25]

$R^1R^2C{=}CH{-}CHO$ + Me S(O) Tol-p (+)-(R) —[LDA/THF, -78°C]→ **45** —[1) NaH, 2.5 equiv., THF, 0°C; 2) MeI, 2.4 equiv., 0°C → r.t.]→ (*R*) - **46**

(3.2.25)

R^1	R^2	**45** Yield [%]	**46** Yield [%]	$[\alpha]_D^{20}$ (acetone)
Me	H	83	68	+224.5
Me	Me	37	69	+216.1
Et	H	92	72	+269.1
Ph	H	82	82	+225.1
$2\text{-}MeOC_6H_4$	H	64	70	+121.2

3.2.3.1.5 Asymmetric Synthesis of α,α-Disubstituted Cyclobutanones[26]

α-Hydroxyalkyl cyclopropyl sulfoxides **47**, which are formed from the anion of cyclopropyl p-tolyl sulfoxide with low diastereoselectivity (diastereomer ratio 3:2), undergo a 1,2-rearrangement to give a mixture of diastereomeric sulfinyl cyclobutenes **50**. It should be added that the major diastereomer formed has always the same configuration on the newly created carbon chiral center irrespective of the diastereomer of **47** applied. The transformation is believed to proceed via the carbonium ion **48**, the degree of asymmetric induction being dependent on the difference in the thermodynamical stability of the intermediates **49a** and **49b**. The sulfinyl cyclobutenes **50** were finally elaborated to cyclobutanones **51** with moderate to good enantiomeric purity.[26]

(R) (100% ee) — 1) BuLi; 2) $R^1C(O)R^2$ → **47** — p-$TolSO_3H$, benzene → [**48** → **49a** + **49b**] → **50** — 1) AcCl; 2) $TiCl_4$, $Pb(OH)_2$, H_2O → **51**

(3.2.26)

R^1	R^2	**47** Yield [%]	**50** Yield from **47**	**51** Yield from **50**	$[\alpha]_D$ (EtOH)	ee [%]
Ph	Me	78	88	68	–9.6	94
Et	Me	72	65		–6.3	73.3

3.2.3.1.6 Syntheses of Five- and Six-Membered Lactones

Addition of the dianion of (+)-(R)-3-(p-toluenesulfinyl)propionic acid **52** to aldehydes gives corresponding hydroxyalkyl derivatives which undergo spontaneous cyclization to γ-lactones. The reaction proceeds with a poor stereoselectivity (diastereomer ratio 3:2), but the diastereomers are readily separable. Bravo et al.[27] and Albinati et al.[28] have used this reaction to obtain enantiopure butenolides **55** and **56** (after thermal elimination of p-toluenesulfenic acid) and γ-butyrolactones **57** and **58** (after desulfurization). Similar reaction has been applied to the synthesis of 2-methylene valerolactones and pentenolides.[29]

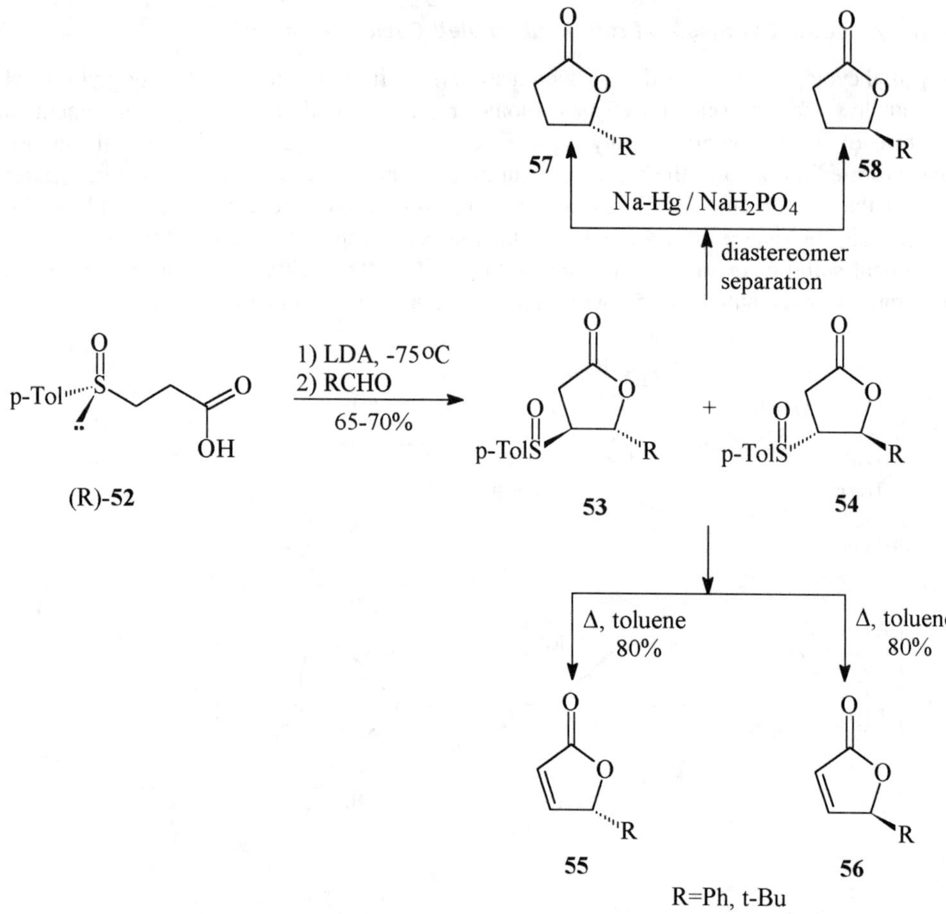

(3.2.27)

R=Ph, t-Bu

3.2.4 Aminoalkylation

Aminoalkylation of α-sulfinyl carbanions takes place when they are treated with compounds having a multiple carbon-nitrogen bond. Authors of some early reports claimed that aminoalkylation of the anions derived from optically active aryl alkyl sulfoxides proceed with a full β-diastereoselection.[30]

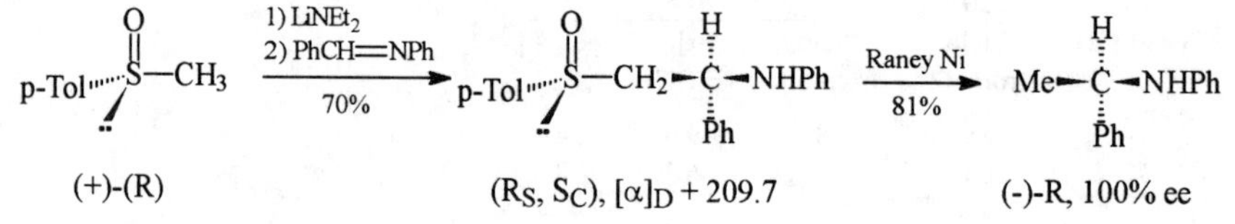

(3.2.28)

Some recent findings have cast doubts on these results.[31] The reaction of α-sulfinyl carbanions with imines has been found to be very sensitive to the conditions applied and thus very capricious. Since several different reports concern reactions performed under various conditions and using differently substituted sulfoxides and imines, the results presented seem inconsistent. Nevertheless, some visible tendencies can be observed (Equation 3.2.29 and Table 3.2.5).

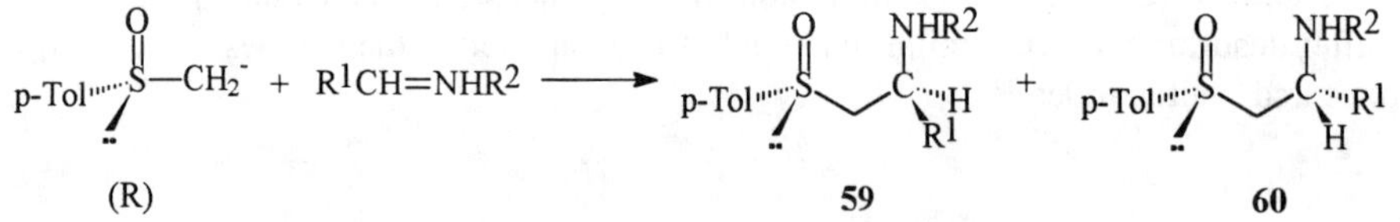

(3.2.29)

Table 3.2.5 Aminoalkylation of Methyl p-Tolyl Sulfoxide

R^1	R^2	Deprotonation temperature [°C]	Reaction temperature [°C]	Reaction time	Ratio 59:60	Yield [%]	Ref.
Ph	Ph	−15	−15	10 min	75:25	98	31
Ph	Ph	0	−78	10 min	92:8	99	31
Ph	Ph	−78	−78 → 0	5 min	86:14	95	32, 33
p-$MeOC_6H_4$	p-$MeOC_6H_4$	0	−78	40 min	95:5	74	31
Ph	Me	−20	−20	10 min	82:18	84	31
Ph	Me	−78	−78 → 0	10 min	91:9	89	32, 33
Ph	Me	−78	−78 → 0	12 h	51:49	89	32, 33
2-Furyl	Ph	−78	−78 → 0	10 min	91:9	96	32, 33
Ph	n-Pr	−20	−20	10 min	90:10	76	31
Ph	i-Pr	−20	−20	10 min	88:12	27	31
i-Pr	Ph	−78	−78 → −45	2 h	81:19	72	32, 33
i-Pr	Ph	−78	−78 → 0	2 h	66:44	62	32, 33
Ph	Cyclopropyl	−20	−20	10 min	84:16	21	31
Ph	n-C_5H_{11}	−20	−20	10 min	87:13	52	31

Thus, the reaction requires strict controlling of the temperature. In some cases the best stereoselectivity is obtained when deprotonation is performed at 0°C and the quench with an imine at −78°C. In general, diastereoselection is better under kinetic control conditions; equilibration usually causes a noticeable drop of the diastereomeric ratio. N-Alkyl imines generally give poorer diastereoselection.

Finally, Pyne and Boche[34] have found that bulky substituents in sulfoxides cause a substantial increase of diastereoselectivity of this reaction. They have reacted the carbanion of t-butyl benzyl sulfoxide with a series of N-phenyl imines and obtained corresponding β-amino sulfoxides with a very high *anti*-diastereoselection.

t-Bu–S(O)–$\bar{C}$HPh + RCH=NPh —(-78°C)→ *anti* - **61** + *syn* - **61**

R	Ratio anti:syn	Yield [%]
Ph	>97:<3	94
2-furyl	97:3	100
(E)-PhCH = CH	>97:<3	52
(E)-PhCH = CH + $ZnCl_2$	>97:<3	82

(3.2.30)

Despite its high sensitiveness to the conditions, the reaction has been successfully applied to the synthesis of enantiopure (+)-(R)-carnegine **65** and (+)-(R)-tetrahydropalmatine **68**. In these cases, the most favorable product diastereoselection (92:8) was unexpectedly observed under equilibrium controlled conditions.[32,33] The major diastereomer **62** was isolated and elaborated to carnegine by N-methylation followed by desulfurization of **64** (cf. Section 3.9.4.1, Scheme 3.9.5). Transformation of **62** into tetrahydropalmatine required reductive alkylation with 2,3-dimethoxybenzaldehyde, intramolecular Pummerer-type cyclization of the sulfoxide **66** and final desulfurization (cf. synthesis of canadine, Section 3.9.4.2, Scheme 3.9.6).

MeO MeO N + p-Tol S(O)—CH2Li 92%

NH H S=O p-Tol 62 + NH H S=O p-Tol 63 92 : 8 (3.2.31)

62 (CH2O)n NaCNBH3 92% NMe H S=O p-Tol 64 Raney Ni NMe Me (-)-(R)-65 (3.2.32)

62 + CHO OMe OMe NaCNBH3 87% N H p-TolSO OMe OMe 66 1) TFAA 2) Δ 82%

67 H p-TolS H OMe OMe Raney Ni 92% (+)-(R)-68 (3.2.33)

The reaction of α-sulfinyl anions with nitrones was reported by Annunziata and Cinquini[35] to proceed with a very high β-stereoselectivity (Equation 3.2.34). In turn, recently Pyne and Hajipour[36] have shown, using racemic methyl phenyl sulfoxide, that the diastereoselectivity of this reaction is only moderate (Equation 3.2.35).

$$\text{p-Tol-S(O)-CH}_2\text{Li} + \text{Ph-CH=N(\to O)-R} \longrightarrow \text{p-Tol-S(O)-CH}_2\text{-}\overset{*}{\text{C}}\text{H(Ph)-N(R)OH} \quad \mathbf{69} \qquad (3.2.34)$$

R	dr	Yield
Me	75:25	85
Ph	82:18	77
t-Bu	100:0	84

$$\text{Ph-S(O)-CH}_2\text{Li} + \text{R}^1\text{-CH=N(\to O)R}^2 \longrightarrow \text{Ph-S(O)-CH}_2\text{-C(H)(N(R}^2\text{)OH)-R}^1\ \mathbf{70a} + \text{Ph-S(O)-CH}_2\text{-C(N(R}^2\text{)OH)(H)-R}^1\ \mathbf{70b} \qquad (3.2.35)$$

R^1	R^2	Ratio **70a:70b**
Ph	Me	67:33
Ph	t-Bu	85:15
Me	t-Bu	50:50

3.2.5 Acylation

Acylation of α-sulfinyl carbanions leads to β-oxosulfoxides. The usual acylating agents are carboxylic esters, acid chlorides, and imidazolides, but in some cases also free acid salts. Optically active β-oxosulfoxides are a very important class of chiral sulfur compounds and therefore their transformations and synthetic applications are separately discussed in Section 3.8. In this section selected examples of their synthesis are presented (Equation 3.2.36, Table 3.2.6).

$$\underset{(R)}{\text{p-Tol-S(O)-CH}_2\text{R}^1} \xrightarrow[\text{2) R}^2\text{C(O)OR}^3]{\text{1) LDA}} \underset{\mathbf{71}}{\text{p-Tol-S(O)-CH(R}^1\text{)-C(O)-R}^2} \qquad (3.2.36)$$

Another interesting example is the acylation of the carbanion derived from the thiane oxide **4**. The acyl group is here introduced exclusively *cis* to the sulfinyl oxygen, which is in accordance with the rules of the α-sulfinyl carbanion reactivity towards electrophiles (cf. Section 3.1). Moreover, the product **72b** has been obtained directly from **4** in 56% ee using a chiral base (cf. alkylation of **4**, Equation 3.2.2).[45]

$$\mathbf{4}\ (\text{OSiBu}^t\text{Ph}_2) \xrightarrow[\text{2) RCO}_2\text{Et}]{\text{1) Base}} \mathbf{72}\ (\text{OSiBu}^t\text{Ph}_2,\ \text{C(O)R}) \qquad (3.2.37)$$

72	R	Yield [%]
a	Me	73
b	Et	86
c	i-Pr	52
d	t-Bu	82
e	Ph	81

Table 3.2.6 Acylation of α-Sulfinyl Carbanions

71	R^1	R^2	R^3	Yield [%]	Absolute config.	$[\alpha]_D$ $CHCl_3$, c = 0.7–2.0	M.p. [°C]	Remarks	Ref.
a	H	Et	Et	60	R	+210	69–70		37
b	H	i-Pr	Et	67	R	+191.3	57–59		37
c	H	t-Bu	Et	82	R	+162	108–108.7		37
d	H	Ph	Et	62	R	+180.9	82–83.5		37
e	H	CH_2F	Et	75	R	+242	120–122		38
f	H	CHF_2	Li	72	R	+251	95–97	Hydrate	38
g	H	CF_2Cl	Li	90	R	+240	120–122	Hydrate	38
g	H	CF_2Cl	Me	82	R				
h	H	CF_3	Li	91	R	+199	95–96	Hydrate	38
h	H	CF_3	Et	90	R	+181.5			40
i	Me	CF_3	Li	80	$R_S(R+S)_C$	+130	100–102	Mixt. of C-1 epimers; hydrate	38
j	H	CF_2Ph	CH_2Ph	92	R	n.r.[a]	n.r.	Mixt. of the keto and hydrated forms	41
k	H	CH_2Cl	Et	90	R	+288	127–129		39
l	H	CH_2Br	Me	83	R	+189	125–127		39
m	H	CMe_2Br	Me	88	R	+217	82–83		39
n	H	CHMeCl	Et	88	R_S,S_C	+7.7	55–57 liq.	Diastereomers	39
n	H	CHMeCl	Et		R_S,R_C	+412		separated by flash chrom.	
o	H	CHMeBr	Et	81	R_S,S_C	−71	77–79	Diastereomers	39
o	H	CHMeBr	Et		R_S,R_C	+382	62–63	separated by flash chrom.	
p	Ph	CH_2Cl	Et	80	R_S	+321	123–125	Single diast.; Conf. at C-1 not determined	39
r	Me	CH_2Cl	Et	88	$R_S(R+S)_C$	+104	44–47	55:45 mixt. of epimers	39
s	H	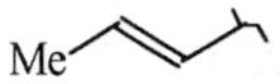 Me	Et	65	(E)-R	+245	n.r.		42
t	H	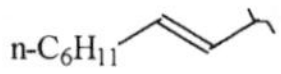 $n\text{-}C_6H_{11}$	Et	61	(E)-R	+177	n.r.		42
u	H	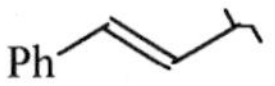 Ph	Et	70	(E)-R	+278	n.r.		42
v	H	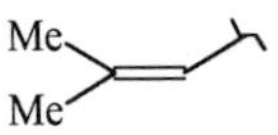 Me, Me	Et	88	R	+278	n.r.		43
w	H	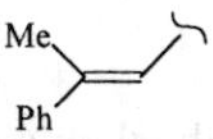 Me, Ph	Me	85	(E)-R	+174	n.r.		43
x	H	$MeC(O)CH_2$	Et	75	R	n.r.	n.r.		44
y	H	$MeCH{=}CHC(O)CH_2$	Et	80	R	n.r.	n.r.		44
z	H	$PhC{\equiv}CC(O)CH_2$	Et	70	R	n.r.	n.r.		44

[a] n.r. = not reported.

3.2.6 Examples of Experimental Procedures

3.2.6.1 Typical Procedure for Asymmetric Alkylation of the Thiane Oxide 4 (See Equation 3.2.2)[2]

A solution of the HCLA base is prepared by treatment of the corresponding amine (0.19 g, 0.74 mmol) in THF (2 mL) under N_2 at −78°C with a solution of n-BuLi (0.46 mL of 1.6 M solution in hexane, 0.74 mmol), followed by warming to RT for 1 h. The resulting yellow solution is recooled to −78°C and added dropwise to a solution of the sulfoxide **4** (0.19 g, 0.51 mmol) in THF (5 mL) at −78°C. The mixture is stirred at −78°C for 1 h and then Me_3SiCl (0.5 mL, 4 mmol) is added.

After stirring for a further 15 min saturated aqueous $NaHCO_3$ (10 mL) is added and the organic product extracted into Et_2O (2 × 20 mL). The organic extract is dried over $MgSO_4$, evaporated under reduced pressure, and subjected to column chromatography (Et_2O as eluent) to give **5b** as a colorless oil (0.206 g, 91%), $[a]_D^{21}$ + 28 (c = 2.05, CH_2Cl_2), 60% ee (from the ^1H NMR experiment in the presence of the shift reagent $Eu[hfc]_3$).

3.2.6.2 Reaction of Lithium and Zinc Derivatives of the Sulfoxides 19 with Carbonyl Compounds: General Procedure (See Equation 3.2.13, Table 3.2.3)[15]

To a solution of *tert*-butyl phenylmethyl sulfoxide **19** (0.186 g, 1.0 mmol) in THF (5 mL) at –78°C is added n-BuLi in hexane (1.1 mmol). After 10 min, a solution of $ZnCl_2$ (1.1 mL of 1.0 M solution in ether) is added (solution turns from a pale yellow color to colorless), and stirring is continued for 10 min. The neat aldehyde (1.2 mmol) or benzophenone (1.2 mmol in 1 mL of THF) is then added dropwise. After 5 min, the solution is quenched rapidly with saturated NH_4Cl (0.5 mL), and then ether (20 mL) and water (20 mL) are added. The layers are separated, and the ether layer dried (K_2CO_3) and concentrated. The diastereoselection of these reactions is determined from ^1H NMR analysis of the crude product. The crude product is purified by column chromatography on silica gel employing AcOEt/hexane as eluent.

3.2.6.3 (2R, R_S)- and (2S, R_S)-1,1,1-Trifluoro-2-Phenyl-(p-Tolylsulfinyl)Propan-2-ol (See Table 3.2.2, entry 5)[12]

A solution of (+)-(R)-methyl p-tolyl sulfoxide (5.78 g, 37.5 mmol) in THF (60 mL) is added dropwise at –78°C and under argon into a stirred solution of LDA (41.2 mmol) in the same solvent (45 mL). After 3 min, 2,2,2-trifluoroacetophenone (5.8 mL, 41.3 mmol) is added at –78°C, stirring is continued for 10 min, and then a saturated aqueous solution of NH_4Cl is added. The aqueous layer is separated and extracted with AcOEt (3 × 100 mL), and the combined organic layers are dried (Na_2SO_4) and evaporated to give a 75:25 mixture of diastereomers (2R, R_S) and (2S, R_S) in nearly quantitative yield. Single pure diastereomers are isolated through flash chromatography (n-hexane/AcOEt 3:1):

(2R,R_S): R_f(n-hexane/AcOEt 3:1) 0.35; $[\alpha]_D^{20}$ + 125.7 (c 1.11, $CHCl_3$); mp 85–87°C; ^1H NMR: δ 2.48 (s, 3H, MeAr), 3.49 (s, 2H, CH_2S), 7.2–7.8 (m, 9H, ArH).

(2S, R_S): R_f 0.31; $[\alpha]_D^{20}$ + 171.7 (c 0.64, $CHCl_3$); mp 76–78°C; ^1H NMR: δ 2.41 (s, 3H, MeAr), 3.13 and 3.40 (m, 1H each, CH_2S), 7.2–7.7 (m, 9H, ArH).

3.2.6.4 Condensation of (+)-(R)-Methyl p-Tolyl Sulfoxide with α,β-Unsaturated Aldehydes; General Procedure[25]

To a solution of diisopropylamine (115 μL, 0.818 mmol) in dry THF (1 mL) cooled to –50°C, is added slowly a solution of BuLi in hexane (1.55 M, 520 μL, 0.799 mmol) under an Ar atmosphere. The mixture is stirred at –50°C for 30 min. Then a solution of (+)-(R) methyl p-tolyl sulfoxide (100 mg, 0.649 mmol) in THF (6.5 mL) cooled to –50°C is added slowly and the resulting mixture stirred at the same temperature for 30 min. Then after cooling down to –78°C, the corresponding aldehyde (1.298 mmol) is added. The resulting solution is stirred at –78°C and the progress of the reaction is monitored by TLC (hexane/EtOAc, 3:7). The mixture is hydrolyzed by addition of a saturated aqueous NH_4Cl solution (10 mL), extracted with CH_2Cl_2 (3 × 10 mL), dried (Na_2SO_4), and the solvent evaporated. The product **45** is purified by flash chromatography (hexane/EtOAc, 3:7).

3.2.6.5 4-Substituted (1E, 3E)-1-[(R)-p-Tolylsulfinyl]-1,3-Butadienes 46; General Procedure[25]

A solution of the corresponding diastereomeric mixture of β-hydroxy sulfoxides **45** in dry THF (10 mL per mmol) is slowly added to a cold (0°C) slurry of NaH (2.5 equiv.) in THF (2.4 mL per mmol). The resulting mixture is stirred for 20 min. Then MeI (2.4 equiv.) is added via a syringe and after 30 min at 0°C, the mixture is allowed to reach room temperature and stirred till the total conversion of the starting product is achieved (TLC, hexane/EtOAc, 7:3). The mixture is diluted with Et_2O and filtered through Celite. The resulting solution is washed twice with a saturated solution of $NaHCO_3$ (2 × 10 mL), dried (Na_2SO_4) and the solvents evaporated. The product **46** is purified by flash chromatography (hexane/EtOAc, 8:2) (Equation 3.2.25).

3.2.6.6 Reaction of the Carbanion of (+)-(R)-Methyl p-Tolyl Sulfoxide with Imines; General Procedure[33]

To a solution of diisopropylamine (121 mg, 1.2 mmol) in dry THF (3 mL) at 0°C is added n-BuLi in hexane (1.1 mmol). After 10 min the solution is cooled to –78°C and then a solution of (+)-(R) methyl p-tolyl sulfoxide (1.0 mmol) in THF (2 mL) is added dropwise. After 1 h at –78°C a solution of the imine (1.3 mmol) in THF (2 mL) is added. After 1 h, the solution is warmed slowly to the temperature specified in Table 3.2.5 for the designated period of time. The reaction is then quenched rapidly by the addition of 10% K_2CO_3 (10 mL) and then extracted with $CHCl_3$ (2 ×). The combined extracts are dried ($MgSO_4$) and evaporated. The diastereoselection of these reactions is determined from ^1H NMR (400 MHz) analysis of the crude reaction product. Purification of the crude product by column chromatography on silica gel using ethyl acetate/hexane as eluent gives the pure product. Yields are reported in Table 3.2.5.

3.2.6.7 Synthesis of β-Oxosulfoxides 71; General Procedure[37]

n-Butyllithium (4 mmol) in n-hexane is added dropwise at –40°C to a stirred solution of diethylamine (4 mmol) in THF (10 mL). The mixture is kept for 30 min below 0°C, cooled to –30°C, and methyl p-tolyl sulfoxide (2 mmol) in THF (10 mL) is added dropwise. The mixture is allowed to reach room temperature, cooled again to –40°C, and the ester (3 mmol) in THF (5 mL) added in one portion. The mixture is kept for 30 min at 0°C, heated at reflux for the appropriate time, and quenched with saturated aqueous ammonium chloride. The organic layer is separated off and the aqueous layer acidified with dilute sulfuric acid to pH 3–4 and extracted with dichloromethane. The combined organic phases are concentrated, and the residue is chromatographed on silica with ether-light petroleum as eluent. Yields, specific rotations, and physical data are reported in Table 3.2.6.

3.2.6.8 3-Fluorinated 2-Oxopropyl Sulfoxides; General Procedure[38]

Method A: a solution of the starting sulfoxide (10 mmol) in dry THF (20 mL) is added dropwise to a stirred solution of the LDA [10.5 mmol; prepared from diisopropylamine (1.49 mL) and a 2.5 normal solution of butyllithium in hexane (4.2 mL)] maintaining the temperature at –75°C and under argon. After 3 min, a suspension of the lithium carboxylate [obtained from the corresponding carboxylic acid (15 mmol) and lithium hydride (20 mmol) in tetrahydrofuran (40 mL)] is added at –40°C and stirring is continued for 15 min.

Method B: a solution of the starting sulfoxide (10 mmol) in dry THF (20 mL) is added dropwise to a stirred solution of LDA [10.5 mmol; prepared from diisopropylamine (1.49 mL) and a 2.5 normal solution of butyllithium in hexane (4.2 mL)] at –75°C under argon. After 3 min, a solution of the fluorinated acetic ester (15 mmol) in THF (40 mL) is added at –75°C and stirring is continued for 5 min.

Workup: the reaction is then quenched by the addition of a saturated aqueous solution of ammonium chloride (60 mL). The pH is adjusted to 7 with dilute hydrochloric acid, the layers are separated, and the aqueous layer is extracted with ethyl acetate (3 × 70 mL). The organic layers are combined and dried with sodium sulfate. Evaporation and column chromatography or crystallization of the residue gives the pure product **71e–i**, Table 3.2.6.

REFERENCES (3.2)

1. Lavielle, S., Bory, S., Moreau, B., Luche, M., Marquet, A., *J. Am. Chem. Soc.*, **100**, 1558, 1978.
2. Cox, P.J., Persad, A., Simpkins, N.S., *Synlett*, 195 and 197, 1992.
3. Bravo, P., Resnati, G., Viani, F., *Tetrahedron Lett.*, **26**, 2913, 1985.
4. Ohta, H., Matsumoto, S., Sugai, T., *Tetrahedron Lett.*, **31**, 2895, 1990.
5. Sato, T., Ito, T., Fujisawa, T., *Tetrahedron Lett.*, **28**, 5677, 1987.
6. Casey, M., Manage, A.C., Nezhat, L., *Tetrahedron Lett.*, **29**, 5821, 1988.
7. Casey, M., Manage, A.C., Gairns, R.S., *Tetrahedron Lett.* **30**, 6919, 1989.
8. Casey, M., Manage, A.C., Murphy, P.J., *Tetrahedron Lett.*, **33**, 965, 1992.
9. Durst, T., Viau, R., Van den Elzen, R., Nguyen, C.H., *J. Chem. Soc., Chem. Commun.*, 1334, 1971.
10. Demailly, G., Greck, C., Solladie, G., *Tetrahedron Lett.*, **25**, 4113, 1985.
11. Braun, M., Hild, W., *Chem. Ber.*, **117**, 413, 1984.
12. Bravo, P., Frigerio, M., Resnati, G., *J. Org. Chem.*, **55**, 4216, 1990.
13. Sakuraba, H., Ushiki, S., *Tetrahedron Lett.*, **31**, 5349, 1990.
14. Farnum, D.G., Veysoglu, T., Carde, A.M., Duhl-Emswiller, B., Pancoast, T.A., Reitz, T.J., Carde, R.T., *Tetrahedron Lett.*, 4009, 1977.
15. Pyne, S.G., Boche, G., *J. Org. Chem.*, **54**, 2663, 1989.
16. Casey, M., Mukherjee, I., Trabsa, H., *Tetrahedron Lett.*, **33**, 127, 1992.
17. Williams, D.R., Phillips, J.G., White, F.H., Huffman, J.C., *Tetrahedron*, **42**, 3003, 1986.
18. Williams, D.R., Phillips, J.G., *J. Org. Chem.*, **46**, 5452, 1981.
19. Williams, D.R., Phillips, J.G., *Tetrahedron*, **42**, 3013, 1986.
20. Stork, G., Paterson, I., Lee, F.K.C., *J. Am. Chem.*, **104**, 4686, 1982.
21. Masamune, S., Ma, P., Okumoto, H., Ellingboe, J.W., Ito, Y., *J. Org. Chem. Soc.*, **49**, 2837, 1984.
22. Kochetkov, N.K., Sviridov, A.F., Ermolenko, M.S., Yashunsky, D.V., Borodkin, V.S., *Tetrahedron*, **45**, 5109, 1989.
23. Alexandre, C., Belkadi, O., Maignan, C., *Synthesis*, 547, 1992.
24. Bonfand, E., Gosselin, P., Maignan, C., *Tetrahedron Lett.*, **33**, 2347, 1992; *Tetrahedron: Asymmetry*, **4**, 1667, 1993.
25. Solladie, G., Ruiz, P., Colobert, F., Carreno, M.C., Garcia-Ruano, J.L., *Synthesis*, 1011, 1991.
26. Hiroi, K., Nakamura, H., Anzai, T., *J. Am. Chem. Soc.*, **109**, 1249, 1987.
27. Bravo, P, Carrera, P., Resnati, G., Ticozzi, C., *J. Chem. Soc., Chem. Commun.*, 19, 1984.
28. Albinati, P., Bravo.P., Ganazzoli, F., Resnati, G., Viani, F., *J. Chem. Soc., Perkin Trans. 1*, 1405, 1986.
29. Bravo, P., De Vita, C., Resnati, G., *Gazz. Chim. Ital.*, **117**, 165, 1987.
30. Tsuchihashi, G., Iriuchijima, S., Maniwa, K., *Tetrahedron Lett.*, 3389, 1973.
31. Ronan, B., Marchalin, S., Samuel, O., Kagan, H.B., *Tetrahedron Lett.*, **29**, 6101, 1988.
32. Pyne, S.G., Dikic, B., *J. Chem. Soc., Chem. Commun.*, 826, 1989.
33. Pyne, S.G., Dikic, B., *J. Org. Chem.*, **55**, 1932, 1990.
34. Pyne, S.G., Boche, G., *J. Org. Chem.*, **54**, 2663, 1989.
35. Annunziata, R., Cinquini, M., *Synthesis*, 929, 1982.
36. Pyne, S.G., Hajipour, A.R., *Tetrahedron*, **48**, 9385, 1992.
37. Annunziata, R., Cinquini, M., Cozzi, F., *J. Chem. Soc., Perkin Trans 1*, 1687, 1979.
38. Bravo, P., Piovosi, E., Resnati, G., *Synthesis*, 579, 1986.
39. Bravo, P., Resnati, G., *Tetrahedron Lett.*, **26**, 5601, 1985.
40. Yamazaki, T., Ishikawa, N., Iwatsubo, H., Kitazume, T., *J. Chem. Soc., Chem. Commun.*, 1340, 1987.
41. Bravo, P., Pregnolato, M., Resnati, G., *J. Org. Chem.*, **57**, 2726, 1992.
42. Solladie, G., Frechou, C., Demailly, G., *Tetrahedron Lett.*, **27**, 2867, 1986.
43. Solladie, G., Maugein, N., Morreno, I., Almario, A., Carreno, M.C., Garcia-Ruano, J.L., *Tetrahedron Lett.*, **33**, 4561, 1992.
44. Solladie, G., Ghiatou, N., *Tetrahedron: Asymmetry*, **3**, 33, 1992.
45. Armer, R., Simpkins, N.S., *Tetrahedron Lett.*, **34**, 363, 1993.

3.3 ALLYL SULFOXIDES

Dialkyl, aryl alkyl, and diaryl sulfoxides are configurationally stable compounds and were found to undergo racemization under rather forced conditions, at temperatures of about 200°C.[1] The estimated activation parameters for racemization of these sulfoxides ($\Delta H^{\ddagger}$ = 35 to 45 kcal/mol, $\Delta S^{\ddagger}$ = –8 to +4 e.u. and $\Delta V^{\ddagger}$ ~ 0 mL/mol) point to the pyramidal inversion mechanism. In contrast to these sulfoxides, allyl sulfoxides undergo a facile thermal racemization already at temperatures from 50 to 70°C. The values of activation energy and entropy for racemization of allyl p-tolyl sulfoxide **1** ($\Delta H^{\ddagger}$ = 23 kcal/mol, $\Delta S^{\ddagger}$ = –4.9 e.u.) as well as the absence of decomposition products indicate a different mechanism for racemization of this and other allyl sulfoxides. A crucial experiment towards understanding the racemization mechanism of allyl sulfoxides has been carried out by Mislow and co-workers.[2] They treated p-toluenesulfenyl chloride with allyl alcohol labeled in the α-position with deuterium and obtained allyl-γ-$^2H_2$ p-tolyl sulfoxide **1** instead of the expected sulfenate (Equation 3.3.1).

$$\text{p-TolSCl} + CH_2{=}CH{-}CD_2{-}OLi \longrightarrow \text{p-Tol}{-}S(=O){-}CH_2{-}CH{=}CD_2 \quad (3.3.1)$$

These and other results have led Mislow et al.[3] to propose a cyclic rearrangement mechanism for the racemization of allyl sulfoxides involving an achiral sulfenate ester as an intermediate. This mechanism is exemplified by racemization of the sulfoxide **1** (Equation 3.3.2).

(-)-(S)-**1** ⇌ **2** ⇌ (+)-(R)-**1** (3.3.2)

The ability of allyl sulfoxides to undergo [2,3]-sigmatropic rearrangement is a unique feature of this class of sulfoxides which has been ingeniously exploited in organic synthesis. Several excellent reviews have been published on this subject.[4-6] Therefore, taking into account a main goal of the book and to keep this chapter to an acceptable length, the discussion herein will be limited to chiral, enantiomeric allyl sulfoxides with emphasis on their most important applications. However, in some cases, when necessary, short comments will be made on the results obtained with racemic allyl sulfoxides.

Like other chiral sulfoxides, allyl p-tolyl sulfoxides can be prepared by the Andersen method using (–)-(S)-menthyl p-toluenesulfinate as a substrate,[2] provided that isolation and purification conditions are carefully selected to avoid racemization (see Chapter 2, Section 2.1). The synthesis of other chiral allylic sulfoxide structures especially designed for synthetic purposes will be discussed below (see also Chapter 3, Section 3.7).

3.3.1 Asymmetric Allylic Sulfoxide-Sulfenate Rearrangements

3.3.1.1 Chirality Transfer from Carbon to Sulfur

The reversible [2,3]-sigmatropic rearrangement of allylic sulfenates to sulfoxides offers a nice opportunity for studies of the transfer of chirality from carbon to sulfur and from sulfur to carbon. The first possibility was investigated by Mislow and co-workers,[2] who observed that chiral (S)-α-methylallyl p-toluenesulfenate **3** rearranges to (–)-(S)-*trans*-crotyl p-tolyl sulfoxide **4** with 37% ee. However, after a short time, it underwent racemization accompanied by isomerization of the carbon-carbon double bond, affording a mixture of racemic E and Z sulfoxides **4** in a 77:33 ratio.

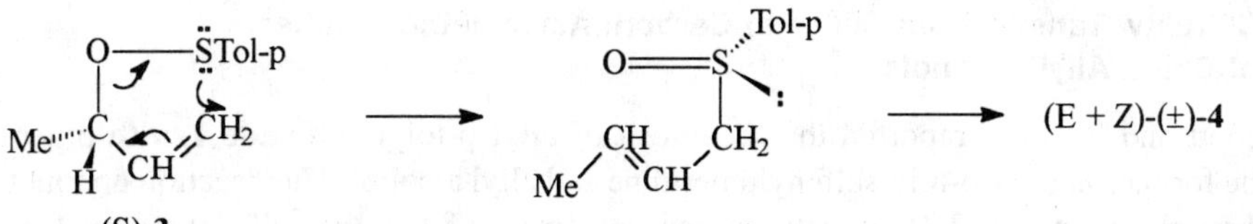

(S)-3 → (E)-(S)-4, ee 37% → (E + Z)-(±)-4 (3.3.3)

Taking into account the concept of the cisoid and transoid transition states originally introduced by Mislow et al.[2] and the *exo* and *endo* transition states proposed by Rautenstrauch,[7] Goldmann et al.[8] have completely rationalized the results mentioned above, i.e., (a) low extent of asymmetric induction in the chirality transfer from C to S, (b) rapid racemization of the sulfoxide **4**, and (c) the E ⇌ Z equilibration. Their interpretation depicted in Scheme 3.3.1 is based on the competitive operation of four readily interconvertible transition states in the reversible sulfenate → sulfoxide rearrangement.

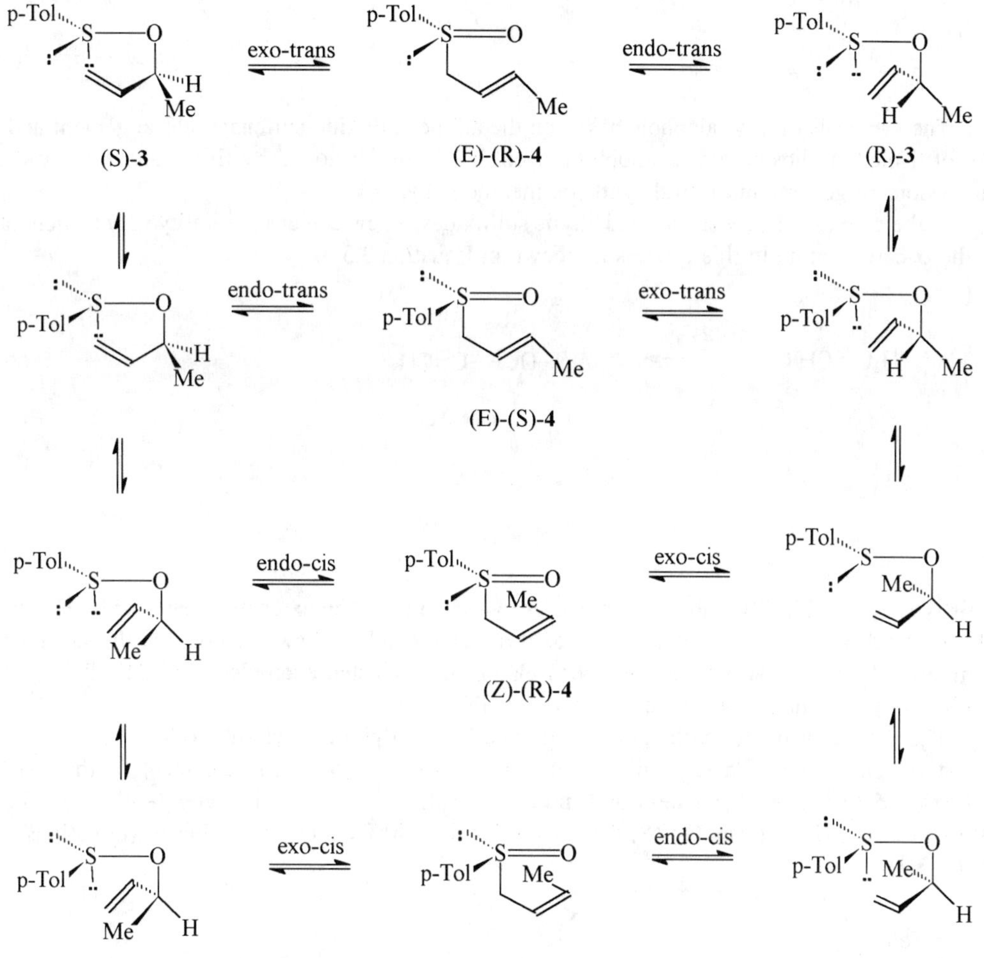

(S)-3 ⇌ (exo-trans) (E)-(R)-4 ⇌ (endo-trans) (R)-3

(endo-trans) (E)-(S)-4 (exo-trans)

(endo-cis) (Z)-(R)-4 (exo-cis)

(exo-cis) (Z)-(S)-4 (endo-cis)

Scheme 3.3.1

(From Braverman, S., in *The Chemistry of Sulphones and Sulphoxides*, Patai, S., Rappoport, Z., Stirling, C. J. M., Eds., John Wiley & Sons, Chichester, 1988, pp. 717–753. With permission.)

3.3.1.2 Chirality Transfer from Sulfur to Carbon: Asymmetric Synthesis of Chiral Allyl Alcohols

In 1969, Abbott and Stirling[9] reported that treatment of allyl p-tolyl sulfoxide **1** with piperidine resulted in the formation of N-p-tolylsulfenylpiperidine and allyl alcohol. This reaction undoubtedly proceeded through the initial [2,3]-sigmatropic rearrangement of **1** to the sulfenate ester **2** which, upon nucleophilic attack of piperidine, afforded the final products (Equation 3.3.4).

p-Tol S(=O)CH₂CH=CH₂ (**1**) ⇌ [p-Tol SOCH₂CH=CH₂ (**2**) + HN(piperidine)] → p-Tol SN(piperidine) + CH₂=CHCH₂OH

(3.3.4)

The synthesis of allyl alcohols based on the allylic sulfoxide-sulfenate rearrangement and the use of trimethyl phosphite as a thiophilic agent has been developed by Evans and Andrews[6] and has become a general and useful synthetic methodology.

In the case of suitably substituted allylic sulfoxides, a new center of chirality may be generated at the α-carbon atom in this process as shown in Equation 3.3.5.

ArS(=O)CH₂C(R¹)=CHR —[2,3]→ ArS–OCH(R)–C(R¹)=CH₂ —P(OMe)₃→ HOCH(R)–C(R¹)=CH₂

(3.3.5)

This type of the chirality transfer from sulfur to carbon has been extensively studied by Hoffmann and co-workers[10] using chiral allyl sulfoxides. As a result, a new method for the asymmetric synthesis of chiral allyl alcohols has been elaborated. Selected examples of chiral allyl alcohols prepared by Hoffmann's group are collected in Table 3.3.1.

The geometry of the carbon-carbon double bond in chiral allyl sulfoxides has an essential effect on the extent of asymmetric induction. For instance, the rearrangement of the (E)-(R)-sulfoxide **5** to (–)-(R)-1-octene-3-ol **6** proceeds with low enantioselectivity (~40%), while the reaction of (Z)-(R)-**5** gives (+)-(S)-**6** with greater than 90% enantiomeric purity (Equations 3.3.6 and 3.3.7).

(E)-(R)-**5** —P(iPr)₃ (":P")→ (-)-(R)-**6** (~40% ee)

(3.3.6)

p-Tol–S(=O)–CH$_2$–CH=CH–C_5H_{11} → (P:) → HO–CH(H)(C_5H_{11})–CH=CH$_2$

(Z)-(R)-**5** (+)-(S)-**6** (>90% ee) (3.3.7)

Table 3.3.1 Asymmetric Synthesis of Chiral Allyl Alcohols by [2,3]-Sigmatropic Rearrangement of Chiral Allyl Sulfoxides

Allyl sulfoxide	Allyl alcohol	Chirality transfer [%]	Transition state exo [%]	Transition state endo [%]
ArS=O	OH	44	72	28
ArS=O	OH	60	80	20
ArS=O	OH	68	84	16
ArS=O	OH	>90	>95	<5
ArS=O	OH	>90	>95	<5

According to Hoffmann, the rearrangement of the (E)- or (Z)-sulfoxide **5** to the corresponding sulfenate can proceed through two diastereomeric transition states, *exo* and *endo*, affording, after treatment with thiophilic agent, enantiomeric forms of allyl alcohol. The extent of asymmetric induction will be determined by the difference in energy between these two transition states which, in turn, is determined by the extent of repulsive non-bonding interaction between the aromatic substituent at sulfur and that at C-3. Thus, for instance, for the conversion of (Z)-(R)-**5** into **6** one can write two diastereomeric transition states depicted below (Scheme 3.3.2).

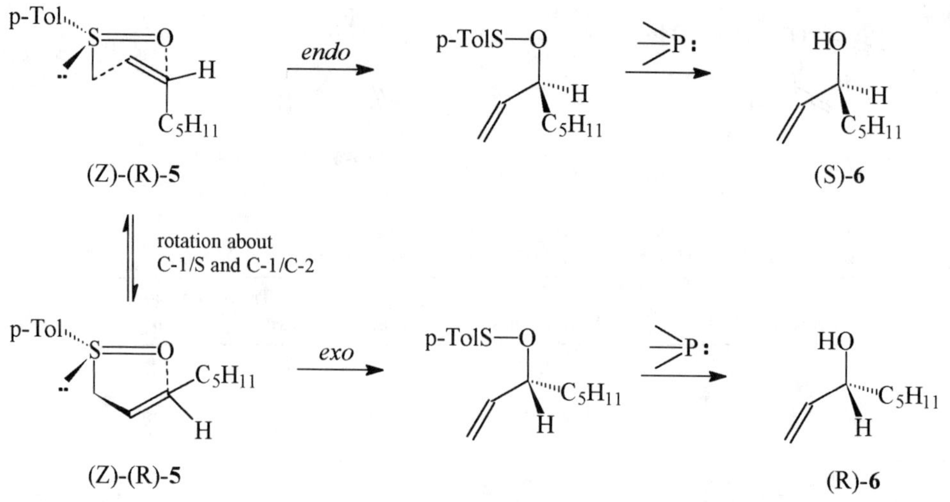

Scheme 3.3.2

A strong preference for the formation of the allyl alcohol (S)-**6** is due to the fact that the *exo* transition state is destabilized by the p-Tol– C_5H_{11} repulsive interactions. Therefore, the rearrangement of (Z)-(R)-**5** to **6** occurs in about 95% via the *endo* transition state and only in 5% via the *exo* transition state. In the case of (E)-(R)-**5**, the energy difference between the *endo* and *exo* transition states is smaller and the corresponding ratio is about 70:30.

The opposite situation is observed with aryl cycloalkenyl methyl sulfoxides (see Table 3.3.1). The rearrangements of these sulfoxides proceed mainly via the *exo* transition state because the presence of substitution at C-2 hinders rearrangements via the *endo* transition state.

Interestingly, when a substituent at C-1 is present in chiral allyl sulfoxides, the stereochemistry of the allyl sulfoxide to allyl alcohol conversion is determined by the chirality at C-1 and not by the chiral sulfur center. This indicates that under the influence of the substituent at C-1, one diastereomeric allyl sulfoxide reacts selectively via the *exo* transition state and the other via the *endo* transition state. Therefore, this particular case represents 1 → 3 chirality transfer from carbon to carbon (Equation 3.3.8).

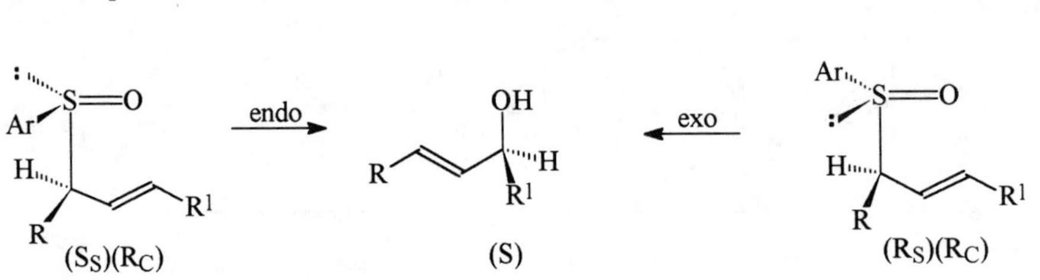

(3.3.8)

3.3.2 Reactions of Allyl Sulfoxide Carbanions

Deprotonation of allyl aryl sulfoxides leads to the corresponding anion which exhibits typical ambident reactivity, i.e., its reactions with electrophiles may occur at the α- or γ-carbon atom.

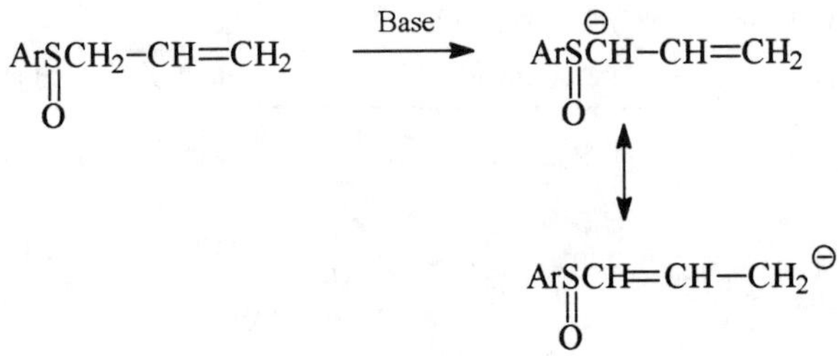

$$\mathrm{ArS(=O)CH_2{-}CH{=}CH_2 \xrightarrow{Base} ArS(=O)\overset{\ominus}{C}H{-}CH{=}CH_2 \longleftrightarrow ArS(=O)CH{=}CH{-}CH_2^{\ominus}} \quad (3.3.9)$$

In fact, addition of these anions to benzaldehyde at –70°C gives a mixture of two adducts **7** and **8** resulting from both α- and γ-attack at the carbonyl group.[11] In the case of the α-attack, all four possible diastereomers are formed, while the γ-adduct **8**, which is a major product, consists of only two diastereomers (Equation 3.3.10). With aliphatic aldehydes at –78°C in the presence of HMPT, the α-addition predominates.[12]

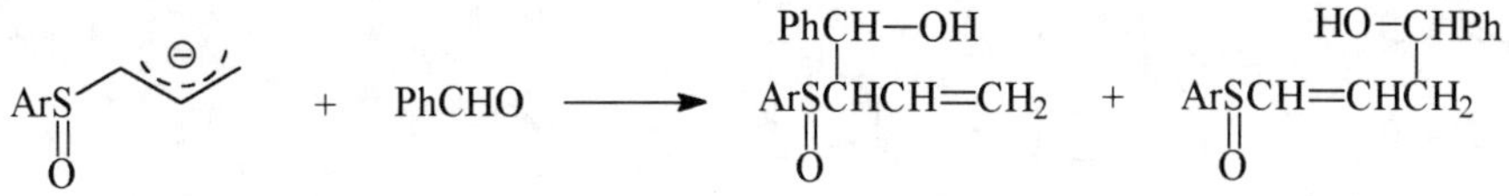

$$\mathrm{ArS(=O)(allyl^{\ominus}) + PhCHO \longrightarrow ArS(=O)CH(CH(OH)Ph)CH{=}CH_2\ (\mathbf{7}) + ArS(=O)CH{=}CHCH_2CH(OH)Ph\ (\mathbf{8})} \quad (3.3.10)$$

Ar	7 dr	8 dr
Ar = Ph	3:6:7:2	3:1
p-Tol	3:3:6:2	3:1
p-$O_2NC_6H_4$		2:1

3.3.2.1 Conjugate Addition of Chiral Allyl Sulfoxide Anions

The Michael addition of allyl sulfoxide anions to α,β-unsaturated carbonyl compounds is more complex because four different adducts may be formed: 1,4α, 1,4γ, 1,2α, and 1,2γ. It has been found, however, that only two of them are actually formed, i.e., 1,4γ and 1,2γ, the former being in most cases prevailing.[13-17] With five-membered cyclic enones, the 1,4γ adduct is the only one which is produced.[18-20] The addition proceeds in a highly diastereoselective way to give exclusively the corresponding (E)-vinyl sulfoxides as single diastereomers. The β-diastereomers arise from (E)-allyl sulfoxides and the α ones from (Z)-sulfoxides. Selected examples collected in Table 3.3.2 illustrate the stereochemical outcome of the Michael addition of allylic sulfoxide anions to cyclopentenones.

Binns and co-workers,[21,22] who did very extensive studies on the reaction under discussion, rationalized these stereochemical results in terms of frontier molecular orbital consideration and in terms of a ten-membered cyclic transition state which is best described as "*trans*-fused chair-like" or "*trans*-decalyl-like" (Figure 3.3.1).

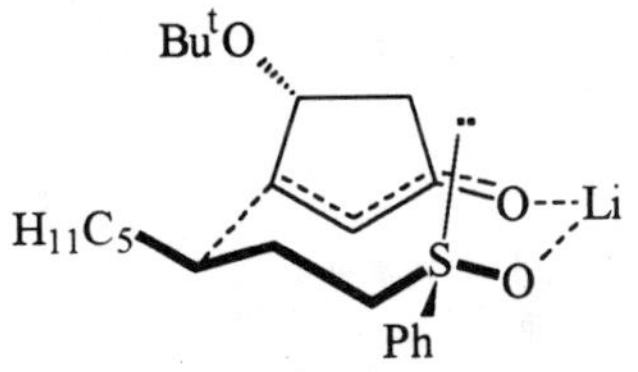

$\mathrm{Bu^tO}$ ... $\mathrm{O{-}{-}{-}Li}$ ... $\mathrm{H_{11}C_5}$ ... S ... O ... Ph

Figure 3.3.1

In this structure, the sulfinyl and carbonyl oxygen atoms are chelated by the lithium cation and the allyl system lies over one face of the cyclic enone structure. It is quite clear, therefore, that chiral allyl sulfoxide anions will react at only one of the enantiotopic faces of the enone to afford enantiomerically pure conjugate addition products.

Table 3.3.2 Michael Addition of Allyl Sulfoxide Anions to Cyclopentenones

Sulfoxide	Enone	Adducts	Ratio β:α	Yield [%]
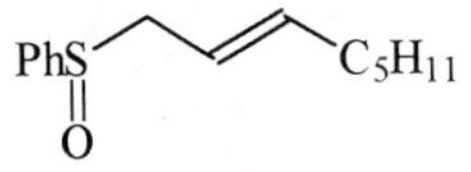 PhS(O) C_5H_{11} E : Z = 85:15	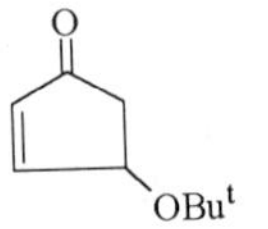 O, OBu^t	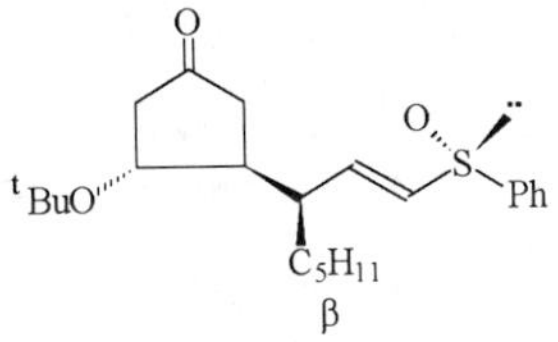 O, ^{t}BuO, O, S, Ph, C_5H_{11} β	83:17	83
E : Z = 17:83		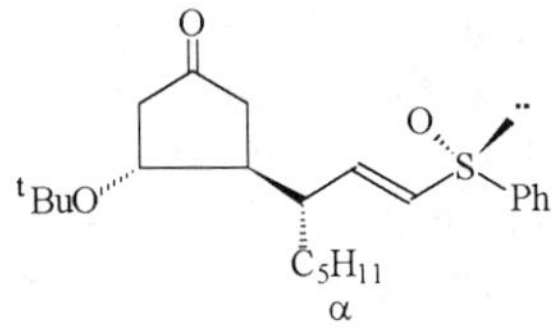 O, ^{t}BuO, O, S, Ph, C_5H_{11} α	21:79	79
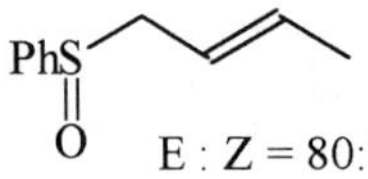 PhS(O) E : Z = 80:20	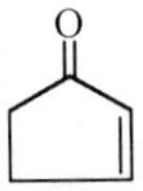 O	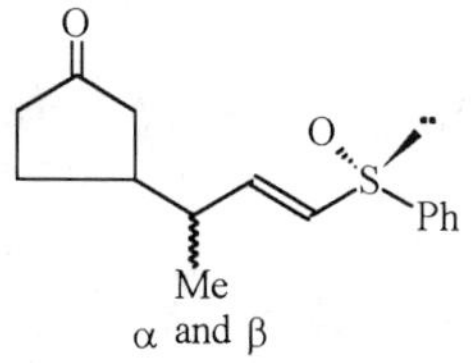 O, O, S, Ph, Me α and β	80:20	64
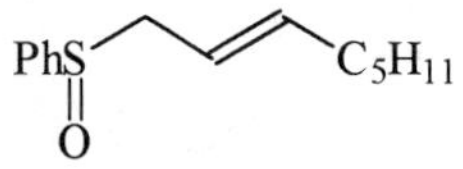 PhS(O) C_5H_{11} E : Z = 85:15	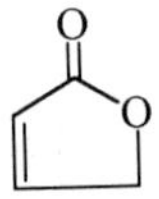 O, O	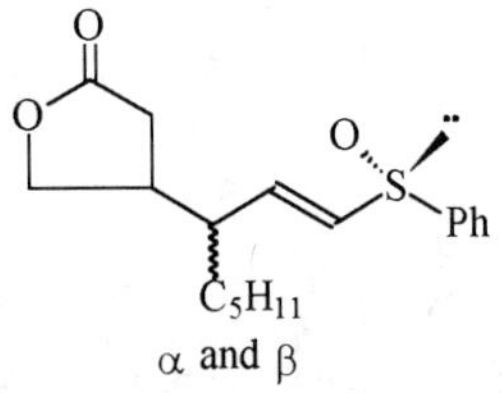 O, O, O, S, Ph, C_5H_{11} α and β	83:17	83

The addition of the anion derived from (+)-(R)-allyl p-tolyl sulfoxide **1** to cyclic enones was found by Hua and co-workers[19] to occur in a highly enantioselective way, especially in the case of cyclopentenones (see Table 3.3.3).

Table 3.3.3 Enantioselective Addition of (+)-(R)-Allyl p-Tolyl Sulfoxide Anion to Cyclic Enones

Enone	1,4-γ Adduct	ee [%]	Yield [%]
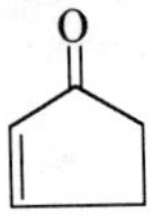 O	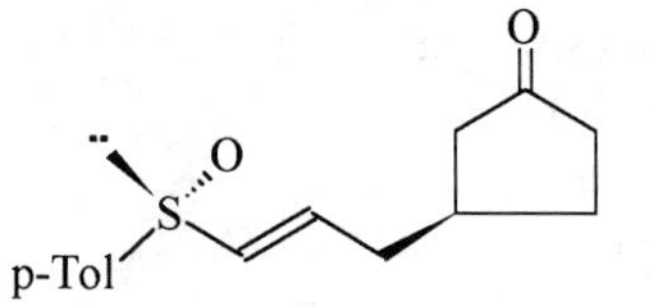 O, S, p-Tol, O	96	91
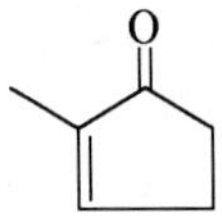 O	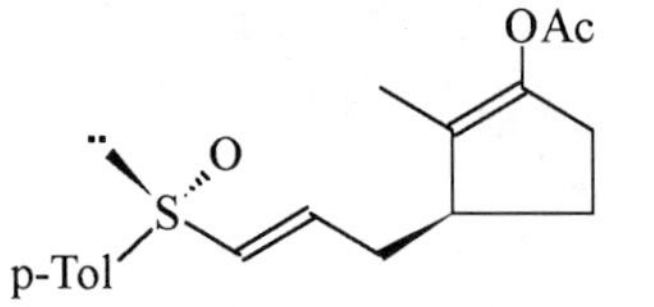 O, S, p-Tol, OAc	95	84[a]
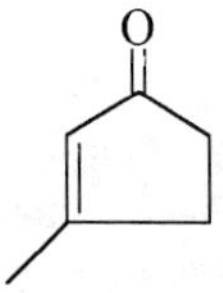 O	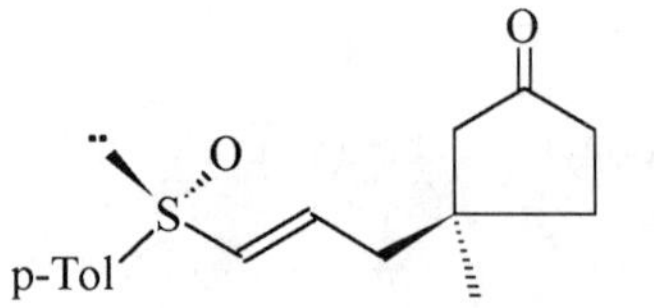 O, S, p-Tol, O	95	80

Table 3.3.3 (continued) Enantioselective Addition of (+)-(R)-Allyl p-Tolyl Sulfoxide Anion to Cyclic Enones

Enone	1,4-γ Adduct	ee [%]	Yield [%]
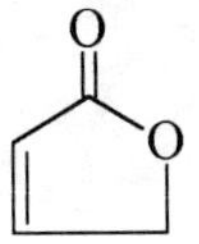	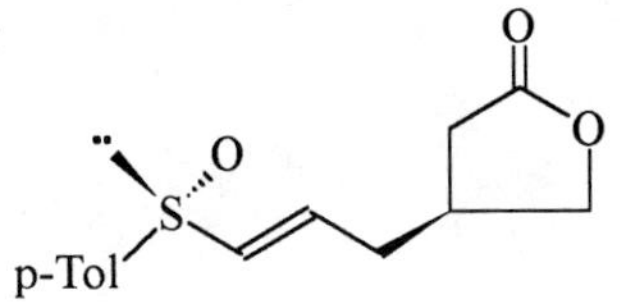 p-Tol	95	70
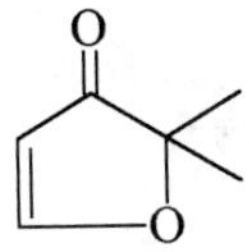	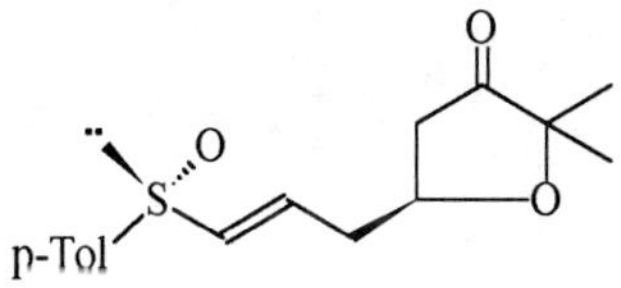 p-Tol	70	82
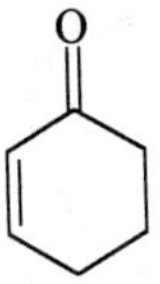	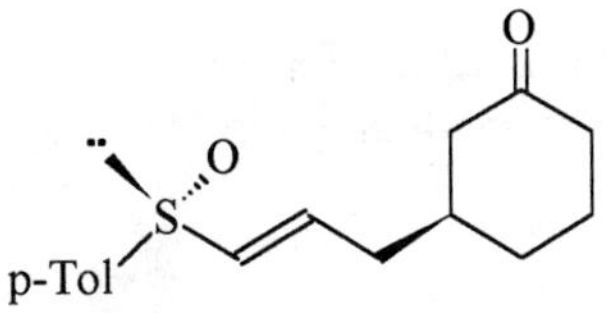 p-Tol	50	79[b]
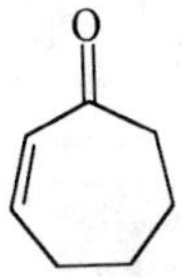	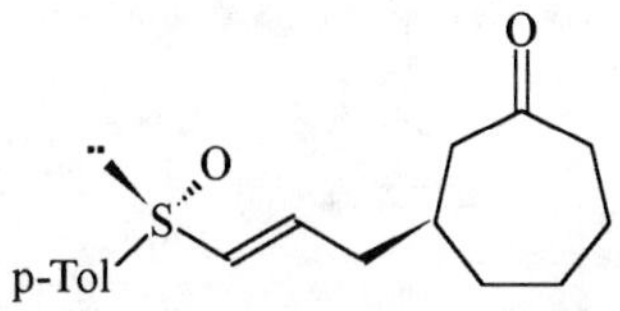 p-Tol	50	25[c]

[a] The resulting enolate anion was trapped with acetyl chloride.
[b] The 1,2-γ adduct was formed in 14% yield and with 50% ee.
[c] The 1,2-γ adduct was formed in 58% yield and with 50% ee.

Since the chiral sulfoxide **1** undergoes ready thermal racemization, a series of the more stable chiral diastereomeric allyl sulfoxides **9** have been synthesized.[23,24] It has turned out that both epimers of each sulfoxide can be obtained by simple thermal epimerization of the (S_S)-diastereomer (Equation 3.3.11).

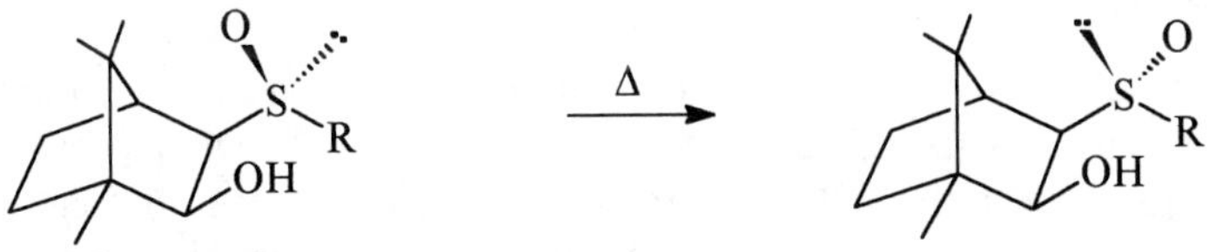

(S_S)-**9** $\xrightarrow{\Delta}$ (R_S)-**9** (3.3.11)

a, R = allyl
b, R = (E)-buten-2-yl
c, R = dimethylallyl
d, R = (E)-octen-2-yl

It was found that the addition of the anions generated from the sulfoxides (S_S)-**9a** and (R_S)-**9a** to cyclopentenone is fully enantioselective and gives the adducts **11** in yields from 50 to 69%.[23]

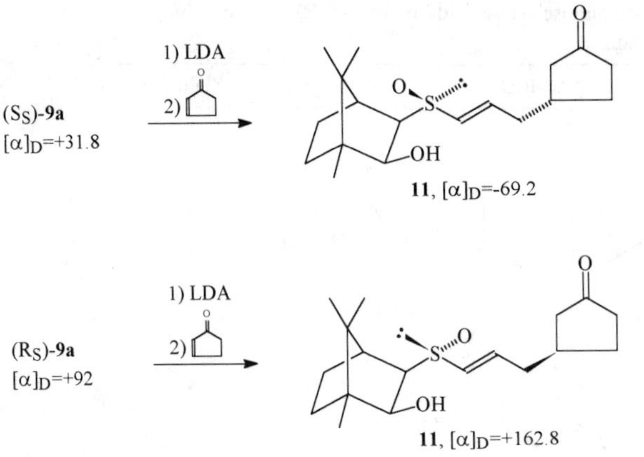

(3.3.12)

Another, strongly hindered allyl sulfoxide **12** prepared by Swindell and co-workers[25] reacted with cyclopentenone also with higher regio- and diastereoselectivity as compared with simple allyl sulfoxides.

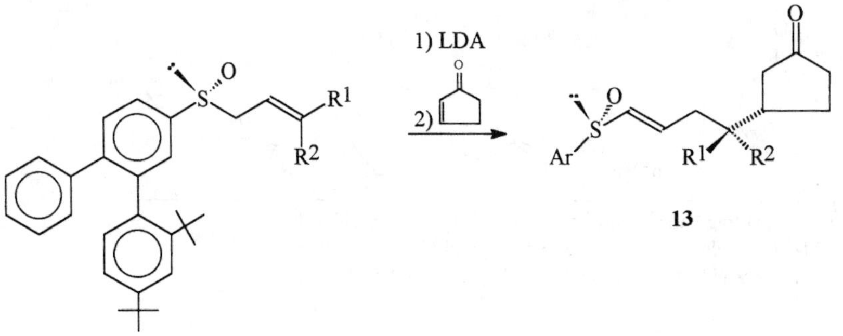

(3.3.13)

R^1	R^2	dr [%]	Yield [%]
H	H	92:8	80
Me	H	85:15	54
Ph	H	90:10	60
Me	Me	93:7	86

The reaction of enantiomeric allyl sulfoxide carbanions with racemic enones has been utilized for kinetic resolution.[19] For example, when two equivalents of racemic 4-[(dimethylbenzyl)oxy]-2-cyclopentenone **14** were allowed to react with the carbanion derived from the allyl sulfoxide (+)-(R)-**1**, only the (S)-enantiomers of enone reacted, indicating that the carbanion approaches 2-cyclopentenone from the *si* face. Similar kinetic resolution was observed with other racemic enones **16**, **17**, and **18**.

(+)-(R)-1 + 2 (±)-14 —LDA→ (3S)-15 (95% ee), yield 68% ; (3R)-15 (90% ee), yield 7%

16 17 18, R = t-BuMe$_2$Si (3.3.14)

The addition reaction of chiral allyl sulfoxide anions with enones has become a crucial step in the asymmetric total syntheses of many natural products. Some examples will be briefly discussed below.

3.3.2.2 Asymmetric Synthesis of (+)-Hirsutene[26,27]

(+)-Hirsutene **19** belongs to a family of sesquiterpenoids isolated from the extract of *Coriolus consors* and is considered to be the biogenetic precursor of coriolin and hirsutic acid. Its total synthesis involves in the first step conjugate addition of the carbanion generated from the sulfoxide (−)-(S)-**1** to 2-methyl-2-cyclopentenone followed by acetylation of the enolate ion formed. The adduct **20** obtained was reduced with Zn-AcOH at room temperature. In the next step, intramolecular cyclization of the enol acetate with the vinylic sulfide moiety took place in the presence of $TiCl_4$ in AcOH to give the hexahydropentalenone **21**. The latter was elaborated into (+)-hirsutene **19**.

Scheme 3.3.3

3.3.2.3 Asymmetric Synthesis of Enantiomers of 12,13-Epoxytrichothec-9-ene[28]

Many of the trichothecenes exhibit interesting biological properties, including antibiotic and antitumor activity. Therefore, there is intense interest in their synthesis. As in the previous synthesis, a key reaction in the total synthesis of the (+)-enantiomer of the title compound **22** was the conjugate addition of the carbanion generated from the chiral cyclohexenyl sulfoxide **23** to an activated enone. The adducts **24** obtained have been used as substrates in the multistep synthesis of (+)- and (–)-**22**.

Scheme 3.3.4

3.3.2.4 Asymmetric Synthesis of (+)-Pentalene[29]

The synthesis of the title compound **25** is based on kinetic resolution which takes place when the carbanion of racemic crotyl sulfoxide **26** is added to enantiomerically pure cyclopentenone **27**. The chiral adduct **28** obtained is then elaborated into the title compound in several steps.

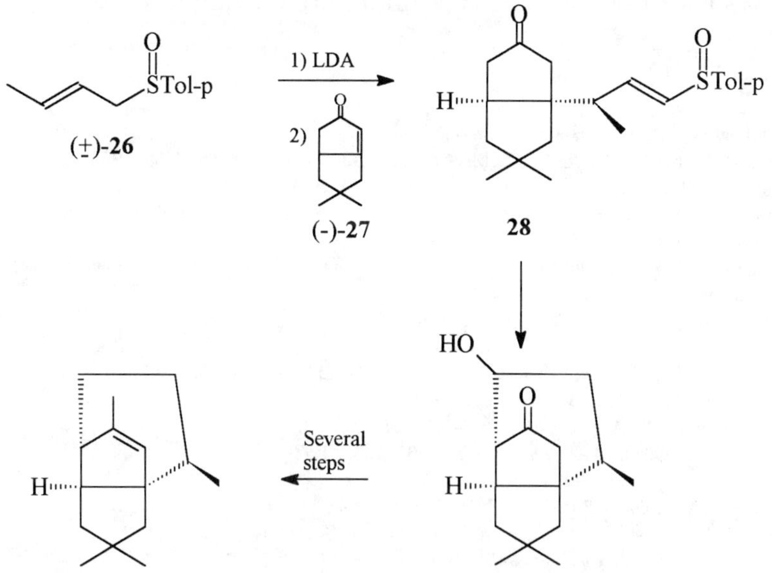

Scheme 3.3.5

3.3.3 Examples of Experimental Procedures

3.3.3.1 Synthesis of (+)-(R) Allyl p-Tolyl Sulfoxide[2]

A solution of allylmagnesium bromide was prepared by dropwise addition of 5.0 g (0.041 mol) of allyl bromide to a stirred suspension of magnesium turnings (0.50 g, 0.020 mol) in 30 mL of anhydrous ether. When most of the magnesium had dissolved, the mixture was added through a glass wool plug to an ethereal solution of (–)-menthyl p-toluenesulfinate, (3.5 g, 0.012 mol). The reaction was immediately quenched by addition of saturated ammonium chloride solution and cooled by addition of ice. The ethereal layer was extracted five times with ice water. The combined aqueous portions were saturated with sodium chloride and extracted five times with a total of 700 mL of chloroform. The cold chloroform solution was dried (magnesium sulfate) and concentrated on a rotary evaporator, care being taken to keep the solution at below 0°C. The residue was chromatographed at 5°C on 25 g of Florisil. The first fractions, eluted with benzene, were discarded, and material eluted with acetate was concentrated on a rotary evaporator at temperatures below 0°C and pumped free of solvent at 0.02 mm. The yellow oil, $[\alpha]_D^{27}$ +212 (c 0.26, ethanol), was obtained. A sample was purified for analysis by kugelrohr distillation at 100°C (0.05 mm).

3.3.3.2 Conjugate Addition of (+)-(R)-Allyl p-Tolyl Sulfoxide 1 to 2-Cyclopentenone; Synthesis of (E)-(3S,S_R)-3-[3-(p-Tolylsulfinyl)-2-propen-1-yl]-cyclopentanone[19]

To a solution of 0.90 g (5.0 mmol) of (+)-(R)-allyl p-tolyl sulfoxide in 16 mL of THF at –78°C under argon was added a cold (-78°C) solution of LDA prepared at –30°C from 0.73 mL (5.2 mmol) of diisopropylamine and 3.3 mL (5.2 mM) of n-butyllithium in 8 mL of THF. The yellow-colored solution obtained was stirred at –78°C for 1 h, and then 0.42 mL (5 mmol) of 2-cyclopentenone was added. After the solution was stirred at –78°C for 5 min, 0.72 g (12 mmol) of acetic acid in 2 mL of ether was added. The reaction mixture was warmed to room temperature, poured into saturated aqueous NH_4Cl solution, and extracted three times with ether. The combined extracts were washed with brine, dried ($MgSO_4$), concentrated, and column chromatographed on silica gel to give 1.205 g (91% yield) of ketone (Table 3.3.3) as an oil: $[\alpha]_D^{21}$ +207 (c 1.23, $CHCl_3$).

REFERENCES (3.3)

1. Mikołajczyk, M., Drabowicz, J., *Top. Stereochem.*, **13**, 406, 1982.
2. Bickart, P., Carson, F. W., Jacobus, I., Miller, E. G., Mislow, K., *J. Am. Chem. Soc.*, **90**, 4869, 1968.
3. Miller, E. G., Rayner, D. R., Mislow, K., *J. Am. Chem. Soc.*, **88**, 3139, 1968.
4. Braverman, S. in *The Chemistry of Sulphones and Sulphoxides*, Patai, S., Rappoport, Z., Stirling, C. J. M., Eds., John Wiley & Sons, Chichester, 1988, pp. 717–757.
5. Oae, S., Uchida, Y., in *The Chemistry of Sulphones and Sulphoxides*, Patai, S., Rappoport, Z., Stirling, C. J. M., Eds., John Wiley & Sons, Chichester, 1988, pp. 583–664.
6. Evans, D. A., Andrews, G. C., *Acc. Chem. Res.*, **7**, 147, 1974.
7. Rautenstrauch, V., *J. Chem. Soc., Chem. Commun.*, 526, 1970.
8. Goldmann, S., Hoffmann, R. W., Maak, N., Geueke, K. J., *Chem. Ber.*, **113**, 831, 1980.
9. Abbott, D. J., Stirling, C. J. M., *J. Chem. Soc., (C)*, 818, 1969.
10. For a review see: Hoffmann, R. W., *Angew. Chem., Int. Ed. Engl.*, **18**, 563, 1979.
11. Antonjuk, D. J., Ridley, D. D., Smal, M. A., *Aust. J. Chem.*, **33**, 2635, 1980.
12. Annunziata, R., Cinquini, M., Cozzi, F., Raimondi, L., *J. Chem. Soc., Chem. Commun.*, 366, 1968.
13. Vasileva, L. L., Melnikova, V. I., Gainullina, E. T., Pivnitsky, K. K., *Zh. Org. Khim.*, **16**, 2618, 1980.
14. Binns, M. R., Haynes, R. K., Houston, T. L., Jackson, W. R., *Aust. J. Chem.*, **34**, 2465, 1981.
15. Nokami, I., Ono, T., Iwao, A., Wakabayashi, S., *Bull. Chem. Soc. Jpn.*, **55**, 3043, 1982.
16. Vasileva, L. L., Melnikova, V. I., Pivnitsky, K. K., *Zh. Org. Khim.*, **19**, 661, 1983.
17. Nokami, J., Ono, T., Wakabayashi, S., Hazato, A., Kurizumi, S., *Tetrahedron Lett.*, **26**, 1985, 1985.
18. Binns, M. R., Haynes, R. K., Katsifis, A. A., Schober, P. A., Vonwiller, S. C., *Tetrahedron Lett.*, **26**, 1565, 1985.
19. Hua, D. H., Venkataraman, S., Coulter, M. J., Sinai-Zingde, G., *J. Org. Chem.*, **52**, 719, 1987.
20. Haynes, R. K., Katsifis, A. A., Vonwiller, S. C., Hambley, T. W., *J. Am. Chem. Soc.*, **110**, 5423, 1988.
21. Binns, M. R., Chai, O. L., Haynes, R. K., Katsifis, A. A., Schober, P. A., Vonwiller, S. C., *Tetrahedron Lett.*, **26**, 1569, 1985.
22. Binns, M. R., Haynes, R. K., Katsifis, A. A., Schober, P. A., Vonwiller, S. C., *J. Am. Chem. Soc.*, **110**, 5411, 1988.
23. Binns, M. R., Goodbridge, R. J., Haynes, R. K., Ridley, D. D., *Tetrahedron Lett.*, **26**, 6381, 1985.
24. Goodbridge, R. J., Hambley, T. W., Haynes, R. K., Ridley, D. D., *J. Org. Chem.*, **53**, 2881, 1988.
25. Swindell, C. S., Blase, F. R., Eggleston, D. S., Krause, J., *Tetrahedron Lett.*, **31**, 5409, 1990.
26. Hua, D. H., Sinai-Zingde, G., Venkataraman, S., *J. Am. Chem. Soc.*, **107**, 4088, 1985.
27. Hua, D. H., Venkataraman, S., Ostrander, R. A., Sinai-Zingde, G., McCann, P. J., Coulter, M. J., Xu, M. R., *J. Org. Chem.*, **53**, 507, 1988.
28. Hua, D. H., Venkataraman, S., Chan-Yu-King, R., Paukstelis, J. V., *J. Am. Chem. Soc.*, **110**, 4741, 1988.
29. Hua, D. H., *J. Am. Chem. Soc.*, **108**, 3839, 1986.

3.4 ALKYL AND ARYLSULFENYLMETHYL SULFOXIDES

The first representatives of alkyl(aryl)sulfenylmethyl sulfoxides (called also alkyl(aryl)thiomethyl sulfoxides or dithioacetal monoxides) were methyl methylthiomethyl sulfoxide **1a** and ethyl ethylthiomethyl sulfoxide **1b**, introduced by Ogura and Tsuchihashi[1] and Schlessinger et al.,[2] respectively. These reagents have an advantage (over the parent dithioacetals) of an easier carbanion generation and removal of the sulfur moiety. The latter is usually readily achieved by a mild acidic hydrolysis, although in certain cases difficulties may be encountered which requires application of a stepwise procedure.

$$R^1S(R^2S{=}O)CH_2 \ (\mathbf{1}) \xrightarrow{1)\ \text{Base};\ 2)\ R^3\text{-}I} R^1S(R^2S{=}O)CHR^3 \xrightarrow{H_3O^+} R^3\text{–CHO}$$

$$R^1S(R^2S{=}O)CHR^3 \xrightarrow{\text{reduction}} R^1S(R^2S)CHR^3 \xrightarrow{\text{hydrolysis}} R^3\text{–CHO} \quad (3.4.1)$$

a: $R^1 = R^2 = Me$
b: $R^1 = R^2 = Et$

In turn, optically active dithioacetal monoxides have been designed for use in asymmetric synthesis as chiral acyl anion equivalents. The anion of the chiral **1** should be capable of inducing formation of new asymmetric centers during its reaction with chiral or prochiral substrates. The following subsections deal with the synthesis of chiral dithioacetal monoxides and their application in the asymmetric carbon-carbon bond formation:

$$R^1S(R^2\overset{*}{S}{=}O)\overset{\ominus}{C}H \equiv H\text{–}\overset{\ominus}{C}{=}O$$

3.4.1 Synthesis

Two main approaches to the synthesis of optically active alkyl(aryl)sulfenylmethyl sulfoxides **1** have been developed. The first one (Method A) consists in the reaction of optically active α-halomethyl sulfoxides with sodium thiolates (Equation 3.4.2).[3,4] The alternative method (B) involves the reaction of metallated alkyl(aryl) methyl sulfides with menthyl p-toluenesulfinate (Equation 3.4.3; cf. Section 2.1).[5,6]

$$R^2\overset{*}{S}(O)\text{–}CH_2X + NaSR^1 \longrightarrow R^1S(R^2\overset{*}{S}{=}O)CH_2 \ (\mathbf{1}) \quad (3.4.2)$$

$$R^1SCH_2M + p\text{-Tol–S(O)–OMen} \longrightarrow R^1SCH_2\text{–S(O)–Tol-p} \quad (3.4.3)$$

(-)-(S_S) or (+)-(R_S) → (S)-**1** / (R)-**1**

An attempt at the synthesis of the sulfoxides **1** by asymmetric oxidation of the parent dithioacetals in the presence of bovine serum albumin (Method C) led to rather disappointing results: the products **1** have been obtained in low yields and with moderate ee values.[7]

$$R^1S-CH_2-SR^2 \xrightarrow[\text{}]{\text{aq. NaIO}_4\text{, Bovine Serum Albumin}} R^1S-CH_2-\overset{*}{S}(=O)R^2$$

$R^1 = R^2$ **1** (3.4.4)

Table 3.4.1 Synthesis of Optically Active Alkyl(Aryl)sulfenymethyl Sulfoxides **1**

					Sulfoxide 1						
Method	**R¹**	**R²**	**X**	**M**	**Symbol**	**Yield [%]**	**M.p. [°C]**	**$[\alpha]_D^{20}$ (acetone)**	**ee [%]**	**Abs. conf.**	**Ref.**
A	Me	p-Tol	Br	—	**c**	90	42–44	+193.3	—	S	4a
A	p-Tol	p-Tol	Br	—	**d**	—	73–74	−77	100	R	3
A	p-Tol	p-Tol	Cl	—	**d**	70	—	−49	—	R	3
A	p-Tol	p-Tol	Br	—	**d**	64	81–82	+76.8	100	S	4b
B	Me	p-Tol	—	Li	**c**	75	44–45	+276	100	S	5
B	Me	p-Tol	—	MgCl	**c**	78	—	+278	100	S	6
B	Ph	p-Tol	—	Li	**e**	76	69–70	+97	100	S	5
B	p-Tol	p-Tol	—	Li	**d**	70	78–79	+76	100	S	5
B	p-Tol	p-Tol	—	MgCl	**d**	81	—	+67.4	88	S	6
C	p-Tol	p-Tol	—	—	**d**	30	—	—	60	S	7
C	Ph	Ph	—	—	**f**	45	—	—	36	S	7

3.4.2 Michael Addition

From the viewpoint of asymmetric synthesis, alkylation of optically active dithioacetal monoxides **1** cannot find any practical application, since removal of the sulfur moieties in the products obtained leads to achiral carbonyl compounds (Equation 3.4.1). However, the use of **1** in the Michael addition can result in an α-, β-, and, in certain cases, γ-stereoselection. Scolastico et al. have used this approach in the synthesis of enantiomerically enriched 3-oxocyclopentanecarboxylic acid **5**[8] and the enantiopure intermediates of 11-deoxy-*ent*-prostaglandin **8** and **9**.[9] The reactions comprised the Michael addition of the anion of (+)-(S)-**1d** to cyclopentenone and 2-(6-methoxycarbonyl)hexylcyclopentenone **6**, respectively. It should be stressed that the use of HMPT as a cosolvent is crucial, since otherwise the 1,2-addition products are exclusively formed. In the former case, the product **2** was obtained as a mixture of four diastereomers and elaborated to the acid (R)-**5** with ca. 35% ee (Equation 3.4.5).

(+)-(S)-**1d** + cyclopentenone → (BuLi, THF, HMPT, 78%) → **2** → (1) $(Me_2N)_3P/I_2/MeCN$; 2) KI, 70%) → **3** → ($I_2/NaHCO_3$, dioxan/H_2O, 70%) → (R)-**4** → (Ag_2O, 52%) → (R)-**5**, 38.4% ee (3.4.5)

In turn, it was shown that the adduct **7** was formed with a poor α-induction (diastereomeric ratio 52:48), but with a full β- and γ-stereoselection, which enabled the authors to obtain the enantiopure dithioacetal **8** and aldehyde **9** (Equation 3.4.6). In conclusion, the substituent at C-2 in cyclopentenone proved to play an important role in the overall stereoselection of the 1,4-addition.[9]

(+)-(S)-**1d** + **6** → (BuLi, THF, HMPT, 45%) → **7** → ($(Me_2N)_3P$, I_2, KI, MeCN, 59%) → (8S,12R)-**8** → ($I_2/NaHCO_3$, dioxan/H_2O) → (8S,12R)-**9** (3.4.6)

3.4.3 Hydroxyalkylation

Reaction of the anion (+)-(S)-**1d** with aldehydes leads to the corresponding α-hydroxyalkyl derivatives, formed as a mixture of three (out of four possible) diastereomers. Methylation of the hydroxy group, followed by a two-step hydrolytic desulfurization of **11** (a direct hydrolysis of the dithioacetal monoxide moiety cannot be achieved) gives α-methoxyaldehydes **13** with ee up to 70%.[10]

(+)-(S)-**1d** + RCHO → (n-BuLi, THF, -78 °C) **10** → (Bu_4NOH, $Me_2SO_4/H_2O/CH_2Cl_2$) **11** → ($NaI/I_2/Ph_3P$ or $(Me_2N)_3P$) **12** → ($I_2/NaHCO_3$, H_2O/dioxan) **13** (3.4.7)

R	10 Yield [%]	11 Yield [%]	12 Yield [%]	12 $[\alpha]_D^{20}$	13 Yield [%]	13 $[\alpha]_D^{20}$	13 ee [%]	13 Abs. conf.
Ph	91	90	68	+45.1	67	–66	70	R
$PhCH_2$	85	76	55	+9.2	43	+28.9	46	R

A full β-stereoselection has been achieved when the magnesium enolate of an enantiopure cyclic dithioacetal monoxide, namely *trans*-2-N,N-diethylacetamide-1,3-dithiolane S-oxide **14**, has been treated with *iso*-butyraldehyde. Note that, unlike **1d**, **14** has two fixed centers of asymmetry. A single diastereomer of **15** has been obtained which is explained in terms of the involvement of a rigid transition state, where the oxygen atoms in **14** and the approaching aldehyde are coordinated by magnesium.[11]

(-)-**14** → (1) t-BuMgCl/THF, -78 °C; 2) Me_2CHCHO; 82%) (R_S, 2R, 1'R)-**15**, $[\alpha]_D^{25}$ -13.3 ($CHCl_3$) (3.4.8)

3.4.4 Acylation

Acylation of the anions of dithioacetal monoxides is usually carried out using acid chlorides[13,14] or esters[12] as acylating agents. α-Stereoselection is usually low and the diastereomers are formed in a thermodynamically controlled ratio. However, they can be easily separated by crystallization,[14] which is important from the viewpoint of their application as substrates in asymmetric reduction (see Section 3.8).

(+)-(S)-**1d** + RCOCl → (BuLi, THF) (2R, S_S)-**16** + (2S, S_S)-**17** (3.4.9)

R	Ratio **16**: **17**
Ph	68:32
t-Bu	36:64
n-C_6H_{13}	59:41

3.4.5 Examples of Experimental Procedures

3.4.5.1 Synthesis of (+)-(S)-p-Tolylthiomethyl p-Tolyl Sulfoxide 1d from (–)-(S)-Menthyl p-Toluenesulfinate (Equation 3.4.3)[5]

A solution of 36 mmol of methyl p-tolyl sulfide and 36 mmol of DABCO in 50 mL of THF at 0°C is treated with 1.61 N n-BuLi in hexane (24.5 mL). After 15 min the resulting mixture, cooled to –78°C, is added to a solution of 18 mmol of (–)-(S)-menthyl p-toluenesulfinate in 60 mL of THF. After 30 min the reaction is quenched with saturated NH_4Cl at –78°C. The usual workup and purification by chromatography (silica gel, n-hexane-ether), and recrystallization (n-pentane-ether) gives 3.5 g of (+)-(S)-**1d** (70% yield). Mp 78–79°C; $[\alpha]_D^{20}$ +76 (c 1, acetone). ^{1}H NMR: δ 2.32 (s, 3H, Me), 2.39 (s, 3H, Me), AB system (2H, $[J_{AB}]$ = 13.5 Hz, δ_A 4.12, δ_B 3.99), 7.00–7.66 (m, 8H, aromatic protons).

3.4.5.2 Synthesis of (+)-(S)-Methylthiomethyl p-Tolyl Sulfoxide 1c using Methylthiomethyl Grignard Reagent[6]

Magnesium (0.6 mol), activated with a small amount of iodine and 1,2-dibromoethane, is suspended in THF (200 mL) under nitrogen, and chloromethyl methyl sulfide (0.29 mol) is added dropwise over a period of 1 h. The temperature of the reaction mixture is kept at 10–20°C by cooling and controlling the speed of addition. After stirring at room temperature for an additional hour, the solution of a Grignard reagent is obtained (concentration 1.3 mol/L, which corresponds to 95% yield). The above solution (38 mL) is added to a THF solution of (–)-(S) menthyl p-toluenesulfinate (45.4 mmol) at 0°C over a 15-min period. After an additional stirring at 0°C for 1.5 h, the usual workup and chromatography on silica gel gives **1c** in 78% yield. Mp 44–45°C[5] $[\alpha]_D^{22}$ +278 (c 1, acetone). ^{1}H NMR[5]: δ 2.19 (s, 3H, Me), 2.42 (s, 3H, Me), AB system (2H, $[J_{AB}]$ = 13.5 Hz, δ_A 3.74, δ_B 3.68), 7.26–7.66 (m, 4H, aromatic protons).

3.4.5.3 Synthesis of (–)-(R)-p-Tolylthiomethyl p-Tolyl Sulfoxide 1d from (–)-(R)-Bromomethyl p-Tolyl Sulfoxide (Equation 3.4.2)[3]

(–)-(R) Bromomethyl p-tolyl sulfoxide (0.92 g, 4 mmol) is added to a stirred solution of sodium p-toluenethiolate (0.58 g, 4 mmol) in methanol (15 mL). The mixture is refluxed under nitrogen for 3 h and then poured into water and extracted with $CHCl_3$. Evaporating and purification by column chromatography (silica gel, ether:light petroleum 1:1) gives (–)-(R)-**1d**. Mp 73–74°C, $[\alpha]_D^{20}$ –77 (c 1, acetone).

3.4.5.4 Hydroxyalkylation of (+)-(S)-1d; Synthesis of α-Methoxyphenylacetaldehyde 13 (Equation 3.4.7)[10b]

a) General Procedure for Metallation of (+)-(S)-1d

A solution of (+)-(S)-**1d** in dry THF (11 mL per 1 mmol of **1d**) is cooled to –40°C and a solution of 1.6 N BuLi in n-hexane (1.1 equiv.) added. The resulting pale yellow solution is stirred at –20°C for 20 min to ensure complete metallation.

b) Reaction of (+)-(S)-1d-Li with Aldehydes

To a solution of the lithium derivative of **1d** (18 mmol) cooled to –78°C, benzaldehyde (3.7 mL, 36 mmol) is added. After 15 min at –78°C the reaction is quenched with aqueous NH_4Cl and extracted with ether. The crude product obtained from the usual workup is chromatographed (silica gel, CH_2Cl_2/MeOH) to give a mixture of three diastereomers of **10** (R = Ph) (3.6 g, 91%).

c) Methylation of 10 (R = Ph)

To a solution of **10** (5.0 g) and Me_2SO_4 (7.5 mL) in CH_2Cl_2 (250 mL), 40% aqueous Bu_4NOH (17.0 g) is added at room temperature. After 3 min, the reaction mixture is acidified by dropwise addition of 2N-HCl, and extracted with CH_2Cl_2. Column chromatography (silica gel, CH_2Cl_2/MeOH) gives a mixture of three diastereomers of **11** (4.7 g, 90%).

d) Reduction of 11(R = Ph)

Ph_3P (27.7 g) and iodine (26.8 g) are stirred together in dry MeCN (200 mL) under nitrogen until a yellow slurry is obtained. A solution of **11** (R = Ph) (3.0 g) in MeCN (35 mL) is added, followed by addition of powdered NaI (27 g). After stirring for 10 min at room temperature the reaction mixture is taken up in ether and then washed with aqueous $Na_2S_2O_3$, $NaHCO_3$, and brine. After the usual workup the residue is taken up in hexane and filtered from most of the Ph_3P and Ph_3PO. After evaporating, the residue is chromatographed on silica gel (hexane-AcOEt) to give **12** (R = Ph) (2.3 g, 68%); $[\alpha]_D^{20}$ +45.1 (c 1, $CHCl_3$).

e) Hydrolysis of 12 and Isolation of Aldehyde 13 (R = Ph)

To a solution of **12** (R = Ph) (0.9 g) in dioxan (15 mL) water (15 mL), I_2 (0.74 g) and $NaHCO_3$ (0.23 g) are added under nitrogen. After 10 min at room temperature a second portion of I_2 (0.39 g) and $NaHCO_3$ (0.12 g) is added and the reaction mixture stirred for another 10 min. Then it is cooled to 0°C and an excess of $Na_2S_2O_3$ added. Extraction with ether, the usual workup, and column chromatography (silica gel, CH_2Cl_2) give 2-methoxy-2-phenylacetaldehyde in 67% yield; $[\alpha]_D^{20}$ –66.0 (c 1, $CHCl_3$).

3.4.5.5 Synthesis of α-Arylthio-β-Oxosulfoxides 16 and 17 (Equation 3.4.9, R = Ph)[14]

Butyllithium (11 mmol) in n-hexane is added at –78°C to a solution of (+)-(S)-**1d** (10 mmol) in dry THF (60 mL). After 10 min benzoyl chloride (5 mmol) is added. After stirring for 20 min, further portions of BuLi (5.5 mmol) and of benzoyl chloride are added following the same procedure. Similar additions are repeated three times. Then the reaction is quenched with a saturated aqueous solution of NH_4Cl and extracted with ether. The crude product obtained from the usual workup is chromatographed (hexane-AcOEt, 1:1) to give a mixture of two diastereomers (70%); **16**:**17** ratio 68:32. The diastereomers are separated by successive crystallization from ether. *No optical rotations are given; the diastereomers are distinguished by their ^1H NMR spectra:*

16 (R = Ph). ^1H NMR: δ 2.29 (s, 3H, MeC_6H_4S), 2.34 (s, 3H, MeC_6H_4S), 5.47 (s, 1H, CH), 6.96–7.94 (m, 13H, Ar).

17 (R = Ph). ^1H NMR: δ 2.28 (s, 3H), 2.41 (s, 3H), 5.39 (s, 1H), 6.97–7.74 (m, 13H).

REFERENCES (3.4)

1. Ogura, K., Tsuchihashi, G., *Tetrahedron Lett.*, 3151, 1971.
2. Richman, J. E., Herrmann, J. L., Schlessinger, R. H., *Tetrahedron Lett.*, 3267, 1973.
3. Cinquini, M., Colonna, S., Fornasier, R., Montanari, F., *J. Chem. Soc., Perkin Trans. 1*, 1886, 1972.
4. a) Numata, T., Oae, S., *Bull. Chem. Soc. Jpn.*, **45**, 2794, 1972.
   b) Kunieda, N., Nokami, J., Kinoshita, M., *Bull. Chem. Soc. Jpn.*, **49**, 256, 1976.
5. Colombo, L., Gennari, C., Narisano, E., *Tetrahedron Lett.*, 3861, 1978.
6. Ogura, K., Fujita, M., Takahashi, K., Iida, H., *Chem. Lett.*, 1697, 1982.
7. Ogura, K., Fujita, M., Iida, H., *Tetrahedron Lett.*, **21**, 2233, 1980.
8. Colombo, L., Gennari, C., Resnati, G., Scolastico, C., *Synthesis*, 74, 1981.

9. Colombo, L., Gennari, C., Resnati, G., Scolastico, C., *J. Chem. Soc. Perkin Trans 1,* 1284, 1981.
10. a) Colombo, L., Gennari, C., Scolastico, C., Guanti, G., Narisano, E., *J. Chem. Soc., Chem. Commun.*, 591, 1979.
    b) Colombo, L., Gennari, C., Scolastico, C., Guanti, G., Narisano, E., *J. Chem. Soc. Perkin Trans 1,* 1278, 1981.
11. Corich, M., Di Furia, F., Licini, G., Modena, G., *Tetrahedron Lett.*, **33**, 3043, 1992.
12. Ogura, K., Fujita, M., Inaba, T., Takahashi, K., Iida, H., *Tetrahedron Lett.*, **24**, 503, 1983.
13. Guanti, G., Narisano, E., Banfi, L., Scolastico, C., *Tetrahedron Lett.*, **24**, 817, 1983.
14. Guanti, G., Narisano, E., Pero, F., Banfi, L., Scolastico, C., *J. Chem. Soc., Perkin Trans. 1*, 189, 1984.

3.5 α-SULFINYL-, α-SULFONYL-, AND α-SULFOXIMINO SULFOXIDES

3.5.1 α-Sulfinyl Sulfoxides

In a search for the most versatile and efficient chiral acyl anion equivalents (cf. Section 3.4) a new class of sulfoxides has been developed, namely, α-sulfinyl sulfoxides. The first example of this class was (S,S)-bis-p-tolylsulfinylmethane **1**, synthesized by Kunieda et al.[1] from menthyl p-toluenesulfinate and the anion of (+)-(R) methyl p-tolyl sulfoxide (Equation 3.5.1).

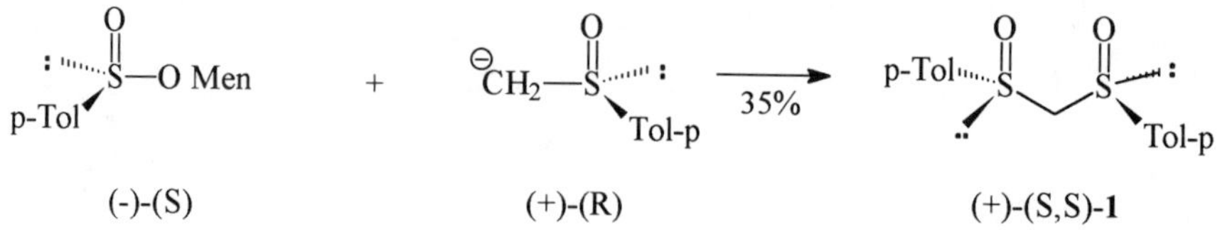

(-)-(S) + (+)-(R) → (35%) (+)-(S,S)-**1** (3.5.1)

A very important feature of the structure of (S,S)-**1** is that it possesses a C_2 axis of symmetry. This not only reduces the number of diastereomeric transition states, but also results in the creation of only one new stereocenter in addition reactions to trigonal electrophiles.[2a] Note that the (R,S) diastereomer is an optically inactive *meso* compound.

The anion of the bis-sulfoxide (S,S)-**1** has been used by Solladie et al.[3] in the reaction with carbonyl compounds to give the corresponding condensation products with moderate to good diastereoselectivity. The use of a magnesium-containing base improves the diastereomer ratio; however, this lowers the yields (Equation 3.5.2, Table 3.5.1).[3]

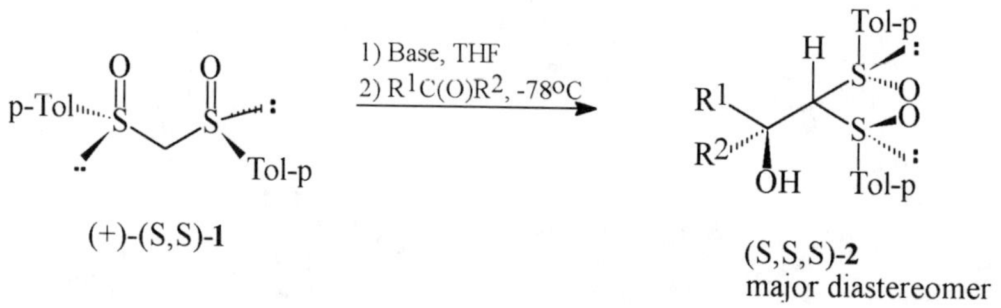

(+)-(S,S)-**1** → 1) Base, THF; 2) $R^1C(O)R^2$, -78°C → (S,S,S)-**2** major diastereomer (3.5.2)

Table 3.5.1 Hydroxylation of the Anion of (+)-(S,S)-**1**

Base	R^1	R^2	Yield [%]	Diastereomer ratio
$(i\text{-}Pr)_2NMgBr$	Ph	H	26	>98:<2
t-BuMgBr	Ph	H	22	>98:<2
BuLi	Ph	H	70	90:10
BuLi	$p\text{-}O_2NC_6H_4$	H	93	85:15
BuLi	$p\text{-}MeOC_6H_4$	H	37	84:16
BuLi	$3,4\text{-}diMeOC_6H_4$	H	No reaction	
BuLi	$p\text{-}MeO_2CC_6H_4$	H	75	85:15
BuLi	C_5H_{11}	H	60	60:40
BuLi	i-Pr	H	56	55:45
BuLi	EtO_2C	Me	60	75:25
BuLi	$p\text{-}O_2NC_6H_4$	Me	No reaction	
BuLi	Ph	Et	No reaction	

Adapted from Solladie, G., Colobert, F., Ruiz, P., Hamdouchi, C., Carreno, M. C., and Garcia Ruano, J. L., *Tetrahedron Lett.*, **32**, 3695, 1991. With kind permission from Elsevier Science Ltd, The Boulevard, Langford Lane, Kidlington OX5 1GB, UK.

When piperidine is applied as a base, the reaction affords γ-hydroxy-α,β-unsaturated sulfoxides **5** as a result of a double bond migration in the initially formed Knoevenagel-type product **3**, followed by the sulfoxide-sulfenate rearrangement and removal of the sulfur moiety from **4** (for a more detailed mechanistic discussion see the "SPAC" reaction, Section 3.8.1.2). The products **5** are formed as mixtures of diastereomers, since both sulfinyl moieties are equally involved in the rearrangement. Their oxidation gives chiral nonracemic (E)-γ-oxo-α,β-unsaturated sulfoxides **6** (Scheme 3.5.1).[4a]

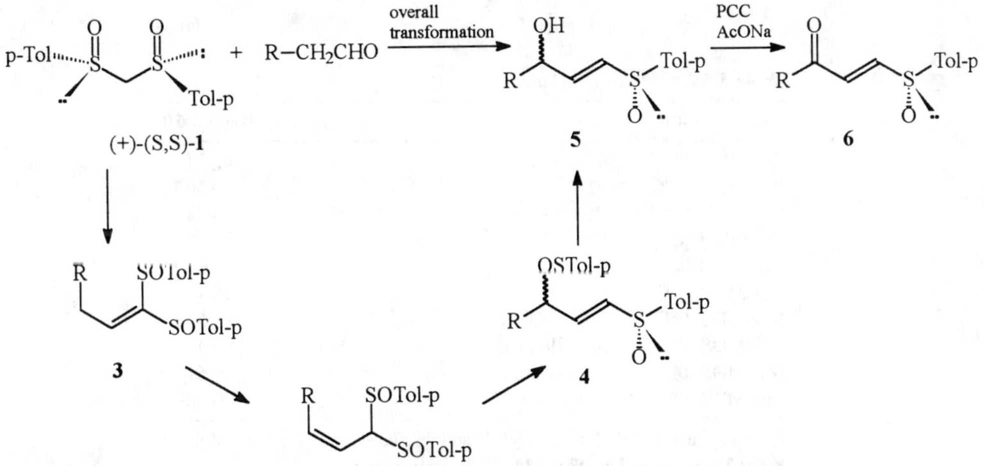

p-Tol–S(O)–CH2–S(O)–Tol-p + R–CH2CHO —overall transformation→ **5** —PCC, AcONa→ **6**

(+)-(S,S)-**1** → **3** → R–CH=CH–CH(SOTol-p)2 → **4** (OSTol-p) → **5**

R	**5**, Yield [%]	**6**, Yield [%]
Et	84	85
n-Pr	98	84
i-Pr	96	86
t-Bu	74	87
n-C_5H_{11}	77	85

Scheme 3.5.1

Application of α,β-unsaturated aldehydes as the carbonyl components in the hydroxyalkylation of the anion of **1** gives in a one-step procedure the corresponding 1,1-bis-(p-tolylsulfinyl)-1,3-butadienes **7**.[3]

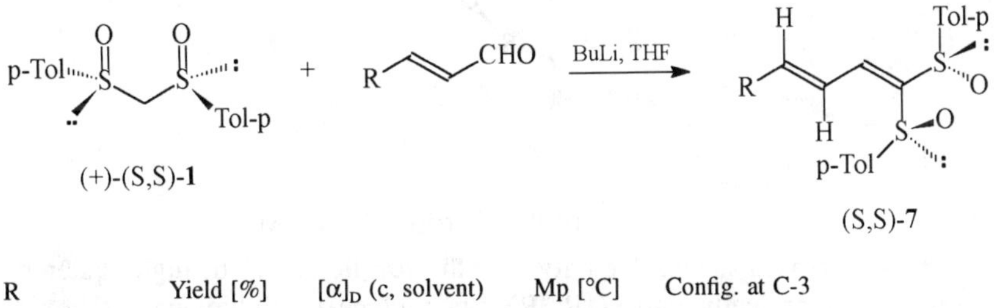

(+)-(S,S)-**1** + R–CH=CH–CHO —BuLi, THF→ (S,S)-**7** (3.5.3)

R	Yield [%]	$[\alpha]_D$ (c, solvent)	Mp [°C]	Config. at C-3
Me	64	–22 (1.9, acetone)	121–123	E
Et	81	–24 (0.7, acetone)	112–114	E
Ph	80	–94 (0.9, $CHCl_3$)	182–183	Not established
o-$MeOC_6H_4$	81	–109 (2.1, $CHCl_3$)	155–157	Not established

Another example of bis-sulfoxide possessing C_2 symmetry is *trans*-1,3-dithiane S,S-dioxide **8**, introduced by Aggarwal et al.[2] The reaction of the anion derived from **8** with aldehydes proved to be very strongly dependent on the conditions and the base applied. Thus, sodium bis-(trimethylsilyl)amide (NaHMDS) gives generally better diastereoselectivity than BuLi/pyridine. The diastereoselectivity is surprisingly low at –78°C and increases substantially at 0°C, when the adducts formed are allowed to equilibrate. Finally, only aromatic aldehydes give products with good diastereoselectivity (Equation 3.5.4, Table 3.5.2).[2]

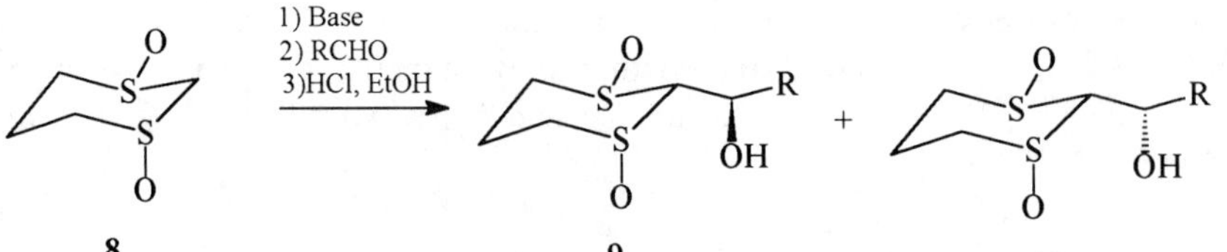

1) Base
2) RCHO
3)HCl, EtOH

8 9 10 (3.5.4)

Table 3.5.2 Hydroxyalkylation of **8**

Base, conditions	R	Time [min]	Ratio 9:10
BuLi/Py, −78°C	n-Bu	1	63:17
BuLi/Py, −78°C	i-Pr	1	50:50
BuLi/Py, −78°C	Ph	1	66:34
BuLi/Py, 0°C	Ph	30	45:55
BuLi/Py, 0°C	Ph	60	36:64
NaHMDS, 0°C	Ph	30	96:4
NaHMDS, 0°C	m-$MeOC_6H_4$	30	95:5
NaHMDS, 0°C	3,4-$(Me_2tBuSi)_2C_6H_3$	30	96:4
NaHMDS, 0°C	p-$O_2NC_6H_4$	30	95:5
NaHMDS, 0°C	i-Pr	30	60:40

From Aggarwal, V.K., Davies, I.W., Maddock, J., Mahon, M.F., and Molloy, K.C., *Tetrahedron Lett.*, **31**, 135, 1990. With permission.

Some chiral ketene equivalents **11**[5], **12**[6], **13**[7], **14**[7], and **15**[8] having C_2 symmetry have been synthesized and used as dienophiles in the Diels-Alder cycloaddition.

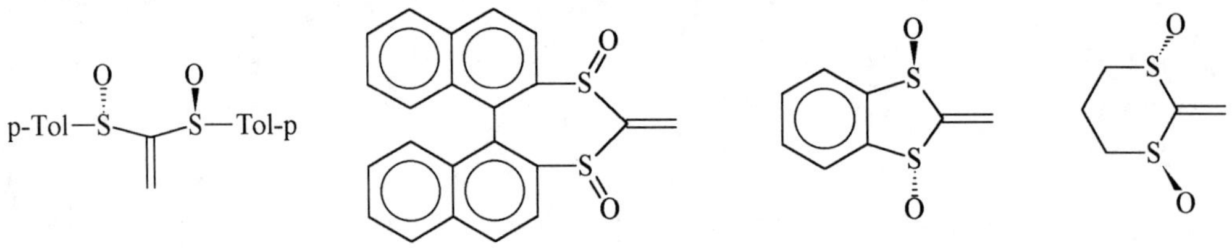

p-Tol—S—S—Tol-p

11 12 13 14

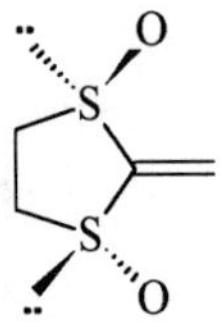

(R,R)-**15**, $[\alpha]_D$ -33 ($CHCl_3$)

The cyclic analogs show increased stereoselectivity over the acyclic ones. The highest stereoselectivity has been observed with optically active (1R,3R)-2-methylene-1,3-dithiolane 1,3-dioxide **15**.[8] Its reaction with a variety of dienophiles proceeds smoothly and gives the adducts as practically single diastereomers (diastereomeric ratio >97:<3).[8]

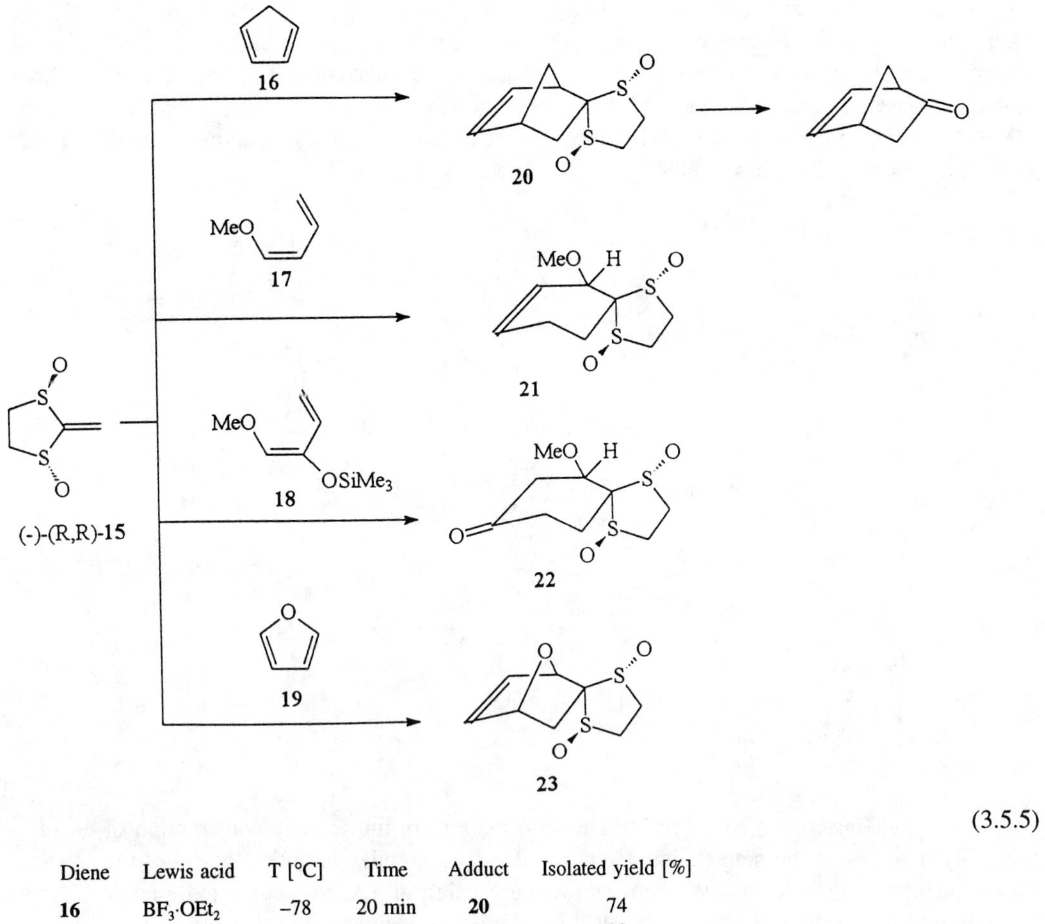

(3.5.5)

Diene	Lewis acid	T [°C]	Time	Adduct	Isolated yield [%]
16	$BF_3 \cdot OEt_2$	–78	20 min	**20**	74
17	—	rt	18 h	**21**	83
18	—	rt	2 h	**22**	90
19	$SnCl_4$	–78	30 min	**23**	65

3.5.2 α-Sulfonyl Sulfoxides

Optically active α-sulfonyl sulfoxides **24** have been easily prepared from the corresponding α-metallated methyl sulfones and (–)-(S) menthyl p-toluenesulfinate.[9,11]

$$RSO_2Me \quad + \quad \text{p-Tol}\overset{O}{\overset{\|}{S}}\text{–O–Men} \xrightarrow{\text{BuLi, THF}} R\text{–}SO_2\text{–}CH_2\text{–}\overset{O}{\overset{\|}{S}}\text{–Tol-p}$$

(-)-(S_S) (S)-**24** (3.5.6)

24	R	$[\alpha]_D^{20}$ ($CHCl_3$)	Mp [°C] (cryst. from $i\text{-}Pr_2O$)	Ref.
a	Ph	+251	71–72	9
b	t-Bu	+143	116–117	11
c	p-Tol	+266	83–84	9
d	$p\text{-}ClC_6H_4$	+266	92–93	9
e	$p\text{-}MeOC_6H_4$	+260	106–107	9

Alkylation of the anion of **24** leads to a mixture of diastereomers in the ratio 95:5.[10] On the other hand, reaction of **24** with aldehydes in the presence of piperidine as a base does not give the corresponding hydroxyalkyl derivatives **25**, but γ-hydroxy-α,β-unsaturated sulfones **26**, the products of their dehydration and subsequent [2,3]-sigmatropic rearrangement (Equation 3.5.7)[12,13] (cf. the "SPAC" reaction, Section 3.8.1.2). The overall stereoselectivity of this reaction is moderate and gives sulfones **26** with ee up to 50%.[14]

piperidine, MeCN, CH_2Cl_2, -35°C; $- H_2O$; proton transfer; 1) [2,3]-shift 2) Thiophile

(S)-**24a** **25** **26**

(3.5.7)

R	Time [h]	Yield [%]	26 ee [%]	Abs. conf.
Me	20	70	34	S
i-Pr	20	76	50	S

Lopez and Carretero[14] have used this transformation for an iterative construction of enantiomerically pure polypropionate chains. However, due to the moderate optical purity of the sulfones **26** obtained, they had to resolve them additionally using an enzyme-promoted enantioselective procedure. The enantiopure sulfone (S)-**26** (R = i-Pr) thus obtained was then transformed via the sequence of reactions in Equation 3.5.8 into the enantiopure aldehyde **27**. The latter was used as a carbonyl component in the next reaction with (S)- and (R)-**24a**.

26 → (Lipase *Pseudomonas cepacia*, AcO–CH=CH$_2$, i-Pr_2O; 70÷88%) → (R)-**26** ee >95% + (S)-**26** ee >98%

(S)-**26** → (1) $(MeO)_2CH_2/P_2O_5$; 2) n-BuLi/THF; 3) TMSCl; 62-84%) → TMS/OMOM vinyl sulfone

→ (1) MeLi, -78°C; 2) KF/MeOH; 85-92%) → $PhSO_2$ OMOM/Me compound, *syn / anti* >98:2, ee >98%

→ (1) n-BuLi; 2) $ClCO_2Et$; 3) Na/Hg; 81%) → EtO_2C OMOM/Me ester

→ (DIBAL, -78°C; 87%) → OHC OMOM/Me aldehyde **27**

(3.5.8)

The sulfonyl sulfoxide **24b** was transformed into (+)-(S)-1-*t*-butylsulfonyl-1-p-tolylsulfinylethene **28** via a Mannich reaction with formaldehyde and dimethylamine followed by quaternization and elimination of the amine moiety.[11] The product **28** was used as a masked chiral ketene equivalent in the Diels-Alder reaction with cyclopentadiene. The stereochemical outcome of this reaction depends on the catalyst used: in the absence of a catalyst no stereoselectivity is observed, while in the presence of a Lewis acid only two out of four possible diastereomers are formed, with one **29** clearly predominating. Its configuration has been determined by chemical correlation with that of known compounds.

(+)-(S)-**24b** (t-$BuSO_2$–CH_2–S(O)–Tol-p) → (1) CH_2O, Me_2NH; 2) MeI, $CaCO_3$, MeCN; 78%) → (S)-**28**

$[\alpha]_D$ +25 ($CHCl_3$) (3.5.9)

(3.5.10)

Lewis acid	Time [h]	29	
		Ratio of diaster.	Isolated yield of predominant diastereomer [%]
No	66	9:23:35:23	—
$ZnBr_2$ (0.8 equivalent)	6	12:0:88:0	58
Eu $(fod)_3$, (1 equivalent)	42	8:0:92:0	62
SiO_2 (without solv.)	48	10:0:84:6	63

3.5.3 α-Sulfoximino Sulfoxides[10]

Synthesis and reactivity of the title compounds is discussed in Chapter 4, Equation 4.3.12.

3.5.4 Examples of Experimental Procedures

3.5.4.1 Synthesis of (S,S)-Bis-p-Tolylsulfinylmethane 1[1]

(+)-(R) Methyl p-tolyl sulfoxide (463 mg, 3 mmol) in 5mL of THF is treated with a solution of lithium diethylamide (prepared from 3.1 mmol of n-BuLi in hexane and 3.1 mmol of Et_2NH) at 0°C under N_2. After 30 min (–)-(S) menthyl p-toluenesulfinate (442 mg, 1.5 mmol) in 5 mL of THF is added and the mixture stirred for 1 h. The solution is acidified (pH~3) with diluted H_2SO_4* and extracted with $CHCl_3$. The organic layer is washed with H_2O and dried over Na_2SO_4. The solution is evaporated and washed with hexane to remove L-menthol. The crude product is obtained in 84% yield. Recrystallization from benzene gives 152 mg of (S,S)-**1** in a diastereomerically pure state; mp 137.5°C, $[\alpha]_D^{20}$ +318 (c 0.19, acetone).

3.5.4.2 Synthesis of α-Sulfonyl Sulfoxides 24; General Procedure[9]

n-BuLi (4 mmol) is added with stirring at –70°C to a solution of aryl methyl sulfone (4 mmol) in anhydrous THF (15 mL). The temperature is allowed to rise to –20°C, then (–)-(S) menthyl p-toluenesulfinate (2 mmol), dissolved in THF (20 mL), is added at –70°C. The mixture is warmed to room temperature and stirred at 20–25°C for 3 h. A saturated solution of aqueous NH_4Cl (25 ml) is added. The organic layer is separated and the aqueous layer extracted with CH_2Cl_2. The combined organic layers are dried with Na_2SO_4 and evaporated. The residue is separated by column chromatography (SiO_2, Et_2O/light petroleum) and the products purified by crystallization from i-Pr_2O (see Equation 3.5.6).

* Originally HCl is used; it is, however, advisable to avoid HCl since the Cl^- anion is known to cause sometimes racemization of sulfoxides.

3.5.4.3 Preparation of γ-Hydroxy α,β-Unsaturated Sulfones 26; General Procedure[12a]

Piperidine (0.13 mL, 1.36 mmol, 2.0 equiv.) and an aldehyde (1.36 mmol, 2.0 equiv.) are added at 0°C to a solution of an α-sulfonyl sulfoxide **24** (0.68 mmol, 1.0 equiv.) in dry MeCN (3 mL). The stirring is continued at 0°C for 1–8 h and then 5% HCl (10 mL) is added. The reaction mixture is extracted with CH_2Cl_2 (2 × 15 mL), the organic extracts are dried with Na_2SO_4 and evaporated. The crude product is purified by flash chromatography (hexane/AcOEt 2 ÷ 5:1) to give pure sulfone **26**.*

REFERENCES (3.5)

1. Kunieda, N., Nokami, J., Kinoshita, M., *Bull. Chem. Soc. Jpn.*, **49**, 256, 1976.
2a. Aggarwal, V. K., Franklin, R. J., Rice, M. J., *Tetrahedron Lett.*, **32**, 7743, 1991.
2b. Aggarwal, V. K., Davies, I. W., Maddock, J., Mahon, M. F., Molloy, K. C., *Tetrahedron Lett.*, **31**, 135, 1990
3. Solladie, G., Colobert, F., Ruiz, P., Hamdouchi, C., Carreno, M. C., Garcia Ruano, J. L., *Tetrahedron Lett.*, **32**, 3695, 1991.
4. Guerrero-de la Rosa, V., Ordonez, M., Alcudia, F., Llera, J. M., *Tetrahedron Lett.*, **31**, 135, 1990.
5. Arai, Y., Kuwayama, S., Takeuchi, Y., Koizumi, T., *Synth. Commun.*, 233, 1986.
6. De Lucchi, O., Fabri, D., Lucchini, V., *Synlett*, 565, 1991.
7. Aggarwal, V. K., Lightowler, M., Lindell, S. D., *Synlett*, 730, 1992.
8. Aggarwal, V. K., Drabowicz, J., Grainger, R. S., Gültekin, Z., Lightowler, M., Spargo, P. L., *J. Org. Chem.*, **60**, 4962, 1995.
9. Annunziata, R., Cinquini, M., Cozzi, F., *Synthesis*, 535, 1979.
10. Annunziata, R., Cinquini, M., *Synthesis*, 767, 1982.
11. Lopez, R., Carretero, J. C., *Tetrahedron: Asymmetry*, **2**, 93, 1991.
12a. Dominguez, E., Carretero, J. C., *Tetrahedron Lett.*, **31**, 2487, 1990.
12b. Dominguez, E., Carretero, J. C., *Tetrahedron*, **46**, 7197, 1990.
13. Trost, B. M., Grese, T. A., *J. Org. Chem.*, **56**, 3189, 1991.
14. Carretero, J. C., Dominguez, E., *J. Org. Chem.*, **58**, 1596, 1993.

3.6 α-HALOGENO SULFOXIDES

Among the α-functionalized sulfoxides, derivatives **1** containing the halogen atom constitute a group of compounds which have recently received increasing attention. From the synthetic viewpoint, especially interesting are α-haloalkyl aryl sulfoxides **2**.

$R-S(=O)-CRR^1(X)$ **1** $\qquad$ $Ar-S(=O)-CH(X)-R$ **2**

Due to the presence of the strongly electron-withdrawing halogen atom in **2**, the α-hydrogen is highly acidic. Therefore, the sulfinyl anions, such as **3a**, can be easily generated by treatment of **2** with a strong nonnucleophilic base, such as LDA, even below –40°C . On the other hand, the sulfinylated carbon centered radicals **3b** can be generated using tributyltin hydride and 2,2′-azobisisobutyronitrile (AIBN) . Finally, as the halogeno anions belong to relatively good leaving groups, α-halogeno sulfoxides **2** may also be considered as a source of the carbocation **3c**.

$Ar-S(=O)-CH(X)-R$ **2** → (LDA, -40 °C) $Ar-S(=O)-C^{\ominus}(X)-R$ **3a**

2 → (Bu_3SnH/AIBN) $Ar-S(=O)-C^{\bullet}(X)-R$ **3b**

2 → ($-X^{\ominus}$) $Ar-S(=O)-C^{\oplus}(H)-R$ **3c** (3.6.1)

Before discussing the synthesis and application of α-halogeno sulfoxides, it should be noted that only α-chloro and α-bromo derivatives have been isolated as optically active species[1] and that a successful use of these compounds as substrates in asymmetric synthesis is limited to α-chloroalkyl p-tolyl sulfoxides.

Chiral α-halogenoalkyl p-tolyl sulfoxides **4** and **5** with high enantiomeric purity were efficiently obtained through halogenation of enantiomerically pure alkyl p-tolyl sulfoxides **6** with electrophilic halogenating agents in the presence of a molar excess of silver (I) nitrate,[2] silica gel,[3] or potassium carbonate[4] (Equation 3.6.2 and Table 3.6.1). As halogenation agents, dichloroiodobenzene **7** and N-halogenosuccinimides **8a** and **b** or bromine were used.

(+)-(R)-6 →([NX], additives) (S)-4,5 or / and (R)-4,5

7, NX = $PhICl_2$
8a, NX = NCS
8b, NX = NBS

4, R = Cl
5, R = Br

(3.6.2)

Table 3.6.1 Halogenation of Optically Active Alkyl p-Tolyl Sulfoxides, R-S(O)Tol-p, **6a–d**, with Electrophilic Halogenation Reagents **7, 8ab**, and Bromine

Sulfoxide 6				Halogenation reagent	α-Halogeno sulfoxide 4 or 5						Ref.
No	R	R^1	$[\alpha]_D$		No	R	R^1	Yield [%]	$[\alpha]_D$	ee [%]	
a	H	H	+144	**7**	**4a**	H	H	75	+92	38.5	2
a	H	H	+144	**7**/$AgNO_3$	**4a**	H	H	69	–106	44.3	2
a	H	H	+138	**8a**/SiO_2	**4a**	H	H	82	–121		3
a	H	H	+149.5	**8a**/K_2CO_3	**4a**	H	H	91	–207.9	87	4
a	H	H	+144	Br_2	**5a**	H	H	68	+153	76.5	2
a	H	H	+144	$Br_2/AgNO_3$	**5a**	H	H	86	–196	98	2
a	H	H	+138	**8b**/SiO_2	**5a**	H	H	79	–142	93	3
b	H	Me	+189	**7**	**4b**	H	Me	78	–7	62	2
b	H	Me	+189	**7**/$AgNO_3$	**4b**	H	Me	68	–153	86	2
b	H	Me	+198	**8a**/SiO_2	**4b**	H	Me	87	–66.8		3
b	H	Me	+202.6	**8a**/K_2CO_3	**4b**	H	Me	94	–154.1	93	4
b	H	Me	+189	Br_2	**5b**	H	Me	76	–83	62	2
b	H	Me	+189	$Br_2/AgNO_3$	**5b**	H	Me	72	–115	86	2
b	H	Me	+189	**8b**/SiO_2	**5b**	H	Me	75	–131		3
c	Me	Me	+178	**7**	**4c**	Me	Me	73	–23	18.8	2
c	Me	Me	+178	**7**/$AgNO_3$	**4c**	Me	Me	66	–119	97	2
c	Me	Me	+172	**8a**/SiO_2	**4c**	Me	Me	75	–86		3
c	Me	Me	+194	**8a**/K_2CO_3	**4c**	Me	Me	92	–115.2	94	4
c	Me	Me	+178	Br_2	**5c**	Me	Me	80	–84	94	2
c	Me	Me	+178	$Br_2/AgNO_3$	**5c**	Me	Me	70	–88	100	2
c	Me	Me	+172	**8b**/SiO_2	**5c**	Me	Me	69	–75.3		3
d	H	CH_2Ph	+119.9	**8a**/K_2CO_3	**4d**	H	CH_2Ph	88	–85.1	87	4

An inspection of the data collected in Table 3.6.1 clearly shows that the conversion of methyl p-tolyl sulfoxide **6a** into chloromethyl and bromomethyl derivatives **4a** and **4b** by dichloroiodobenzene or bromine, respectively, is accompanied by retention of configuration at the sulfur atom, but the same reaction proceeds with inversion of configuration in the presence of a molar excess of silver (I) nitrate.[2] Predominant inversion of configuration is also observed when N-halogenosuccinimides **8ab** are used as halogenation agents and reactions are carried out in the presence of SiO_2 or K_2CO_3.[3,4] When the methyl group in alkyl p-tolyl sulfoxide **6** is replaced by the ethyl, isopropyl, or benzyl group, α-halogenation is always accompanied by predominant inversion of configuration.[2]

Halogenation of ethyl p-tolyl sulfoxide **6b** is accompanied by asymmetric induction at the α-carbon atom and gives the corresponding α-halogeno derivatives **4b** and **5b** as a mixture of diastereomers. This is evident from the fact that the sulfones **7a** or**7b** obtained from them are always optically active.[2]

p-Tol—S(=O)—CH_2CH_3 —[X]→ p-Tol—S(=O)—CH(X)—CH_3

(+)-(R)-**6b**, $[\alpha]_D$ +189 → **4b** or **5b**

X = $PhICl_2$ → **4b**, X = Cl, $[\alpha]_D$ -7
X = $PhICl_2$ + $AgNO_3$ → **4b**, X = Cl, $[\alpha]_D$ -153
X = Br_2 → **5b**, X = Br, $[\alpha]_D$ -83
X = Br_2 + $AgNO_3$ → **5b**, X = Br, $[\alpha]_D$ -115

(3.6.3)

p-Tol—S(=O)—CH(X)—CH_3 —[O]→ p-Tol—S(=O)(=O)—CH(X)—CH_3

4b or **5b**

4b, X = Cl, $[\alpha]_D$ -7 → **7a**, X = Cl, $[\alpha]_D$ -0.5
4b, X = Cl, $[\alpha]_D$ -153 → **7a**, X = Cl, $[\alpha]_D$ -6.9
5b, X = Br, $[\alpha]_D$ -115 → **7b**, X = Br, $[\alpha]_D$ -14.9

(3.6.4)

It was shown that halogenation of (+)-(R)-**6b** in the presence of $AgNO_3$ gave almost pure diastereomers **4b** and **5b** (de values above 90%).[2] Later on, a similar induction was observed in the α-chlorination of this sulfoxide with NCS in the presence of K_2CO_3.[4]

Bromination of (–)-(S)-benzyl methyl sulfoxide **8** with bromine in pyridine was found to give a mixture of two regioisomers: enantiomeric α-bromomethyl benzyl sulfoxide **9** and diastereomeric α-bromobenzyl methyl sulfoxide **10** in a ratio 3:2. Oxidation of the latter gave the corresponding sulfone **11** whose absolute configuration was determined by X-ray analysis (Equation 3.6.5).[5a]

(-)-(S)-**8** —Br_2 / Pyr→ (-)-(R)-**9** (CH_2Br, CH_2Ph) + (R_S,S_C)-**10** (Me, C(H)(Ph)Br) —[O]→ (-)-(S)-**11** ($MeSO_2$–C(H)(Ph)(Br))

(3.6.5)

Interestingly, in the α-halogenation reaction of sulfoxides a rare case of asymmetric induction caused by isotopic substitution was observed. When optically active (+)-(R)-α,α-dideuteriodibenzyl sulfoxide **12** was chlorinated with dichloroiodobenzene in pyridine, α,α-dideuteriobenzyl α′-chlorobenzyl sulfoxide **13** was obtained as a major regioisomer with at least 78% isotopic purity.[5b] A high stereoselectivity of this conversion was indicated by the formation of essentially only one of the possible diastereomers, which after oxidation gave the sulfone **15** having relatively high optical rotation (Equation 3.6.6).

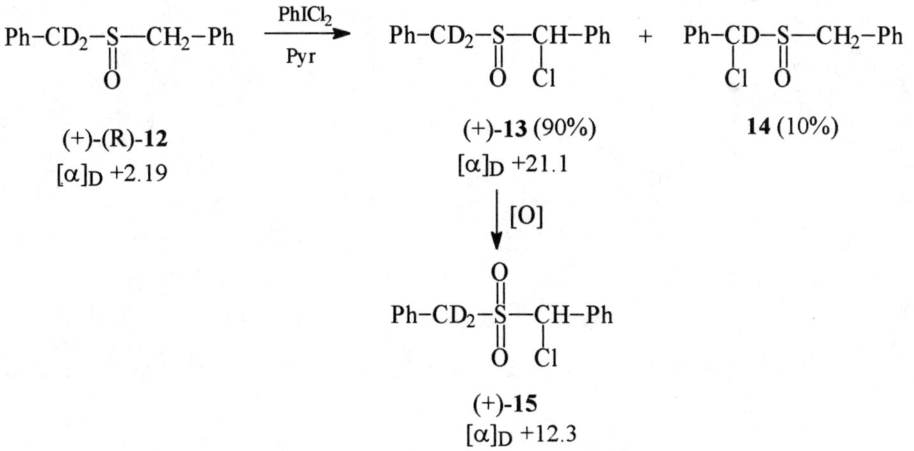

$$Ph{-}CD_2{-}\underset{\underset{O}{\|}}{S}{-}CH_2{-}Ph \xrightarrow[Pyr]{PhICl_2} Ph{-}CD_2{-}\underset{\underset{O}{\|}}{S}{-}\underset{\underset{Cl}{|}}{CH}{-}Ph + Ph{-}\underset{\underset{Cl}{|}}{CD}{-}\underset{\underset{O}{\|}}{S}{-}CH_2{-}Ph$$

(+)-(R)-**12** $[\alpha]_D$ +2.19 → (+)-**13** (90%) $[\alpha]_D$ +21.1 + **14** (10%)

(+)-**13** —[O]→ $Ph{-}CD_2{-}SO_2{-}CHCl{-}Ph$

(+)-**15** $[\alpha]_D$ +12.3 (3.6.6)

α-Chloroalkyl aryl sulfoxides were found to be the key substrates for the synthesis of various kinds of α-functionalized carbonyl compounds, epoxides, allylic alcohols, and aziridines. Very recently, this methodology was extended to asymmetric synthesis of these classes of compounds using enantiomeric α-chloroalkyl p-tolyl sulfoxides. Thus, starting from chiral α-chloromethyl p-tolyl sulfoxide (–)-**4a**, the synthesis of optically active epoxides **16** was accomplished by the sequence of reactions shown in Scheme 3.6.1.[6]

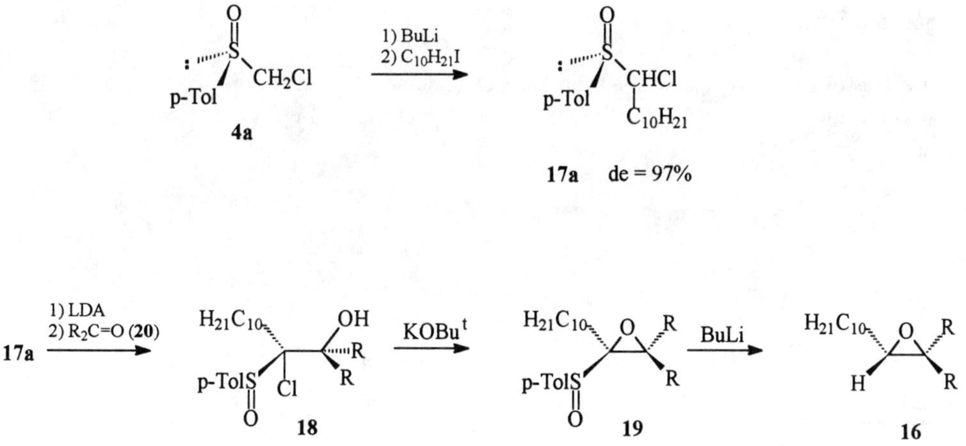

4a (p-Tol–S(O)–CH_2Cl) —1) BuLi; 2) $C_{10}H_{21}I$→ **17a** (p-Tol–S(O)–$CHCl$–$C_{10}H_{21}$) de = 97%

17a —1) LDA; 2) $R_2C{=}O$ (**20**)→ **18** —$KOBu^t$→ **19** —BuLi→ **16**

No	**20** R_2	**18** Yield[a] [%]	**19** Yield[a] [%]	$[\alpha]_D$[b]	**16** Yield[a] [%]	$[\alpha]_D$[b]
a	Me_2	100	98	+8.9	73	–13.7
b	Ph_2	—[c]	99	–91.7	89	–28.6
c	$-(CH_2)_5-$	93	95	+13.7	85	–14.0
d	$-(CH_2)_5-$	84	86	+18.5	61	–13.7

[a]Isolated yield.
[b]Measured in CCl_4 at 25°C.
[c]Not isolated.

Scheme 3.6.1

In a similar manner, alkylation of **4a** with 1-iododecane gave diastereomerically pure α-chloro sulfoxide **17a**, which upon deprotonation with LDA and treatment with 6-methylheptanal afforded an easily separable mixture of the chlorohydrines **21** and **22** in good yield. The diastereomerically pure **22** was used as a substrate for the synthesis of optically pure (+)-disparlure **23**, the sex attractant of the female gypsy moth (Scheme 3.6.2).[6]

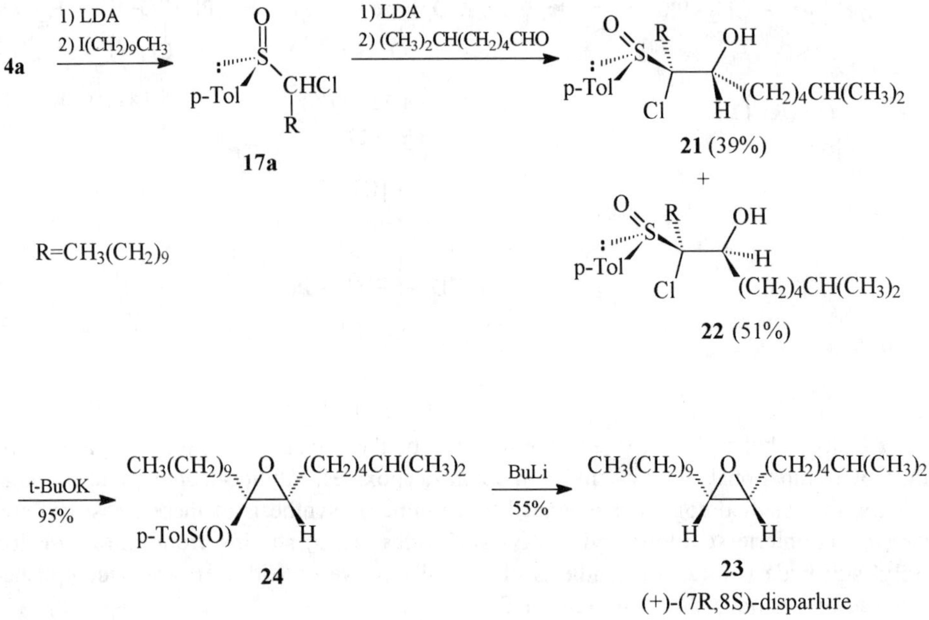

Scheme 3.6.2

A series of optically active α-amino ketones (**25** and **26**) and α-amino aldehydes **29–32** were prepared starting from (–)-**4a** via the corresponding chiral α,β-epoxy sulfoxides **27, 28, 33, 34** (Equations 3.6.7–3.6.9).

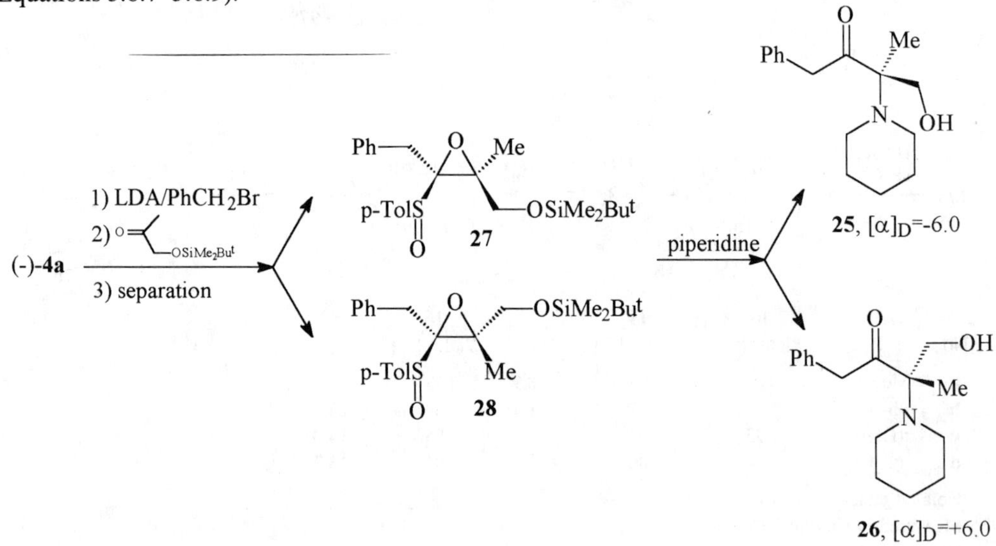

(3.6.7)

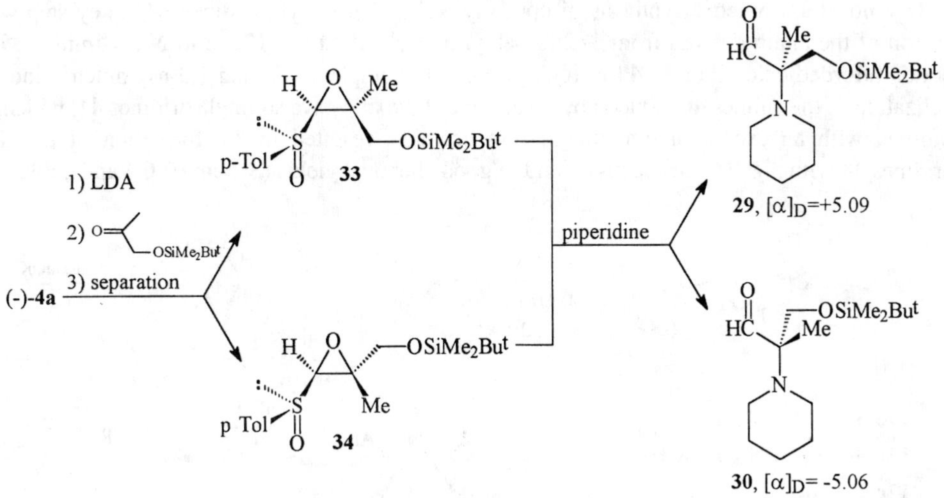

O
Me
HC
OSiMe2But
N

H O Me
p-Tol S
O
OSiMe2But
33

1) LDA

2) O=C(Me)–CH2OSiMe2But

3) separation

(-)-**4a**

piperidine

29, $[\alpha]_D$=+5.09

O
HC
OSiMe2But
Me
N

H O
OSiMe2But
p Tol S
O
Me
34

30, $[\alpha]_D$= -5.06

(3.6.8)

O
Me
HC
OSiMe2But
PhNH

(-)-**4a**

33

34

$PhNH_2$

31, $[\alpha]_D$=+36.1

O
OSiMe2But
HC
Me
PhNH

32, $[\alpha]_D$=-35.8

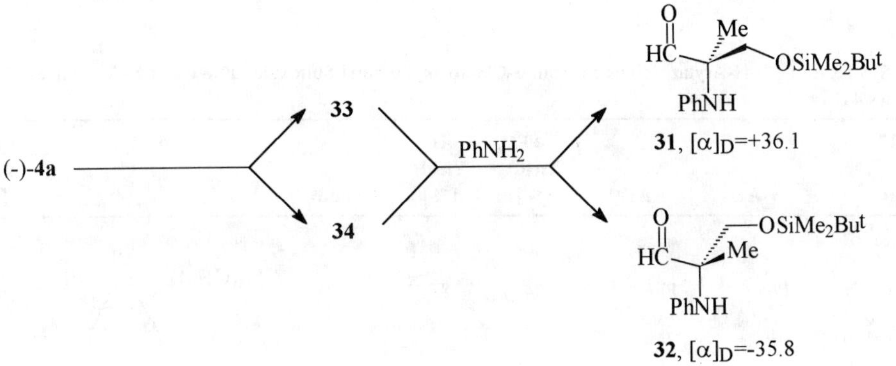

(3.6.9)

The epoxy sulfoxides **27** and **28** were also successfully used for the preparation of optically active allylic alcohols. The α,β-epoxy sulfoxide **27** gave upon treatment with 3 eq of BuLi the desired allylic alcohol **35** ($[\alpha]_D$ = +3.0; ee = 96%) and epoxy silane **36** in 21 and 63% yields, respectively. The same reaction with **28** afforded the allylic alcohol **37** ($[\alpha]_D$ = –3.3; ee > 96%) in 90% yield (Scheme 3.6.3).[6]

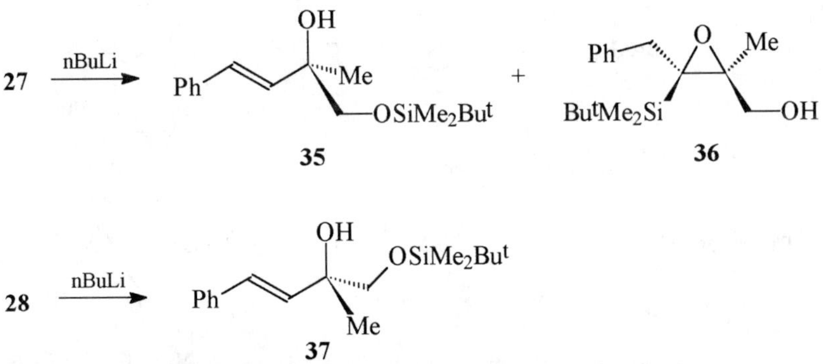

27 —nBuLi→ Ph
OH
Me
OSiMe2But
35

+

Ph O Me
ButMe2Si
OH
36

28 —nBuLi→ Ph
OH
OSiMe2But
Me
37

Scheme 3.6.3

In a novel asymmetric synthesis of optically active (Z)-N-arylaziridines **38**, a key step was the addition of the anion derived from 1-chloroalkyl p-tolyl sulfoxides **17a–c** to N-arylimines **39**. This reaction afforded chloroamines **40** in high yields with complete 1,2- and 1,3-asymmetric induction. Cyclization of the amines **40** induced by potassium t-butoxide gave sulfinylaziridines **41**, which, upon treatment with an excess of ethylmagnesium bromide, resulted in the formation of the desired aziridines **38** with high stereoselectivity and in good chemical yields (Scheme 3.6.4 and Table 3.6.2).[7]

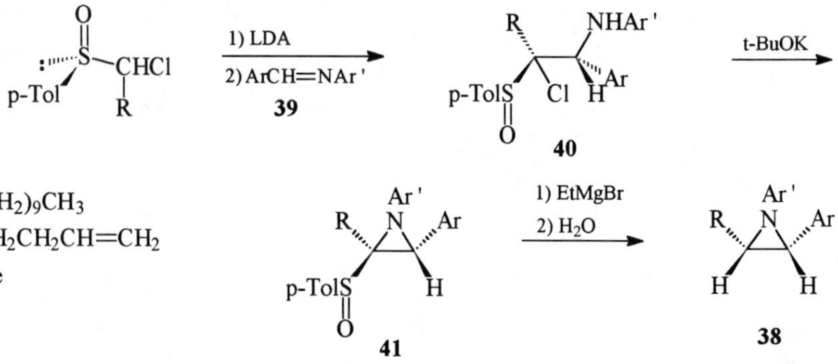

17a, R = $(CH_2)_9CH_3$
17b, R = $CH_2CH_2CH{=}CH_2$
17c, R = Me

Scheme 3.6.4

Table 3.6.2 Synthesis of (Z)-N-Arylaziridines **38** from α-Chloroalkyl p-Tolyl Sulfoxides **17a–c** and N-Arylimines **39** via Sulfinylaziridines **41**[a]

	17			40	41	38		
No	R	Ar	Ar′	Yield[b] [%]	Yield[b] [%]	Condn[c]		Yield[b] [%]
a	$CH_3(CH_2)_9$	Ph	Ph	94	87	A	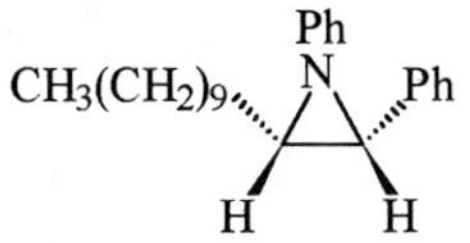	95
		p-ClC_6H_4	Ph	84	92	B	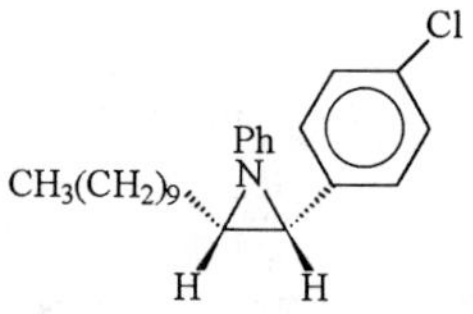	88
		Ph	p-BrC_6H_4	88	98	C	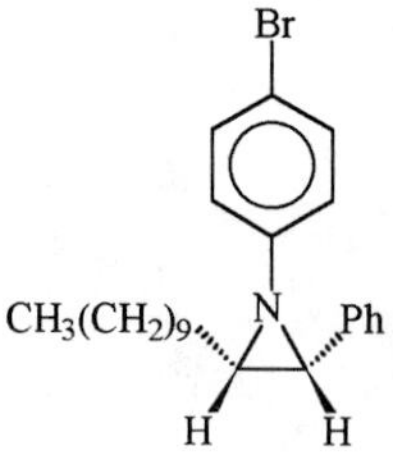	85
b	$CH_2{=}CH(CH_2)_2$	Ph	Ph	76	83	D	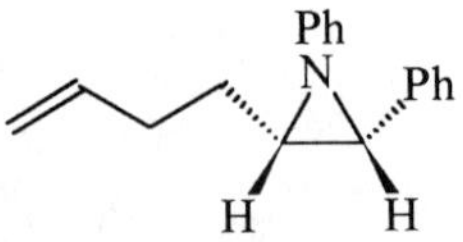	90
c	Me	Ph	Ph	91	92	E	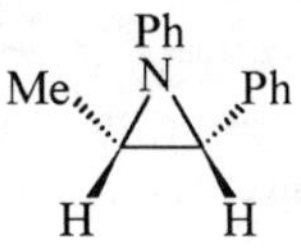	89

Table 3.6.2 (continued) Synthesis of (Z)-N-Arylaziridines **38** from α-Chloroalkyl p-Tolyl Sulfoxides **17a–c** and N-Arylimines **39** via Sulfinylaziridines **41**[a]

No	17 R	Ar	Ar′	40 Yield[b] [%]	41 Yield[b] [%]	38 Condn[c]	38	38 Yield[b] [%]
		p-ClC$_6$H$_4$	Ph	74	90	F	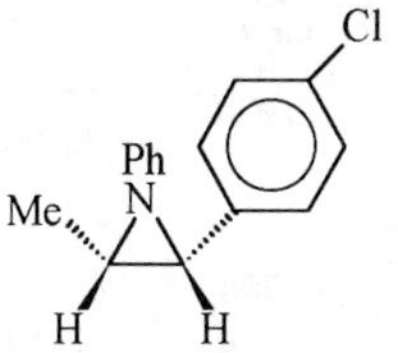	91

[a] Taken from Reference 7.
[b] Isolated yield.
[c] All reactions were conducted with EtMgBr in THF. A: 3.5 equiv, –55 to –35°C, 2 h. B: 5 equiv, –60 to –50°C, 1 h, then room temperature for 20 min. C: 3.5 equiv, –55 to –20°C, then room temperature for 10 min. D: 5 equiv, –55 to –30°C, then room temperature for 10 min. E: 5 equiv, –55 to –40°C, 1 h. F: 3.5 equiv, –55 to –25°C, 2 h.

Similarly, the asymmetric synthesis of both enantiomers of α-hydroxy acids, esters, amides, and propargylic alcohols was accomplished using the anion derived from (–)-chloroalkyl p-tolyl sulfoxides **17d,e**. Thus, addition of **17d** to aldehydes **48** in the presence of LDA gave the adducts **42** and **43** in a quantitative yield. These adducts were separated by silica gel chromatography, and after thermal elimination of the sulfinyl group gave unsaturated chlorohydrines **44** and **45**, respectively, in quantitative yields. They were converted into both enantiomers of α-hydroxy acids, esters, and amides **46** and **47** via ozonolysis followed by treatment with the appropriate nucleophile.[8]

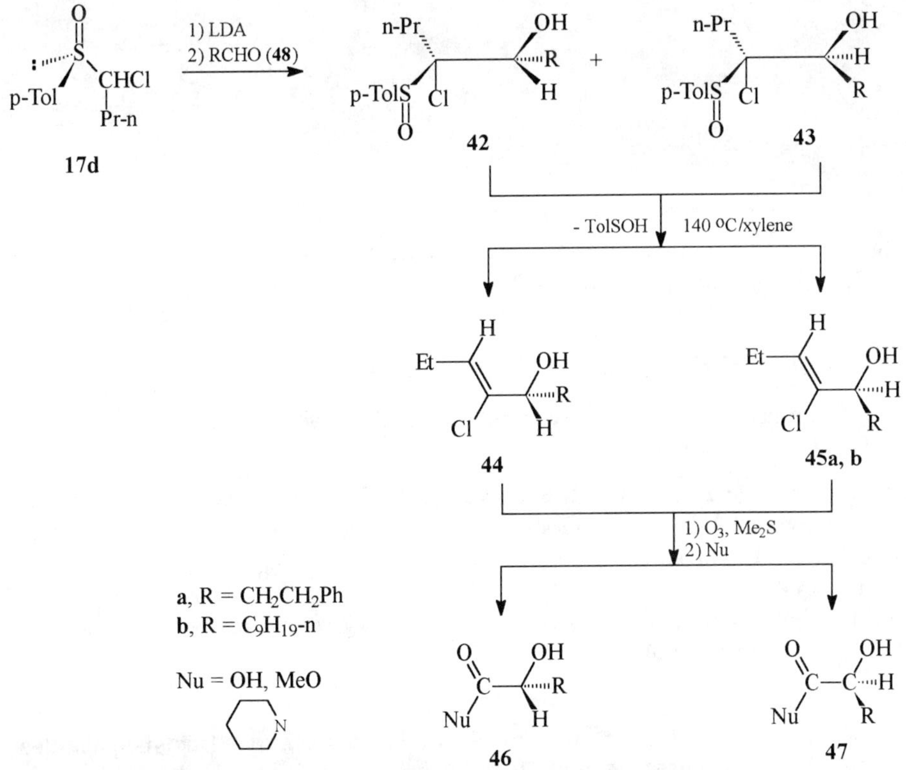

1) LDA
2) RCHO (**48**)

17d → **42** + **43**

- TolSOH 140 °C/xylene

44 **45a, b**

1) O$_3$, Me$_2$S
2) Nu

46 **47**

a, R = CH$_2$CH$_2$Ph
b, R = C$_9$H$_{19}$-n

Nu = OH, MeO, N (piperidine)

Scheme 3.6.5

Both enantiomers of the propargylic alcohol **51a** and **51b** were synthesized from **49a** and **49b** via the vinyl chlorides **50a** and **50b** (Scheme 3.6.6).[8]

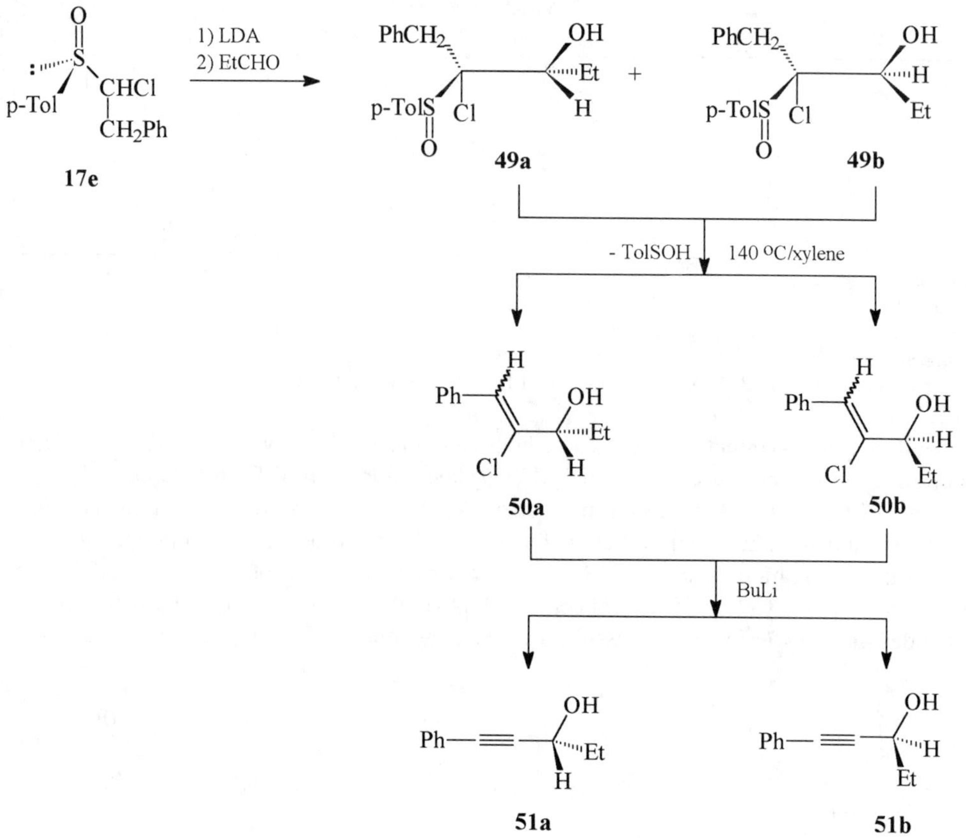

Scheme 3.6.6

Another interesting application of chiral 1-chloroalkyl p-tolyl sulfoxides concerns the asymmetric Favorskii rearrangement of optically active α-chloro α-sulfonyl ketones.[9]

3.6.1 Examples of Experimental Procedures

3.6.1.1 α-Bromination of Chiral Alkyl p-Tolyl Sulfoxides with Bromine; General Procedure[2b]

A solution of bromine (3.2 g, 0.02 mol) in anhydrous MeCN (20 mL) cooled to –20°C is added dropwise to a stirred solution of the sulfoxide (0.01 mol) and anhydrous pyridine/MeCN (25 ml, 1:4 v/v) at –40°C. After 1 h at –40°C, the mixture is kept at room temperature for 20 h. MeCN is evaporated, the residue dissolved in $CHCl_3$ (50 mL), washed with an aqueous solution of $Na_2S_2O_3$, and then with aqueous sulfuric acid. Evaporation of $CHCl_3$ gives the α-bromosulfoxide which is purified by column chromatography.

3.6.1.2 α-Halogenation of Chiral Alkyl p-Tolyl Sulfoxides 6 with N-Halosuccinimides on Silica Gel Plates; General Procedure[3]

A Merck silica GF_{254} TLC plate (20 × 5 cm) is covered with a solution of sulfoxide **6** (0.5–1 mml) and the N-halosuccinimide (10% excess) in acetonitrile (1 mL) and kept at room temperature for an appropriate time. After the reaction is completed the silica gel is removed from the glass plate and the product washed out from the silica gel with ether (~50 mL). The crude racemic products obtained by evaporating the ether under reduced pressure are purified by column chromatography on silica gel. Optically active products are isolated by preparative TLC on silica gel.

3.6.1.3 α-Chlorination of Chiral Alkyl p-Tolyl Sulfoxides with N-Chlorosuccinimide in the presence of K_2CO_3; General Procedure[4,9]

To a solution of **6** (10 mmol) in 10 mL of dry CH_2Cl_2 was added K_2CO_3 (800 mg) followed by NCS (2.64 g, 20 mmol). The suspension was stirred at room temperature for 40 h. The reaction mixture was diluted with ether (50 mL), and the solution was washed with 4% NaI (50 mL) followed by 10% $Na_2S_2O_3$ (50 mL). The organic layer was dried over Na_2SO_4, and the solvent was evaporated to leave a residue, which was purified by silica gel column chromatography and recrystallization from AcOEt.

REFERENCES (3.6)

1. Mikołajczyk, M., Drabowicz, J., *Top. Stereochem.*, **13**, 283, 1982.
2. a) Calzavara, P., Cinquini, M., Colonna, S., Fornasier, R., Montanari, F., *J. Am. Chem. Soc.*, **95**, 7431, 1973.
   b) Cinquini, M., Colonna, S., Fornasier, R., Montanari, F., *J. Chem. Soc., Perkin Trans 1*, 1883, 1972.
   c) Cinquini, M., Colonna, S., Montanari, F., *J. Chem. Soc., Chem. Commun.*, 607, 1969.
   d) Cinquini, M., Colonna, S., Montanari, F., *J. Chem. Soc., Chem. Commun.*, 1441, 1970.
3. Drabowicz, J., *Synthesis*, 831, 1986.
4. Satoh, T., Oohara, T., Ueda, Y., Yamakawa, K., *Tetrahedron Lett.*, **29**, 313, 1988.
5. a) Cinquini, M., Colonna, S., Montanari, F., *J. Chem. Soc., Perkin Trans 1*, 1719, 1974.
   b) Cinquini, M., Colonna, S., *J. Chem. Soc., Chem. Commun.*, 769, 1974.
6. Satoh, T., Oohara, T., Ueda, Y., Yamakawa, K., *J. Org. Chem.*, **54**, 3130, 1989.
7. Satoh, T., Sato, T., Oohara, T., Yamakawa, K., *J. Org. Chem.*, **54**, 3973, 1989.
8. Satoh, T., Onda, K., Yamakawa, K., *Tetrahedron Lett.*, **31**, 3567, 1990.
9. Satoh, T., Motohashi, S., Kimura, S., Yamakawa, K., *Tetrahedron Lett.*, **34**, 4823, 1993.

3.7 α-PHOSPHORYL SULFOXIDES AND α-SULFINYL PHOSPHONIUM YLIDES

Among α-phosphoryl substituted organosulfur compounds,[1] α-phosphoryl sulfoxides of the general structure **1** deserve special attention as useful synthetic reagents and model compounds in stereochemical studies. A diverse reactivity of this class of sulfoxides is caused by the presence of an additional reactive moiety in the structure of **1**, namely the phosphonate grouping. Both these strongly electron-withdrawing phosphoryl and sulfinyl groups enhance the acidity of the α-methylene hydrogens in **1**, and proton elimination readily occurs upon treatment with a base yielding the corresponding α-carbanions. They have been found to react with a variety of electrophilic reagents. However, the most important reaction of these α-carbanions is that with carbonyl compounds (Horner-Wittig reaction), resulting in the formation of α,β-unsaturated sulfoxides.

α-Phosphoryl sulfoxides **1**, like other unsymmetrical sulfoxides, are chiral due to the presence of an asymmetric center at the sulfur atom. Besides the chiral sulfinyl center, the phosphonate moiety may also be chiral provided that two substituents (R^1 and R^2) connected with the phosphorus atom are different. Moreover, substitution of one of the two diastereotopic hydrogens of the methylene group by another substituent creates a third chiral center at the α-carbon atom. Such a situation is depicted below.

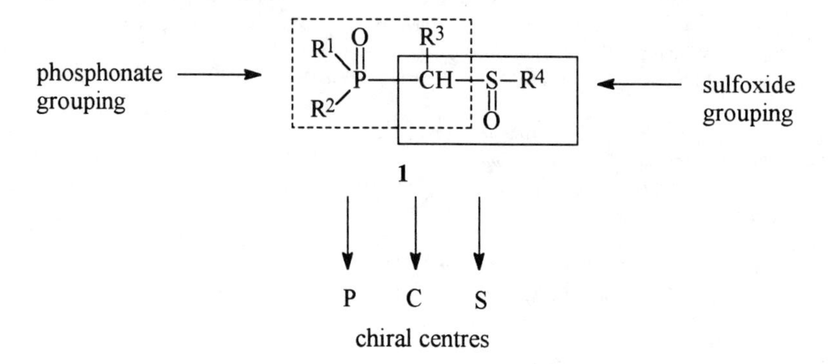

In contrast to the well-known, chiral α-phosphoryl sulfoxides **1**, α-phosphonium substituted sulfoxides **2** have been obtained only recently.

$$R_3\overset{+}{P}CH_2S(O)R^1 \;(\mathbf{2}) \xrightarrow{\text{Base}} R_3\overset{+}{P}-\overset{-}{C}HS(O)R^1 \longleftrightarrow R_3P{=}CHS(O)R^1 \;(\mathbf{3}) \qquad (3.7.1)$$

The sulfinyl ylides **3** generated from **2** represent a new class of useful Wittig olefination reagents.

3.7.1 Synthesis of Chiral α-Phosphoryl Sulfoxides

Racemic α-phosphoryl sulfoxides can be prepared on three different ways which comprise: (a) oxidation of α-phosphoryl sulfides, (b) reaction between phosphonate α-carbanions and aromatic sulfinic esters, and (c) condensation of α-halogeno sulfoxides with dialkyl phosphite anions.[2-5]

$$(RO)_2P(O)CH_2SR^1 \xrightarrow{[O]} (RO)_2P(O)CH_2S(O)R^1 \quad (3.7.2)$$

$$(RO)_2P(O)CH_2^- + ArS(O)OR^1 \longrightarrow (RO)_2P(O)CH_2S(O)Ar \quad (3.7.3)$$

$$(RO)_2P(O)^- + XCH_2S(O)R^1 \longrightarrow (RO)_2P(O)CH_2S(O)R^1 \quad (3.7.4)$$

Utilizing the reaction shown in Equation 3.7.3, racemic dimethoxyphosphorylmethyl p-tolyl sulfoxide **4** was obtained in 80% yield. It was found[6] to undergo readily alkaline hydrolysis to the corresponding phosphonic acid **5** which formed a crystalline salt with quinine. Separation of the diastereomeric quininium salts of the acid **5** by fractional crystallization and subsequent methylation of the tetramethylammonium salts of the resulting enantiomers of **5** gave both enantiomeric forms of the α-phosphoryl sulfoxide **4**. The (–)-(R)-**4** was obtained in this way with 100% enantiomeric purity, whereas the (+)-(S)- enantiomer was obtained in approximately 70% enantiomeric purity. The resolution of the α-phosphoryl sulfoxide **4**, described briefly above, is shown in Scheme 3.7.1.

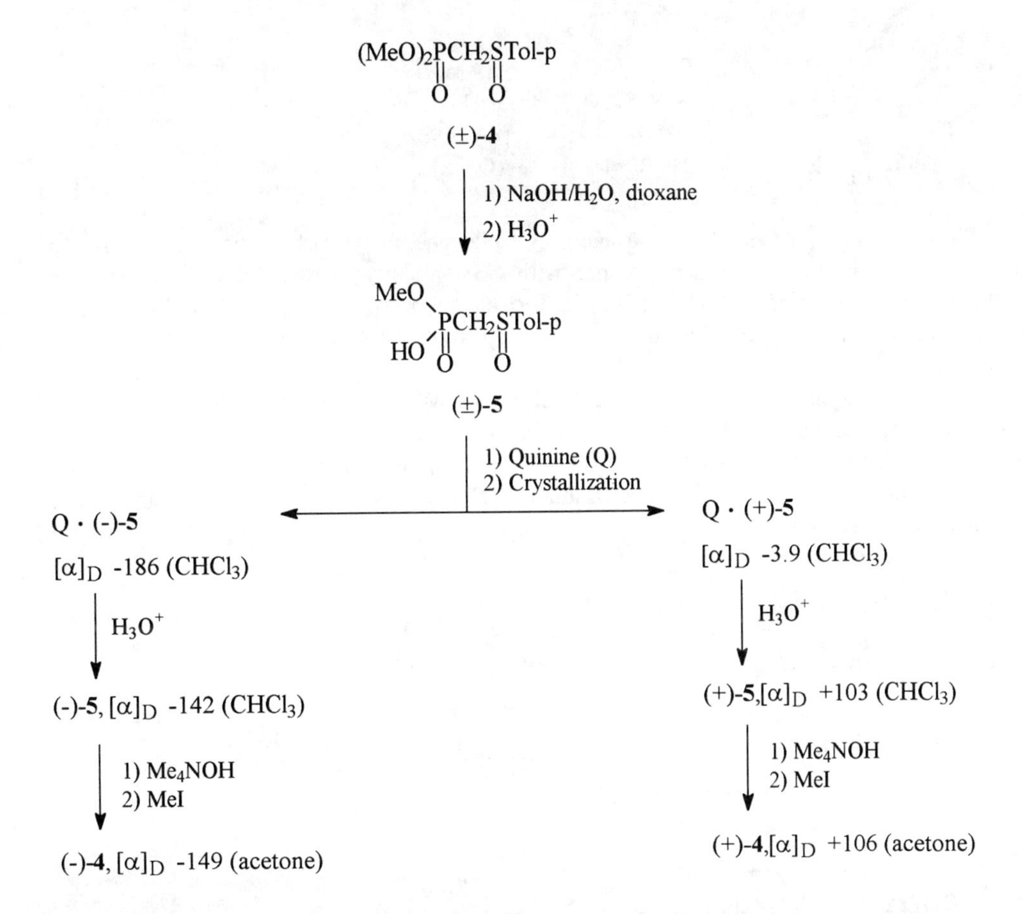

Scheme 3.7.1

A more convenient and efficient approach to chiral α-phosphoryl sulfoxide **4** is based on a stereoselective synthesis involving treatment of (–)-(S)-menthyl p-toluenesulfinate **6** with dimethoxyphosphorylmethyllithium used in a molar excess (Equation 3.7.5).[6] This reaction leads to the formation of the sulfoxide (+)-**4** in 70% yield and with almost full enantiomeric purity.

$(MeO)_2P(O)CH_2Li$ + (-)-(S)-**6** (p-Tol–S(O)–OMen) → (+)-(S)-**4** ($(MeO)_2P(O)CH_2$–S(O)–p-Tol), $[\alpha]_D$ +144 (acetone) (3.7.5)

The reaction described above is a typical nucleophilic substitution at sulfinyl sulfur and takes place with inversion of configuration at sulfur. This enabled the assignment of the chirality at sulfur in the sulfoxide (+)-**4** as S. The correctness of this assignment was later confirmed by CD measurements.[7]

From the point of view of the synthesis of enantiomeric α-phosphoryl sulfoxides **4**, it is noteworthy that both methods discussed above, that is, resolution and stereoselective synthesis, are complementary and allow both pure enantiomers of this sulfoxide to be prepared.

The stereoselective synthesis of α-phosphoryl sulfoxides has a general character, and other α-phosphoryl sulfoxides and their structural analogs **7–10** have been prepared starting from (–)-**6** and the appropriate phosphonate α-carbanions.[7]

7-10: $R_2P(X)CH_2$–S(O)–p-Tol

7, R = MeO, X = S, $[\alpha]_D$ +132
8, R = EtO, X = O, $[\alpha]_D$ +105
9, R = Ph, X = O, $[\alpha]_D$ +93
10, R = Me_2N, X = O, $[\alpha]_D$ +134

Similarly, starting from (–)-(S)-**6** and chiral phosphonate α-carbanions it is also possible to synthesize chiral α-phosphoryl sulfoxides with two optically active centers at phosphorus and sulfur.[8] The synthesis of this type of P,S-chiral α-phosphoryl sulfoxides will be discussed later in connection with their application in the stereoselective conversion of chiral thiophosphonic acids into chiral thiophosphoric acids.[8]

A different synthesis of α-phosphoryl sulfoxides with two chiral centers at phosphorus and sulfur has been developed by Koizumi and co-workers.[9] They used as a starting material the diastereomerically pure α-phosphoryl sulfide **11** which, upon oxidation, yielded a separable mixture of two diastereomeric sulfoxides **12a** (Equation 3.7.6). Their acid-catalyzed alcoholysis affords chiral P,S-sulfoxides **12b**.

(+)-(S)-**11** (EtO–P(O)(PrOOEt)–CH_2SPh) —MCPBA→ (S_S,S_P)-**12a** + (R_S,S_P)-**12a** (EtO–P(O)(PrOOEt)–$CH_2S(O)Ph$)

PrOOEt = EtO_2C-substituted pyrrolidin-1-yl (N)

12b: (EtO)(RO)P(O)$CH_2S(O)Ph$ (3.7.6)

3.7.2 Synthesis of Chiral α-Sulfinyl Phosphonium Salts and Ylides

As in the case of chiral α-phosphoryl sulfoxides **2**, the chiral title compounds may be prepared also from (–)-(S)-menthyl p-toluenesulfinate **6** upon treatments with the appropriate phosphonium ylides. Scheme 3.7.2 shows the synthesis of (S)-p-tolylsulfinylmethyltriphenylphosphonium ylide **13** and (+)-(S)-p-tolylsulfinylmethyltriphenylphosphonium iodide **14**.[10] The reaction of triphenylphosphonium ylide **15** with (–)-(S)-**6** used in a 2:1 molar ratio results in the formation of a mixture of the desired ylide **13** and triphenylphosphonium iodide **16** from which the ylide **15** was generated. Quenching this mixture with hydroiodic acid and removal of the salt **16** by filtration gives the crude α-sulfinyl phosphonium iodide **14** which was purified by column chromatography and obtained in 70% yield (calculated in respect to the sulfinate **6**).

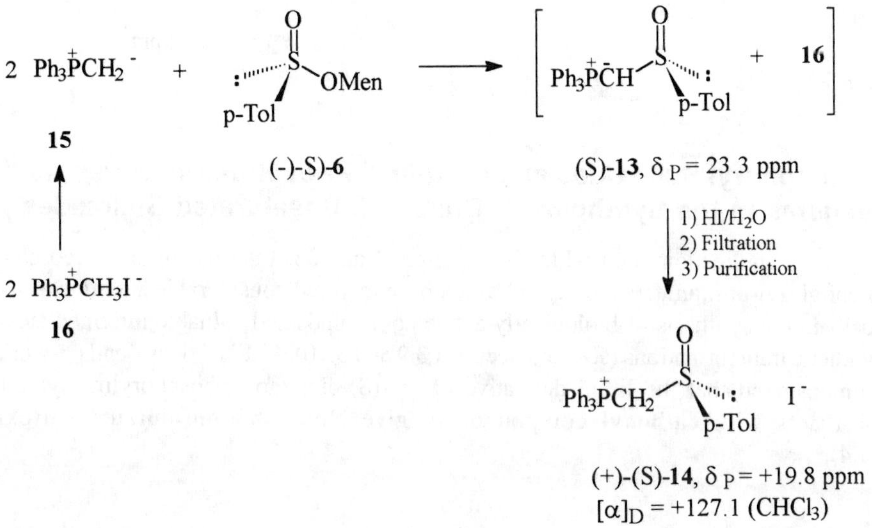

2 $Ph_3\overset{+}{P}CH_2^-$ + (-)-(S)-**6** (p-Tol–S(O)–OMen) → [$Ph_3\overset{+}{P}\overset{-}{C}H$–S(O)–p-Tol + **16**]

15 ← 2 $Ph_3\overset{+}{P}CH_3I^-$ **16**

(S)-**13**, δ_P = 23.3 ppm

1) HI/H_2O
2) Filtration
3) Purification

$Ph_3\overset{+}{P}CH_2$–S(O)–p-Tol I^-

(+)-(S)-**14**, δ_P = +19.8 ppm
$[\alpha]_D$ = +127.1 ($CHCl_3$)

Scheme 3.7.2

Although the synthesis of the ylide **13** and the α-sulfinyl phosphonium iodide **14** has been accomplished as shown above, this method has one drawback connected with the formation of the ylide **13**, together with the starting phosphonium salt **16**. This is due to a fast proton migration from the primarily formed salt **14** to the ylide **15**.

To overcome this problem, it is advantageous to use the diylide **17** (generated from the phosphonium salt **18**) instead of the monoylide **15** in the reaction with (–)-(S)-**6**. This condensation gives only the corresponding α-sulfinyl ylide **19** which can be used *in situ* for the Wittig reaction with carbonyl compounds.[10]

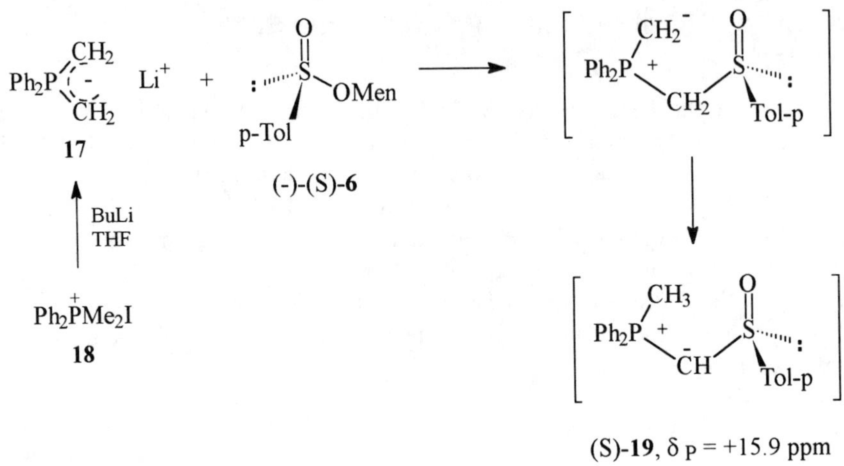

(-)-(S)-**6**

(S)-**19**, $\delta_P = +15.9$ ppm

Scheme 3.7.3

3.7.3 α-Phosphoryl Sulfoxides and α-Sulfinyl Phosphonium Ylides As Key Reagents in the Synthesis of Chiral α,β-Unsaturated Sulfoxides

The most important application of chiral title compounds **1** and **3** is their use in the stereoselective synthesis of chiral α,β-unsaturated sulfoxides, which have received considerable attention as useful building blocks in the synthesis of biologically active compounds and valuable intermediates in a variety of synthetic transformations (see also Sections 3.9 and 3.10).[11] Mikołajczyk and co-workers[6] have first demonstrated that the lithio derivative of (+)-(S)-dimethoxyphosphorylmethyl p-tolyl sulfoxide **4** reacts with carbonyl compounds to give chiral α,β-unsaturated sulfoxides (Scheme 3.7.4).[6]

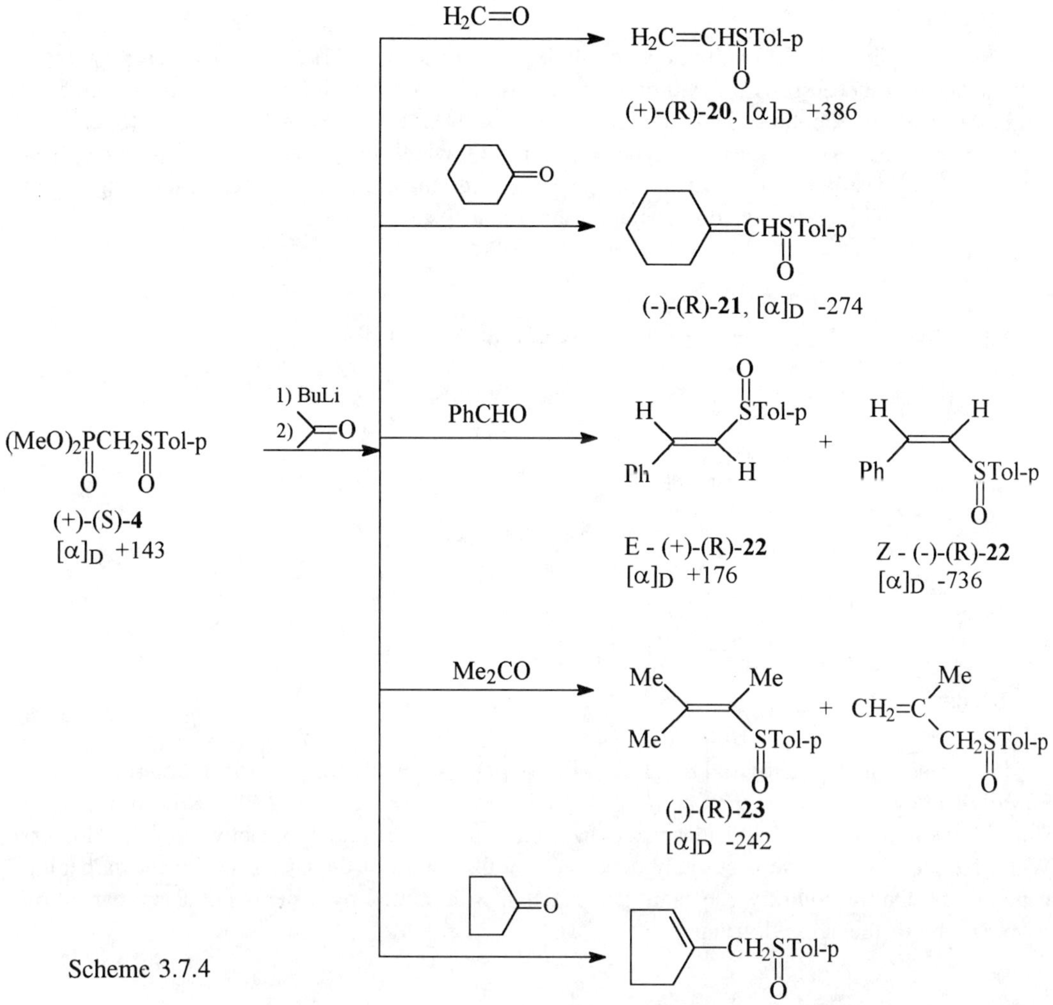

$(MeO)_2P(O)CH_2S(O)Tol\text{-}p$

(+)-(S)-**4**
$[\alpha]_D$ +143

1) BuLi
2) >=O

$H_2C{=}O$ → $H_2C{=}CHS(O)Tol\text{-}p$
(+)-(R)-**20**, $[\alpha]_D$ +386

Cyclohexanone → cyclohexylidene=CHS(O)Tol-p
(-)-(R)-**21**, $[\alpha]_D$ -274

PhCHO → E - (+)-(R)-**22** + Z - (-)-(R)-**22**
E - (+)-(R)-**22**, $[\alpha]_D$ +176
Z - (-)-(R)-**22**, $[\alpha]_D$ -736

Me_2CO → $Me_2C{=}CHS(O)Tol\text{-}p$ + $CH_2{=}C(Me)CH_2S(O)Tol\text{-}p$
(-)-(R)-**23**
$[\alpha]_D$ -242

Cyclopentanone → cyclopentenyl-$CH_2S(O)Tol\text{-}p$

Scheme 3.7.4

At least three characteristic features of this Horner-Wittig reaction should be discussed. First of all, the reaction essentially occurs without racemization of the chiral sulfoxide moiety. In some cases, when the carbonyl component contains a hydrogen atom in the α-position, highly racemized allylic sulfoxides are formed. Most probably, the vinylic sulfoxides first formed rearrange under basic reaction conditions to the corresponding allylic sulfoxides which, in turn, undergo fast racemization by a [2,3]-sigmatropic process involving the achiral sulfenate ester as an intermediate. With regard to the geometry of α,β-unsaturated sulfoxides, the Horner-Wittig reaction is nonstereoselective and affords mixtures of (E) and (Z) isomers. This is best illustrated by the reaction of (+)-(S)-**4** with benzaldehyde resulting in a separable mixture of (E) and (Z)-styryl p-tolyl sulfoxides **22** in a 75:25 ratio. It is noteworthy that these isomers having the same chirality at sulfur exhibit opposite rotation signs, and the (Z)-isomer of **22** has a very high negative rotation. A similar relationship was observed for other (E) and (Z) vinylic sulfoxides prepared by the Horner-Wittig reaction.[12,13]

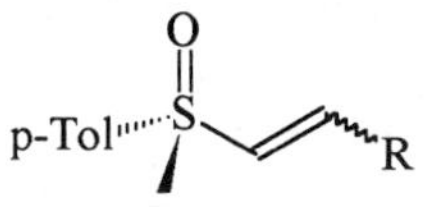

p-Tol–S(O)–CH=CH–R

24, R = n-C_5H_{11}, E : $[\alpha]_D$ +148, Z : $[\alpha]_D$ -357

25, R = t-Bu, E : $[\alpha]_D$ +33, Z : $[\alpha]_D$ -240

29, R = CO_2Me, E : $[\alpha]_D$ +421, Z : $[\alpha]_D$ -438

The conversion of vinyl sulfoxides to allylic sulfoxides and [2,3]-sigmatropic rearrangement of the latter was successfully utilized by Hoffmann and Maak[14] and Hoffmann[15] in the asymmetric synthesis of allylic alcohols. Thus, the chiral vinyl sulfoxides **26** prepared from (+)-(S)-**4** were converted by means of a base into allylic sulfoxides **27**, which underwent [2,3]-sigmatropic rearrangement and desulfurization to chiral alcohols **28** with enantiomeric purities in the range from 40 to 95% (for more extensive discussion on this subject see Section 3.3).

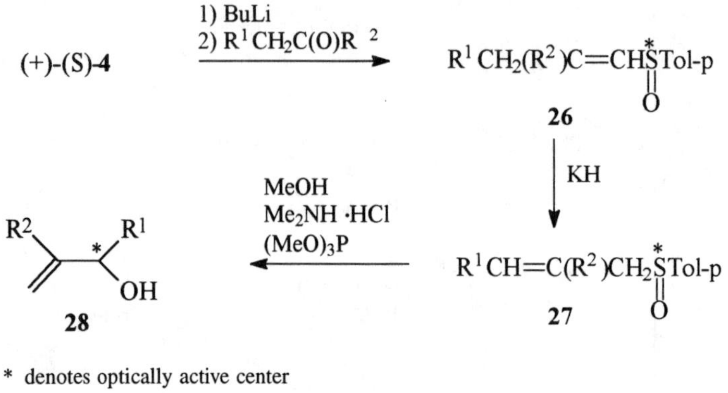

Scheme 3.7.5

The reaction of methyl glyoxalate with (+)-(S)-**4** results in the formation of methyl 3-p-tolylsulfinylpropenoate **29** as an (E) + (Z) mixture. The (Z)-isomer of **29** has been used as a chiral dienophile in the Diels-Alder reaction.[16] Interestingly, the enantioselectivity of the Horner-Wittig reaction in this case is strongly dependent on the nature of the base used for the carbanion generation.[13] On the contrary, the analogous reaction with methyl pyruvate is not accompanied by racemization of the sulfinyl group.

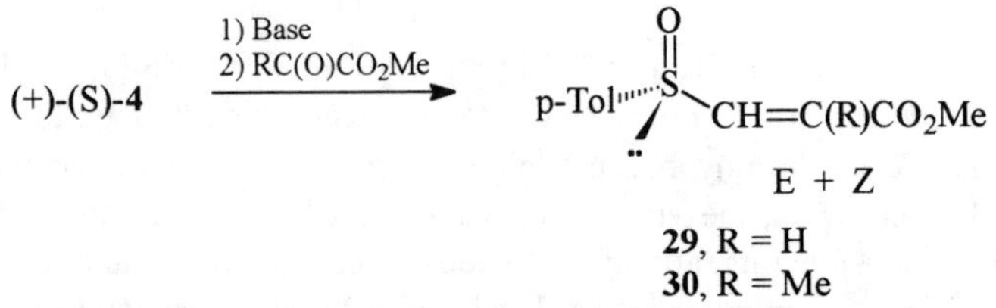

(3.7.7)

The Horner-Wittig reaction of α-phosphoryl sulfoxides was of importance in the stereoselective conversion of chiral phosphonothioic acids into phosphorothioic acids.[8] This reaction cycle (Scheme 3.7.6) consists of three steps. The first is a well-known synthesis of the phosphonothionates **32** from chiral O-isopropyl methanephosphonothioic acid **31**. The second reaction shows the approach to the synthesis of the sulfoxide **33** with two chiral centers at phosphorus and sulfur. In the last step the chiral P,S-sulfoxide **33** was converted into chiral O-alkyl O-isopropyl phosphorothioic acid **34** by means of the Horner-Wittig reaction. Since the stereochemistry at phosphorus of each step is known, this reaction cycle was used for the correlation of the absolute configuration between the chiral acids **31** and **34**.

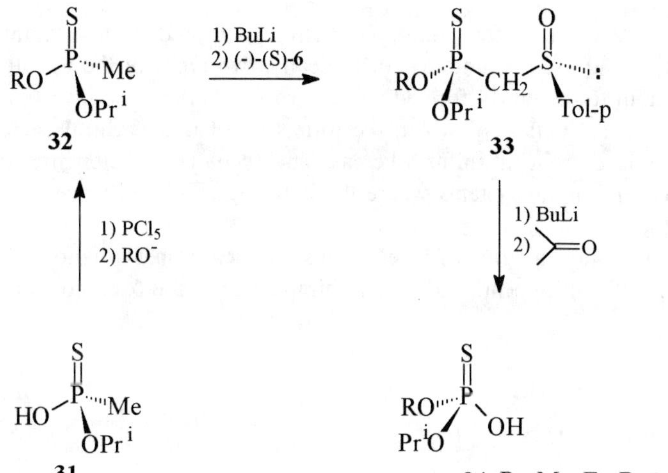

1) BuLi
2) (-)-(S)-6

32 → **33**

1) PCl_5
2) RO^-

31 → **32**

1) BuLi
2) acetone

33 → **34**, R = Me, Et, Bu^n

Scheme 3.7.6

The synthesis of chiral vinyl sulfoxides using (+)-(S)-dimethoxyphosphorylmethyl p-tolyl sulfoxide **4** as a key reagent was an important step in the total synthesis of some natural products like α-tocopherol,[17,18] carnegine,[19] and fragolide[20] as well as of α-sulfinyl δ-lactones.[21]

Finally, to close a discussion on the Horner-Wittig reaction of (+)-(S)-**4**, it should be added that chiral vinyl sulfoxides can be prepared in a one-pot reaction starting from dimethyl methane-phosphonate by sequential addition of butyllithium, (–)-(S)-**6**, and carbonyl compounds.[22]

In contrast to the synthesis of vinyl sulfoxides via the Horner reaction, the Wittig reaction of the α-sulfinyl phosphonium ylides (S)-**13** and (S)-**19** with carbonyl compounds is fully or almost fully (E)-stereoselective.[10] Thus, when the ylide (S)-**13** generated as shown in Scheme 3.7.2 is treated with aldehydes, the chiral E-vinyl sulfoxides are formed in yields exceeding 70%. Their optical rotations, enantiomeric purities (ep), and E-isomer contents are given below.

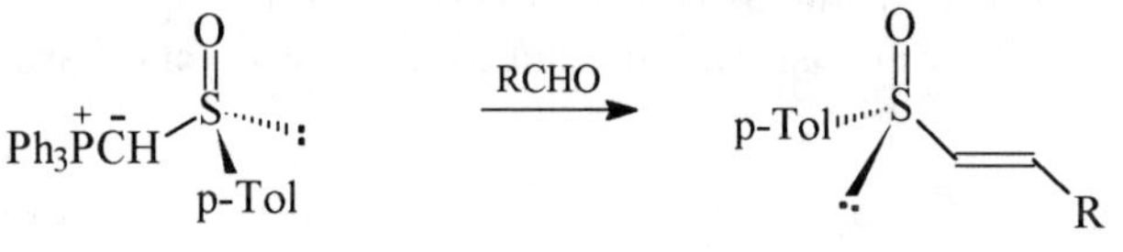

$Ph_3\overset{+}{P}\overset{-}{C}H$–S(O)–p-Tol (S)-**13** —RCHO→ p-Tol–S(O)–CH=CH–R (3.7.8)

20, R = H, $[\alpha]_D$ +389.7 (EtOH), ee 54%
22, R = Ph, $[\alpha]_D$ +159.5 ($CHCl_3$), ee 100%, E 97%
35, R = p-$MeOC_6H_4$, $[\alpha]_D$ +100.0 ($CHCl_3$), ee 96%, E 100%
36, R = CH_2=CH, $[\alpha]_D$ +283.4 ($CHCl_3$), ee 100%, E 100%
37, R = (E)-MeCH=CH, $[\alpha]_D$ +158.8 ($CHCl_3$), ee 100%, E 100%
38, R = (E)-PhCH=CH, $[\alpha]_D$ +168.2 ($CHCl_3$), ee 100%, E 98%

Similar results have been obtained with the sulfinyl ylide (S)-**19**.

3.7.4 Asymmetric Reactions of α-Phosphoryl Sulfoxides

The structure of α-phosphoryl sulfoxides **1** offers many possibilities for studies of asymmetric induction, especially on the methylene α-carbon atom. Theoretically, two types of the transfer of chirality from the chiral sulfur atom to the newly formed chiral carbon atom may be investigated. The first type includes reactions in the course of which the formation of a new chiral center is accompanied by disappearance of the chirality at sulfur, whereas to the second type belong reactions resulting in the formation of diastereomeric systems where the inducing chiral sulfur center preserves fully or partly its optical activity.

The Pummerer reaction[23] shown in Equation 3.7.9 represents a typical transformation of sulfoxides in which the chirality at sulfur disappears and a new chiral center at the α-carbon atom is created.

$$RS(=O)-CH_2R^1 \xrightarrow{Ac_2O} RS-CHR^1(OAc) \qquad (3.7.9)$$

In fact, the Pummerer reaction of (+)-(S)-dimethoxyphosphorylmethyl p-tolyl sulfoxide **4** with acetic anhydride under reflux was found to occur in an asymmetric way and afforded optically active α-acetoxy-α-(dimethoxyphosphoryl)methyl p-tolyl sulfide **39** having $[\alpha]_D = -4$.[24] When the Pummerer reaction of (+)-(S)-**4** was carried out in the presence of DCC, the extent of asymmetric induction was even higher (Equation 3.7.10).[25]

$$(MeO)_2P(=O)CH_2S(=O)Tol\text{-}p \xrightarrow{Ac_2O/DCC/reflux} (MeO)_2P(=O)CH(OAc)STol\text{-}p \qquad (3.7.10)$$

(+)-(S)-**4**
$[\alpha]_D$ +144

39, $[\alpha]_D$ -8.5 (45.3 % ep)

The observation of the chirality transfer from sulfur to carbon in the Pummerer reaction of (+)-(S)-**4** and the complementary results of experiments with the ^{18}O-labeled sulfoxide **4** shown in Equation 3.7.11 strongly support the view that migration of the acetoxy group from sulfur to carbon in the intermediate acetoxy sulfur ylide proceeds via five- and three-membered cyclic transition states **40a** and **40b**.

$$(MeO)_2P(=O)CH_2S(={}^{18}O)Tol\text{-}p \xrightarrow{Ac_2O} (MeO)_2P(=O)CH({}^{18}O\text{-}C(={}^{18}O)Me)STol\text{-}p$$

4 → **39** (^{18}O 0.23 on ester O, 0.77 on carbonyl O)

^{18}O - content
4, 28.4%
39, 21.2%

40a: $>P(=O)-CH\cdots{}^{\delta+}STol\text{-}p$, five-membered cyclic O···C(Me)···O bridge (δ^-)

40b: $>P(=O)-CH\cdots{}^{\delta+}STol\text{-}p$, three-membered ring via O; O–C(Me)=O (δ^-)

(3.7.11)

Similarly, a substantial extent of asymmetric induction was observed in the reaction of (+)-(S)-**4** with titanium tetrachloride, which produced the corresponding α-chloro sulfide **41** as the Pummerer reaction type product with 56% enantiomeric purity.[25]

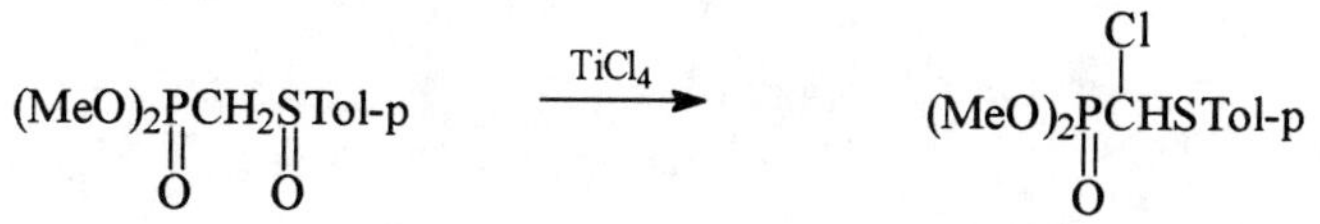

$$(MeO)_2P(O)CH_2S(O)Tol\text{-}p \xrightarrow{TiCl_4} (MeO)_2P(O)CH(Cl)STol\text{-}p$$

(+)-(S)-**4**, $[\alpha]_D$ +144 → **41**, $[\alpha]_D$ +108 (56% ee)

(3.7.12)

In addition to the synthetically important Horner-Wittig reaction of α-phosphoryl sulfoxides, the lithiated sulfoxide (+)-(S)-**4** reacts with alkyl halides. For example, methylation results in the formation of two diastereomeric sulfoxides **42** in a 3:1 ratio.[26] The major diastereomer isolated in a pure state has an $[\alpha]_D = +157$ (acetone) and (S_C, S_S)-configuration as established by X-ray analysis. Interestingly, a mixture of the same diastereomers of **42** (2.4:1) is formed when the lithium salt of dimethyl ethanephosphonate is reacted with (–)-(S)-**6**. The configurational relationship between **4**, **6**, and **42** is depicted in Scheme 3.7.7.

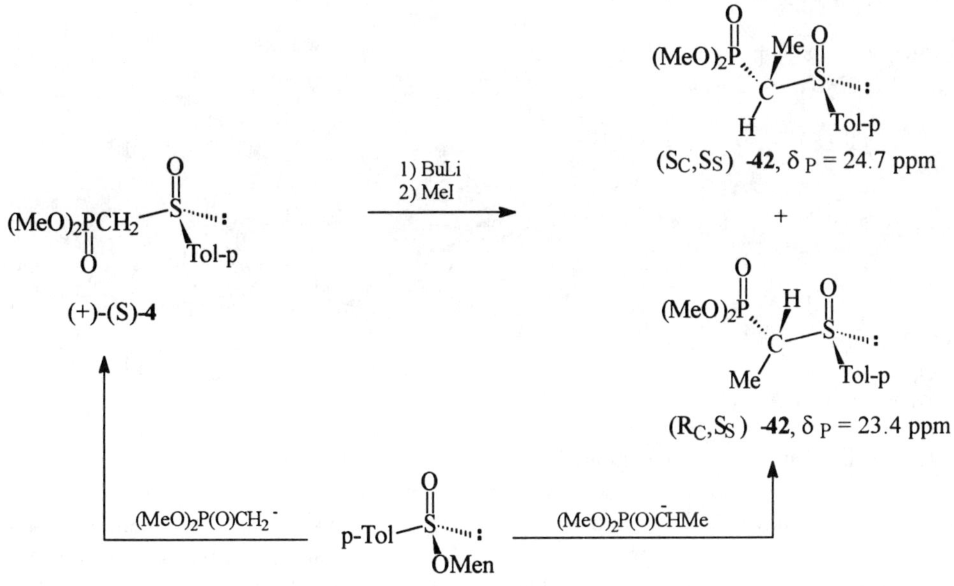

(MeO)₂P(O)CH₂–S(O)–Tol-p, (+)-(S)-**4** —1) BuLi, 2) MeI→ (S_C,S_S) **-42**, $\delta_P = 24.7$ ppm + (R_C,S_S) **-42**, $\delta_P = 23.4$ ppm

p-Tol–S(O)–OMen, (-)-(S)-**6**: with $(MeO)_2P(O)CH_2^-$ → (+)-(S)-**4**; with $(MeO)_2P(O)\bar{C}HMe$ → **42**

Scheme 3.7.7

The asymmetric course of methylation of (+)-(S)-**4** was rationalized by assuming preferential abstraction of the pro-R-hydrogen in **4** and inversion at carbon during the α-carbanion methylation.[26]

A mixture of diastereomeric sulfoxides **43** obtained by methylation has been used as a substrate in the synthesis of (+)-(S)-diethoxyphosphorylvinyl p-tolyl sulfoxide **44**, which is a new chiral Michael acceptor and dienophile.[27]

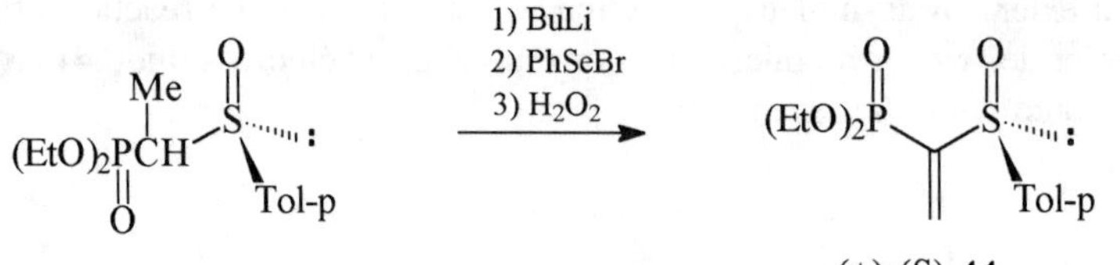

1) BuLi
2) PhSeBr
3) H_2O_2

(+)-(S)-**43** → (+)-(S)-**44** $[\alpha]_D$ +157 (acetone) (3.7.13)

The reactivity of this sulfoxide was exemplified by the diastereoselective addition of ethanethiol, tandem Michael addition/intramolecular Horner-Wittig reaction with 2-formyl pyrrole, and cycloaddition with cyclopentadiene (see Scheme 3.7.8). The latter reaction, when carried out in the presence of zinc chloride, produces only two of four possible diastereomeric cycloadducts.

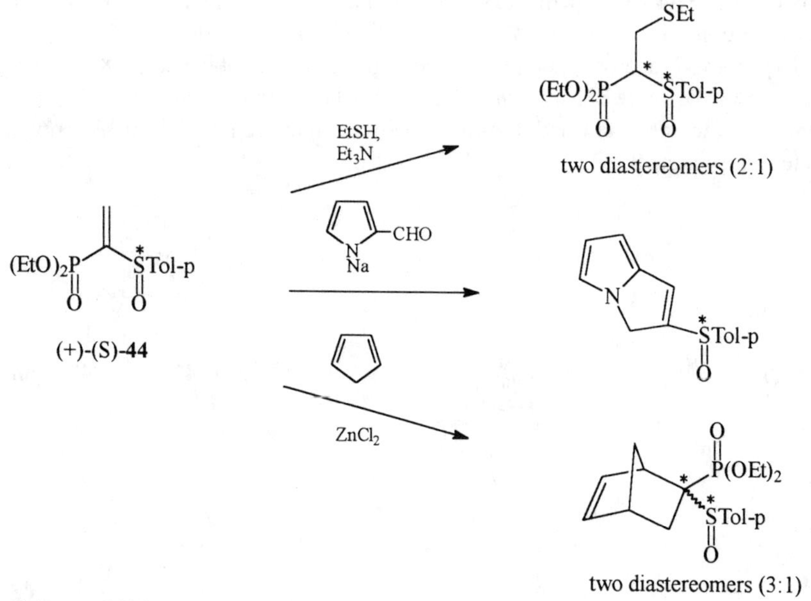

(+)-(S)-**44**

EtSH, Et_3N → two diastereomers (2:1)

CHO, N Na →

ZnCl$_2$ → two diastereomers (3:1)

Scheme 3.7.8

A closely related reaction to methylation of the sulfoxide (+)-(S)-**4** is its chlorination, which results in the formation of the corresponding α-chloro-α-phosphoryl sulfoxide **45** as a mixture of two diastereomers having a different configuration at the α-carbon atom.[28] The extent of asymmetric induction at the α-carbon atom and racemization of the chiral sulfinyl center observed in this case depend mainly on the nature of the chlorinating agent as illustrated by the data listed in Equation 3.7.14.

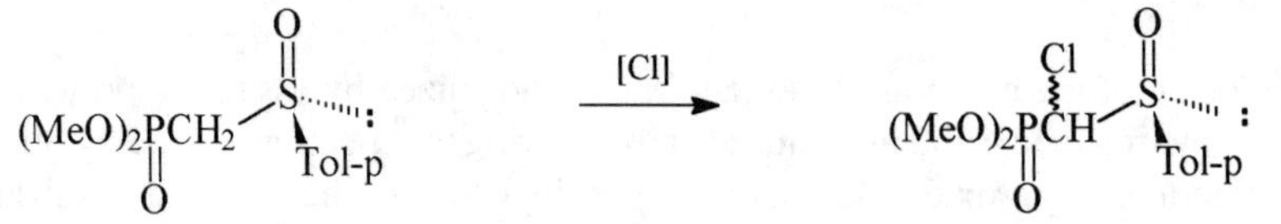

(+)-(S)-**4** →[Cl] (S_C,S_S)-**45** + (R_C,S_S)-**45** (3.7.14)

[Cl]	d.r.	$[\alpha]_D$
$PhICl_2$/Py	3.3:1	+63
$PhICl_2$	2.8:1	+131
SO_2Cl_2/Py	2.8:1	+180
SO_2Cl_2	3.3:1	+185

The structure of the major diastereomer, (S_C, S_S)-**45**, $[\alpha]_D = +243$ ($CHCl_3$), formed in this reaction was established by X-ray analysis. All the experimental data on the chlorination of chiral (+)-(S)-**4** point to retention of the configuration at both chiral centers (C and S), and this stereochemistry can be rationalized assuming an addition-elimination mechanism involving the participation of positively charged ylides as intermediates.[28]

3.7.5 Examples of Experimental Procedures

3.7.5.1 Preparation of (+)-(S)-Dimethoxyphosphorylmethyl p-Tolyl Sulfoxide 4 from (–)-(S)-Menthyl p-Toluenesulfinate 6[6]

To a solution of dimethyl methanephosphonate (2.48 g, 0.02 mol) in THF (30 mL) a solution of n-butyllithium (16 mL, 0.022 mol) in hexane was added at –78°C under a nitrogen atmosphere. The reaction mixture was stirred at this temperature for 0.5 h and then a solution of (–)-(S)-menthyl p-toluenesulfinate **6** (2.94 g, 0.01 mol), $[\alpha]_D = -202$ (c 1.2, acetone) in 20 mL of THF was added at –78°C. After 15 min the reaction mixture was warmed to –20°C and quenched with aqueous ammonium chloride. After evaporating of the organic solvents (THF, hexane) the aqueous layer was extracted with petroleum ether (to remove menthol) and then with chloroform (3 × 25 mL). The chloroform solution was dried and evaporated. Careful removal of dimethyl methanephosphonate under reduced pressure gave 1.87 g (72%) of (+)-(S)-**4**. The analytically pure sample of this sulfoxide, $[\alpha]_D = +144$ (c 1.0, acetone), was obtained after column chromatography using benzene-acetone (5:1) as the eluent.

3.7.5.2 General Procedure for Synthesis of Chiral α,β-Unsaturated Sulfoxides from (+)-(S)-4[4,6]

To a solution of (+)-(S)-**4** prepared as above (0.015 mol) in 25 mL of THF, a solution of n-butyllithium (0.016 mol) in ether was added at –78°C under a nitrogen atmosphere. After 1 h a clear and colorless solution of the lithium derivative was obtained. A solution of the carbonyl compound (0.015 mol) in 20 mL of THF was then added dropwise at –78°C and the reaction mixture was stirred for 30 min at this temperature. This mixture was warmed slowly to room temperature and stirred for an additional 2 h. At –20°C the reaction mixture becomes turbid and in some cases an appearance of the yellow-orange color is observed. After removal of solvents the residue was treated with water (50 mL) and extracted with chloroform (3 × 25 mL). The chloroform solution was washed with water (25 mL) and evaporated to afford the crude vinyl sulfoxides.

3.7.5.3 General, One-Pot Procedure for Synthesis of Chiral α,β-Unsaturated Sulfoxides from Methyltriphenylphosphonium Iodide 16[10]

To a solution of methyltriphenylphosphonium iodide **16** (0.81 g, 2 mmol) in dry benzene (50 mL), a solution of n-butyllithium (2.1 mmol) in hexane was added at room temperature under argon. The reaction mixture was stirred for 1 h. Then, to a solution of **15** so generated (–)-(S)-menthyl p-toluenesulfinate **6** (0.294 g, 1 mmol) was added and the mixture was stirred for 1 h. A solution of a freshly distilled aldehyde (1.5 mmol) was added and the mixture was refluxed for 10 h. After refluxing the reaction was cooled to room temperature and quenched by addition of hydrochloric acid (0.2 N) (pH of solution ca. 3 ÷ 4) and water (10 mL). The layers were separated. The water layer was extracted with $CHCl_3$ (2 × 15 mL) and the combined organic layers were dried over sodium sulfate. Evaporation of solvents afforded the crude vinyl sulfoxides which were purified by column chromatography on silica gel using hexane-dichloromethane as the eluent.

REFERENCES (3.7)

1. Mikołajczyk, M., Bałczewski, P., in *Advances in Sulfur Chemistry*, Block, E., ed., JAI Press, 1994, vol. 1, pp. 41–96.
2. Mikołajczyk, M., Grzejszczak, S., Zatorski, A., in *Organic Sulfur Chemistry*, Stirling, C. J. M., Ed., Butterworth & Co., Ltd., London, 1975, p. 413.
3. Mikołajczyk, M., Zatorski, A., *Synthesis*, 669, 1973.
4. Mikołajczyk, M., Grzejszczak, S., Zatorski, A., *J. Org. Chem.*, **40**, 1979, 1975.
5. Mikołajczyk, M., Zatorski, A., *Gazz. Chim. Ital.*, **125**, 323, 1995.
6. Mikołajczyk, M., Midura, W., Grzejszczak, S., Zatorski, A., Chefczyńska, A., *J. Org. Chem.*, **43**, 473, 1978.
7. Mikołajczyk, M., Midura, W., Kajtar, M., *Phosphorus Sulfur*, **36**, 79, 1988.
8. Mikołajczyk, M., Grzejszczak, S., Midura, W., Popielarczyk, M., Omelańczuk, J., in *Phosphorus Chemistry*, ACS Symposium Series, **171**, 55, 1981.
9. Koizumi, T., Iwata, M., Tanaka, N., Yoshii, E., *Chem. Pharm. Bull.*, **31**, 4198, 1983.
10. Mikołajczyk, M., Perlikowska, W., Omelańczuk, J., Cristau, H.-J., Perraud-Darcy, A., *Synlett*, 913, 1991; and unpublished results.
11. Koizumi, T., *Phosphorus, Sulfur, Silicon*, **58**, 111, 1991.
12. Chaigne, F., Gotteland, J.-P., Malacria, M., *Tetrahedron Lett.*, **30**, 1803, 1989.
13. Cardellicchio, C., Iacuone, A., Naso, F., *Tetrahedron Lett.*, **36**, 6563, 1995.
14. Hoffmann, R. W., Maak, N., *Tetrahedron Lett.*, **26**, 2237, 1976.
15. For a review see Hoffmann, R. W., *Angew. Chem. Int., Ed. Engl.*, **18**, 563, 1979.
16. Maignan, Ch., Guessous, A., Rouessac, F., *Tetrahedron Lett.*, **25**, 1727, 1984.
17. Akkerman, J. M., De Koning, H., Huisman, H. O., *Heterocycles*, **15**, 797, 1981.
18. Solladie, G., Moine, G., *J. Am. Chem. Soc.*, **106**, 6097, 1984.
19. Pyne, S. G., Chapman, S. L., *Chem. Commun.*, 1688, 1986.
20. Burke, S. D., Shankaron, K., Jones Helber, M., *Tetrahedron Lett.*, **32**, 4655, 1991.
21. Holton, R. A., Kim, H.-B., *Tetrahedron Lett.*, **27**, 2191, 1986.
22. Craig, D., Daniels, K., Marsh, A., Reinford, D., Smith, A. M., *Synlett*, 531, 1990.
23. Pummerer, R., *Ber.*, **43**, 1401, 1910.
24. Mikołajczyk, M., Zatorski, A., Grzejszczak, S., Costisella, B., Midura, W., *J. Org. Chem.*, **43**, 2518, 1978.
25. Mikołajczyk, M., Midura, W., unpublished results.
26. Mikołajczyk, M., Midura, W., Miller, A., Wieczorek, M. W., *Tetrahedron*, **43**, 2967, 1987.
27. Mikołajczyk, M., Midura, W., *Tetrahedron: Asymmetry*, **3**, 1515, 1992.
28. Mikołajczyk, M., Midura, W., Grzejszczak, S., Montanari, F., Cinquini, M., Wieczorek, M. W., Karolak-Wojciechowska, J., *Tetrahedron*, **50**, 8053, 1994.

3.8 β-OXO SULFOXIDES AND RELATED DERIVATIVES

This section deals with the synthesis and applications of the following types of sulfoxides:

a. α-Sulfinylcarboxylates A,
b. β-Oxosulfoxides B, called also β-ketosulfoxides or sometimes α-ketosulfoxides,
c. Analogs of the above in which the ester or keto function is replaced by another heteroatom containing substituent, e.g., C and D.

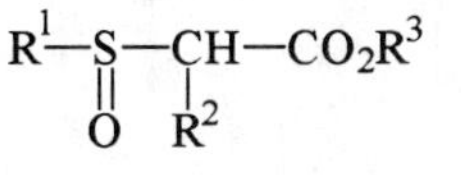

$$R^1-S(=O)-CH(R^2)-CO_2R^3 \quad \text{(A)} \qquad R^1-S(=O)-CH(R^2)-C(=O)R^3 \quad \text{(B)}$$

$$R^1-S(=O)-CH(R^2)-C(=NR^4)R^3 \quad \text{(C)}, \qquad R^1-S(=O)-CH(R^2)-C(=S)R^3 \quad \text{(D)}$$

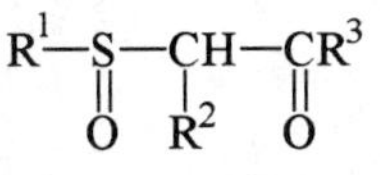

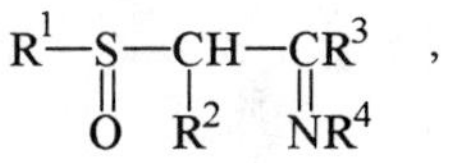

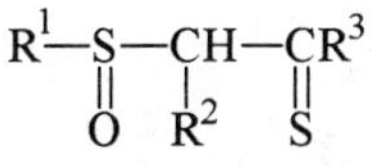

Since each class of these sulfoxides shows different reactivity and is applied in a different way, their syntheses and utility will be discussed in separate subsections.

3.8.1 α-Sulfinylcarboxylates

3.8.1.1 Synthesis

Optically active α-sulfinylcarboxylates are synthesized in the reaction of magnesium enolates of the corresponding esters with menthyl p-toluenesulfinate, e.g., Equation 3.8.1.[1] Another approach utilizes α-anions of optically active methyl sulfoxides which are acylated using corresponding carbonates; an example is shown in Equation 3.8.2.[2]

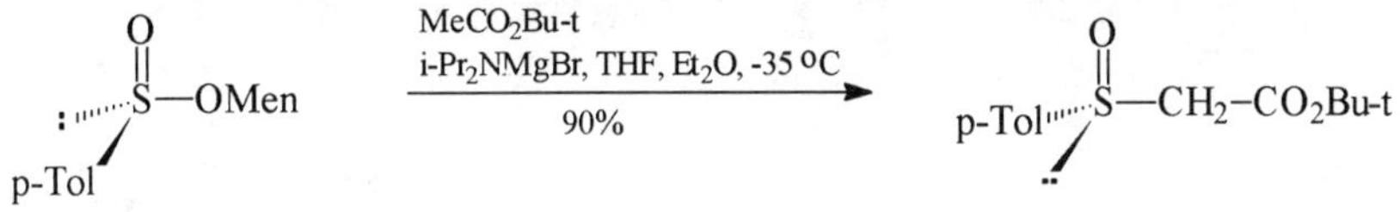

$$\underset{(-)\text{-}S}{p\text{-Tol}-S(=O)-OMen} \xrightarrow[90\%]{MeCO_2Bu\text{-}t,\ i\text{-}Pr_2NMgBr,\ THF,\ Et_2O,\ -35\ ^oC} \underset{(+)\text{-}R}{p\text{-Tol}-S(=O)-CH_2-CO_2Bu\text{-}t} \qquad (3.81)$$

$$p\text{-Tol}-S(=O)-Me \xrightarrow[91\%]{1)\ LDA,\ THF,\ -78\ ^oC;\ 2)\ (t\text{-}BuO)_2C=O} p\text{-Tol}-S(=O)-CH_2-CO_2Bu\text{-}t \qquad (3.8.2)$$

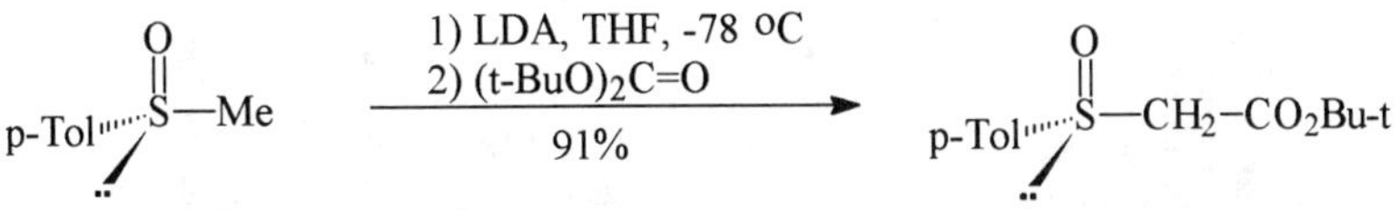

Recently, a new method of the synthesis has been developed which allows preparation of a great variety of different β- and γ-oxosulfoxides. It is based on the enzyme-promoted hydrolysis of sulfinylcarboxylates[3-7] or sulfinyldiacetates.[8,9] The former one consists in a kinetic resolution of racemic substrates, which means that the reaction is stopped after ca. 50% conversion and both enantiomerically enriched unreacted ester and acid are isolated (Equation 3.8.3, Table 3.8.1). Hydrolysis of sulfinyldiacetates **1** consists in an asymmetric synthesis in which prochiral substrates are transformed into enantiomerically enriched products (Equation 3.8.4, Table 3.8.2).

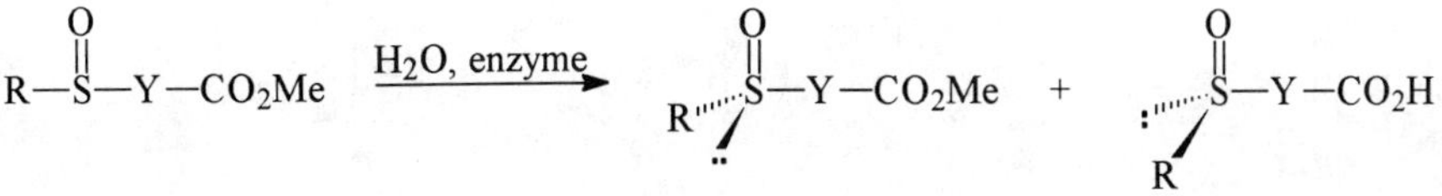

$$R-S(=O)-Y-CO_2Me \xrightarrow{H_2O,\ enzyme} \underset{\text{'recovered ester'}}{R-S(=O)-Y-CO_2Me} + \underset{\text{'acid'}}{R-S(=O)-Y-CO_2H} \qquad (3.8.3)$$

Table 3.8.1 Enzyme-Promoted Hydrolytic Resolution of Racemic Methyl Sulfinylcarboxylates

				Recovered ester				Acid				
R	Y	Enzyme, Conditions	Time [h]	Yield [%]	$[\alpha]_D$	ee [%]	Abs. conf.	Yield [%]	$[\alpha]_D$	ee [%]	Abs. conf.	Ref.
Ph	CH_2	Corynebacterium equi. IFO 3730, 30°C	24	43	+155	90	R	n.i.				3
p-Tol	CH_2		24	30	+181	90	R	n.i.				3
p-ClC_6H_4	CH_2		24	30	+193	97	R	n.i.				3
Ph	CH_2CH_2		24	22	+87	96	R	n.i.				3
p-ClC_6H_4	CH_2	Pseudomonas K-10, pH 7.5, H_2O/toluene, 25°C	25	48	+201	>95[a]	R	38	(−) n.r.	91	S	5
p-$O_2NC_6H_4$	CH_2		96	33	+179	>95[a]	R	22[b]	−174	>95[a]	S	5
Ph	CH_2		55	48	(+) n.r.	>95[a]	R	17	(−) n.r.	92	S	5
p-$MeOC_6H_4$	CH_2		48	48	+126	>95[a]	R	34[b]	−117	88	S	5
2-Naph	CH_2		27	45	+124	>95[a]	R	35	(−) n.r.	80	S	5
n-Bu	CH_2		96	33	(+) n.r.	>95[a]	R	n.i.				
c-C_6H_{11}	CH_2		96	49	(+) n.r.	>95[a]	R	18	(−) n.r.	>95[a]	S	5
p-ClC_6H_4	CH_2CH_2		37	30	+143	>95[a]	R	29	(−) n.r.	63	S	5
p-$O_2NC_6H_4$	CH_2CH_2		66	44	+64	>95[a]	R	35	(−) n.r.	82	S	5
Ph	CH_2CH_2		38	48	+99	>95[a]	R	24	(−) n.r.	91	S	5
p-ClC_6H_4	$CH_2CH_2CH_2$		54	No react.				No react.				5
p-Tol	CH=CH	OF 360 lipase phosphate buffer	48	34	−370	85	R	66		44	S	7

Note: n.i. = not isolated; n.r. = not reported.

[a] None of the other enantiomers was detected.

[b] These acids were converted to the corresponding methyl esters (using CH_2N_2) prior to isolation from the crude reaction mixture.

$$ROC(=O)CH_2S(=O)CH_2C(=O)OR \ (\mathbf{1}) \xrightarrow[\text{enzyme}]{\text{Buffer, NaOH}} ROC(=O)CH_2S(=O)CH_2COOH \ (\mathbf{2}) \quad (3.8.4)$$

Table 3.8.2 Enzymatic Hydrolysis of Prochiral Sulfinyldiacetates **1**

1	R	Enzyme	pH	Time [h]	Temp. [°C]	2 Yield [%]	$[\alpha]_D$(MeOH, c 1)	ee [%]	Abs. conf.	Ref.
a	Me	PLE	7.2	16	r.t.	70	+15.8	79	S	8
a	Me	PLE	7.5	0.5	40	74[a]	n.r.	86	S	9
a	Me	PPL	7.2	72	r.t.	35	–6.9	35	R	8
a	Me	PPL	8.0	24	r.t.	86[a]	n.r.	91	R	9
a	Me	α-CT	7.5	16	r.t.	63	–18.3	92	R	8
b	Et	PLE	7.5	0.5	40	74[a]	n.r.	82	S	9
b	Et	PPL	8.0	24	r.t.	92[a]	n.r.	83	R	9
c	i-Pr	PLE	7.5	1	40	74[a]	n.r.	76	S	9
c	i-Pr	PPL	8.0	72	40	46[a]	n.r.	4	S	9

Note: r.t. = room temperature; n.r. = not reported.

[a] The monoacid **2** transformed into methyl benzoylmethyl diester, PLE = porcine liver esterase; α-CT = α-chymotrypsin; PPL = porcine pancreatic lipase.

3.8.1.2 Reactions of the Carbanion of α-Sulfinylcarboxylates with Electrophiles

Alkylation[10] and Michael addition[11] of the carbanions of α-sulfinylcarboxylates proceed with a poor diastereoselection. On the other hand, hydroxyalkylation gives the corresponding derivatives in high yields and with a good to excellent β-diastereoselection, the degree of which depends on the nature of the substituents at the carbonyl group.[12] The use of t-butylmagnesium bromide as a base is essential for good results, since no hydroxyalkylation products can be obtained using alkyllithium or sodium hydride. Desulfurization gives optically active β-hydroxy carboxylic acids.

$$p\text{-Tol}S(O)CH_2CO_2Bu\text{-}t \ ((+)\text{-}R) + R^1C(=O)R^2 \xrightarrow{\text{t-BuMgBr, THF, -78°C}} R^1R^2C(OH)CH(S(O)p\text{-Tol})CO_2Bu\text{-}t \ (\mathbf{3}) \xrightarrow{\text{Al–Hg}} R^1R^2C(OH)CH_2CO_2Bu\text{-}t \ (\mathbf{4}) \quad (3.8.5)$$

R^1	R^2	3 Symbol	3 Yield [%]	4 ee [%]	4 Abs. conf.
H	Ph	**a**	85	91	S
Me	Ph	**b**	75	68	S
Ph	CF_3	**c**	75	20	R
H	n-C_7H_{15}	**d**	80	86	R
Me	c-C_6H_{11}	**e**	88	95	S

Application of this chemistry includes the preparation of several optically active lactones, among them insect pheromones (+)-(R)-δ-n-hexadecanolactone **6** and (+)-(R)-γ-n-dodecanolactone **8**, obtained with ee > 80%.[13]

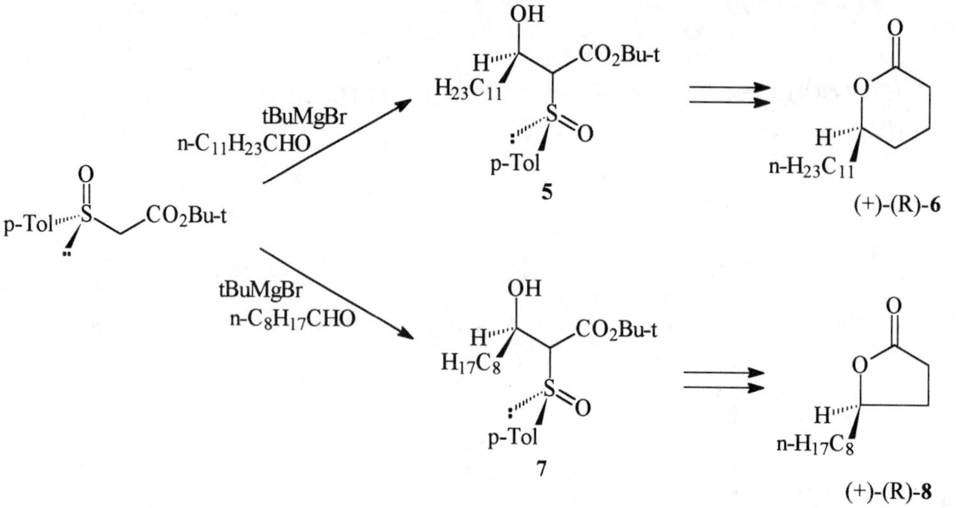

tBuMgBr, n-$C_{11}H_{23}CHO$ → 5 ⇒ (+)-(R)-6

tBuMgBr, n-$C_8H_{17}CHO$ → 7 ⇒ (+)-(R)-8

(3.8.6)

Pyrolytic elimination of the sulfinyl moiety from the initial adducts produces α-hydroxyalkylacrylates **10** in 25% overall yield and with 75% ee.[14]

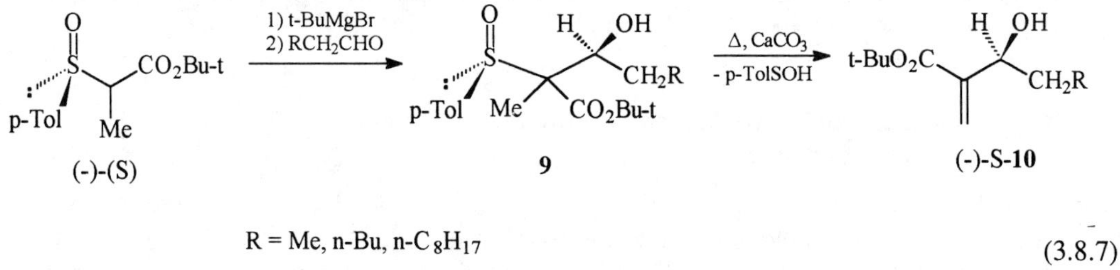

(-)-(S) → 1) t-BuMgBr, 2) RCH_2CHO → **9** → Δ, $CaCO_3$, - p-TolSOH → (-)-S-**10**

R = Me, n-Bu, n-C_8H_{17}

(3.8.7)

When a similar reaction is performed under the Knoevenagel condensation conditions, i.e., using piperidine as a base, and with the use of α-unsubstituted aldehydes, enantiomerically enriched (E)-γ-hydroxy-α,β-unsaturated esters **14** are formed as final products. This type of transformation is in some papers called the "SPAC" reaction (an abbreviation from **S**ulfoxide **P**iperidine **A**nd **C**arbonyl[15,16] or **S**ulfoxide **P**iperidine **A**ldehyde **C**ondensation[17]). The "SPAC" reaction involves several mechanistic steps (Scheme 3.8.1).

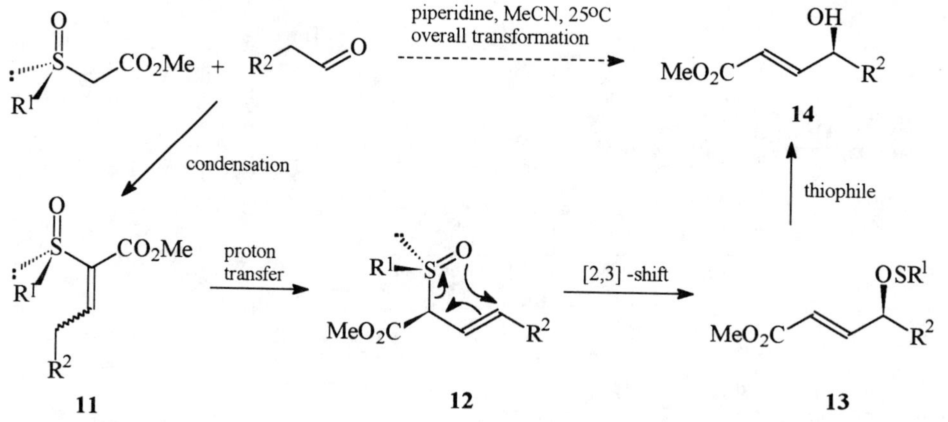

piperidine, MeCN, 25°C, overall transformation → **14**

condensation → **11** → proton transfer → **12** → [2,3]-shift → **13** → thiophile → **14**

Scheme 3.8.1

In the first step the condensation product **11** is formed which subsequently undergoes a carbon-carbon double bond shift. This step is considered as the crucial one in determining the configuration of the alcohol **14**, since the asymmetric induction is believed to arise from conjugate deprotonation by piperidine and asymmetric α-protonation by the piperidinium ion. The sulfoxide chirality is significant in this step, but does not affect the stereochemical outcome of the next transformation, i.e., the [2,3]-sigmatropic rearrangement to **13**, "because the sulfoxide rearranges on the face of the double bond which corresponds to minimum 1,3-allylic strain in the transition state"[15] [for a detailed discussion on the mechanism of the (2.3)-sigmatropic rearrangement see Section 3.3.1]. The final step consists in a simple removal of the sulfenyl moiety by a thiophile (which is here again piperidine). Enantiomeric purity of the products **14** varies from low to moderate and depends on the kind of substituents at the sulfinyl sulfur atom and in the aldehyde. In general, large aliphatic substituents at sulfur give poor optical yields of **14** and so does the phenyl group in the aldehyde (Table 3.8.3). The sulfinylacetates having the spatial arrangements as shown in Scheme 3.8.1 (the absolute configuration R or S depending on the substituent R^1) give rise to the formation of (S)-allylic alcohol **14**.

Table 3.8.3 The "SPAC" Reaction

R^1	R^2	Reaction time [h]	Allylic alcohol 14 Yield [%]	Allylic alcohol 14 ee [%]	Ref.
n-Bu	n-Pr	20	69	46	4
c-C_6H_{11}	n-Pr	20	74	16	4
p-$MeOC_6H_4$	n-Pr	23	52	60	4
Ph	n-Pr	4	68	61	4
p-$O_2NC_6H_4$	n-Pr	8	42	79	4
p-ClC_6H_4	n-Pr	3	65	75	15
p-ClC_6H_4	i-$PrCH_2$	24	54	74	15
p-ClC_6H_4	Me	19	71	64	15
p-ClC_6H_4	Et	23	86	60	15
p-ClC_6H_4	i-Pr	21	94	50	15
p-ClC_6H_4	Ph	24	16	0	15
HO_2CCH_2	n-Pr	12	n.r.	~75	8

Analogs of sulfinylacetates in which the ester group is replaced by its derivative having also electron-withdrawing properties, (e.g., the cyano or oxazoline group) behave similarly and give the corresponding allylic alcohols with similar enantioselection.[18,19]

(R) Ar–S(O)–CH$_2$–R^1 $\xrightarrow[\text{piperidine}]{R^2CH_2CHO}$ R^1–CH=CH–CH(OH)–R^2 **15**

Ar	R^1	R^2	Solvent	15 Yield [%]	15 ee [%]	Ref.
p-Tol	CN	n-Bu	MeOH	66	47	18
p-ClC_6H_4	CN	n-Bu	MeCN	77	75	19
p-Tol	CN	i-Pr	MeOH	42	50	18
p-Tol	4,4-dimethyloxazolin-2-yl (N, O ring; Me, Me)	i-Pr	MeCN	66	60	18
p-ClC_6H_4	CN	$MeO_2C(CH_2)_2$	MeCN	71	76	19
p-ClC_6H_4	CN	$THPOCH_2C{\equiv}C(CH_2)_2$	MeCN	54	80	19

(3.8.8)

Since the stereoselectivity of the SPAC reaction does not exceed 75 ÷ 80%, some attempts have been made to improve it by a double diastereoselection. Thus, Burgess and Henderson[16,20] used for this purpose optically pure sulfinylacetates bearing chiral auxiliaries in the ester functionality. The stereochemical outcome of the reaction was found to depend on the use of "matched" or "mismatched" combinations of the absolute configurations at the sulfinyl center and of the auxiliary. For example, the combination of the (S)-sulfoxide and the (–)-phenylmenthyl auxiliary is a "mismatched" one and gives practically no stereoselection, while that of the (R)-sulfoxide and the same auxiliary is a "matched" one and gives an 88:12 diastereomer ratio (Equation 3.8.9). In general, introduction of a chiral auxiliary only slightly enhances stereoselectivity of the "SPAC" reaction.

p-ClC$_6$H$_4$S(O)CH$_2$CO$_2$R* (**16**) + RCH$_2$CH$_2$CHO —(piperidine, MeCN, 25°C)→ R*O$_2$C–CH=CH–CH(OH)–R (**17**, C4 = carbinol carbon) (3.8.9)

Sulfinylacetate 16			Allylic alcohol 17	
Configuration of sulfoxide	R*	R	Yield [%]	(R):(S) ratio at C4
S	(–) PhMen	n-Pr	82	42:58
R	(–) PhMen	n-Pr	98	88:12
S	(–)-Cam	Me	83	25:75
R	(–)-Cam	Me	69	88:12
R	(–)-Cam	Et	89	87:13
R	(–)-Cam	n-Pr	93	91:9

(-) PheMen = [cyclohexane chair bearing Me and Me$_2$PhC–]

(-) Cam = [bornane skeleton with two Me groups and CH$_2$SO$_2$NCy$_2$]

In turn, Trost and Mallart[21] investigated a double diastereodifferentiation during the SPAC reaction using optically active sulfinylacetates and an optically active aldehyde, (S)-citronellal. Again, "matched" and "mismatched" pairs were found, and in the former case the stereoselection was slightly higher.

(3.8.10)

Ar	Conf. at SO	Ratio **18:19**
p-ClC_6H_4	R	75:25
p-$O_2NC_6H_4$	R	88:12
p-$PhSO_2C_6H_4$	R	88:12
p-$PhSO_2C_6H_4$	S	18:82

On the basis of the results obtained, the authors deduced that the stereochemistry of the sulfoxide in the [2,3]-sigmatropic rearrangement may play a role in double distereodifferentiation,[21] which is in contrast to the conclusion of Burgess et al.[15] presented earlier (see discussion on Scheme 3.8.1).

3.8.2 β-Oxosulfoxides

Optically active β-oxosulfoxides are readily available from α-sulfinyl carbanions and derivatives of carboxylic acids which is described in detail in Section 3.2.5. Reactivity of their α-carbanions is similar to that of α-sulfinylacetates and has been a subject of a rather limited interest.

The most important transformation of β-oxosulfoxides, which has become a commonly used methodology during the last 15 years, is their reduction to β-hydroxysulfoxides. The reduction leads to diastereomeric products, the ratio of which strongly depends on the reducing agent used.

(R_S) [H] → (R_C,R_S) + (S_C,R_S) (3.8.11)

Recent detailed studies on the stereoselectivity of the reduction of acyclic and cyclic β-oxosulfoxides performed by a Spanish-French team[22] (Equation 3.8.11, Table 3.8.4) allowed them to draw some general mechanistic conclusions.

Table 3.8.4 Reduction of Acyclic ß-Oxosulfoxides with Various Hydrides.

		Product ratio (R_C, R_S):(S_C, R_S)			
Ar	**R**	**$NaBH_4$**	**$LiAlH_4$**	**DIBAL**	**DIBAL/$ZnCl_2$**
Ph	p-Tol	41:59	84:16	5:95	95:5
Ph	Me	45:55	64:36	16:84	80:20
2-Pyr	p-Tol	44:56	56:44	0:100	44:56
2-Pyr	p-Tol	52:48	63:37	0:100	50:50

From Carreno, M. C., Garcia Ruano, J. L., Martin, A. M., et al., *J. Org. Chem.*, 55, 2120, 1990.

1. The poor stereoselectivity observed in reduction of acyclic substrates with $NaBH_4$ is believed to be due to the fact that both the stereoelectronic repulsion of the sulfur lone pair and the steric effects are simultaneously responsible for the control of the hydride approach to the carbonyl.
2. When diisobutylaluminum hydride (DIBAL) is used as a reducing agent, an Al-O=S associate is pre-formed; in this adduct the intramolecular hydride transfer takes place through a chair-like transition state **B**, whose relative stability is responsible for the high stereoselectivity (Scheme 3.8.2).
3. In the case of $LiAlH_4$ and DIBAL/$ZnCl_2$, which exert the same diastereoselection opposite to that of DIBAL alone (Table 3.8.4), the high stereoselectivity observed for the substrates with Ar=Ph is attributed to the involvement of a metal chelate. For $LiAlH_4$ this is a lithium chelate, to which hydride is intramolecularly transferred from associated AlH_4^-, while for DIBAL/$ZnCl_2$ this is a zinc chelate, and the hydride transfer occurs here intermolecularly in the most stable half-chair conformation (Scheme 3.8.2). A poorer stereoselectivity observed for the substrates with Ar=pyridyl is attributed to the competing pyridyl nitrogen towards association with $ZnCl_2$, resulting in an incomplete chelation.[22]

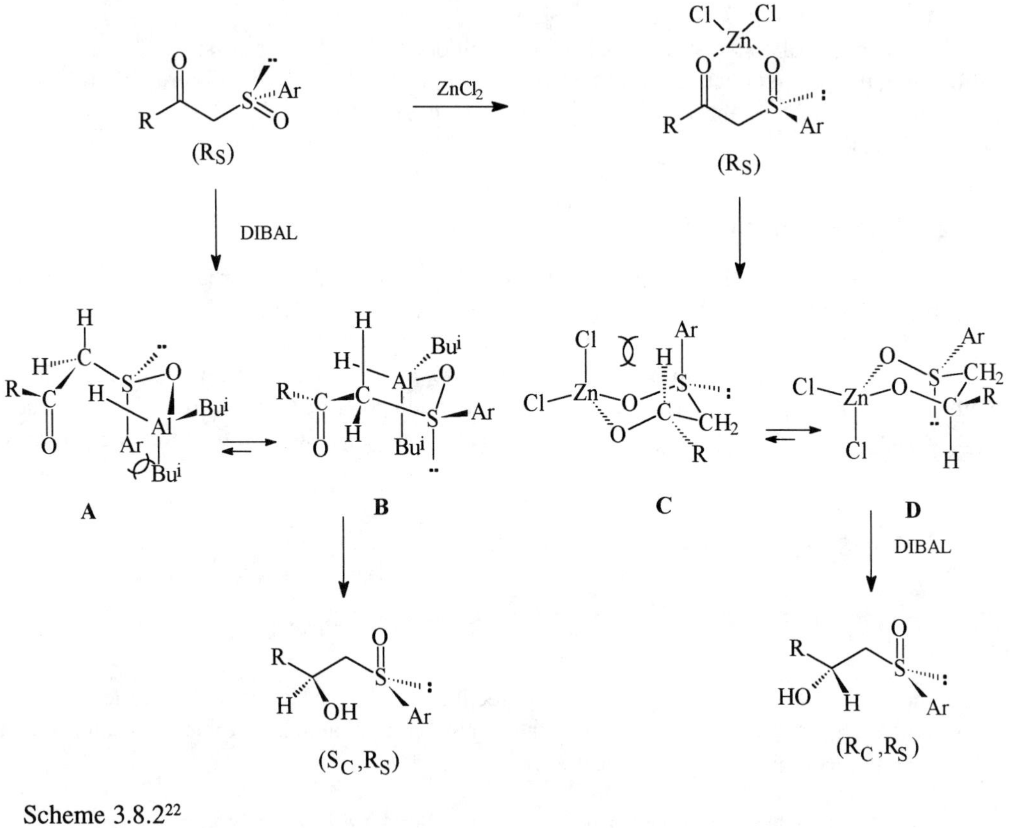

Scheme 3.8.2[22]

It should be stressed that DIBAL and the DIBAL/$ZnCl_2$ system, introduced simultaneously by Solladie et al.[23] and Kosugi et al.,[24] are presently the most commonly used reducing agents, since a simple modification of the reagent allows for a highly stereoselective synthesis of both epimeric β-hydroxysulfoxides from the same substrate (for a similar approach see Sections 3.9.6 and 3.10.3). This methodology has been widely used in the asymmetric synthesis of hydroxyl compounds, and some selected representative examples are presented below (for a review see Reference 25).

3.8.2.1 Asymmetric Synthesis of Both Enantiomers of 4-Hydroxy-2-Cyclohexenone[26]

A pure epimer of the ketosulfoxide **20** was reduced with DIBAL and with DIBAL/$ZnCl_2$ to give the *trans*-hydroxysulfoxide **21** and a mixture of *trans*-**21** and *cis*-hydroxysulfoxide **22** in a ratio 30:70, respectively. In the latter case the pure *cis*-**22** was obtained by crystallization. Elimination of the sulfinyl group in each product was performed using acidic silica gel and gave (S)- and (R)-4-hydroxy-2-cyclohexanones **23** in moderate yields, but with ee > 95%.

20 → DIBAL, 95% → *trans* - **21**

trans - **21** → SiO_2, H_2SO_4, 40% → (S)-**23**

20 → 1) DIBAL / $ZnCl_2$ 2) crystallization, 70% → *cis* - **22**

cis - **22** → SiO_2, H_2SO_4, 41% → (R)-**23**

(3.8.12)

3.8.2.2 Synthesis of Optically Active Epoxides[23]

Reduction of the β-hydroxysulfoxide **24** either with DIBAL and with DIBAL/$ZnCl_2$ led to each epimeric β-hydrosulfoxide **25** with high stereoselectivity. Each of them was transformed into the epoxide **27** via a reduction of the sulfoxide to sulfide, alkylation, and intramolecular substitution of the sulfonium moiety in the sulfonium salt **26**.

(R)-**24** → DIBAL → (R_S, S_C)-**25** → 1) $LiAlH_4$ 2) Me_3OBF_4 3) NaOH → (S)-**27**

(R)-**24** → DIBAL, $ZnCl_2$ → (R_S, R_C)-**25** → $LiAlH_4$ → p-TolS… → Me_3OBF_4 → **26** → NaOH → (R)-**27**

(3.8.13)

R	Reducing agent	**25** (R_S, R_C):(R_S, S_C)	**25** Yield [%]	**27** Yield [%]	**27** (R):(S)
Ph	DIBAL	<5:>95	95	60	5:95
Ph	DIBAL/$ZnCl_2$	>95:<5	90	60	95:5
n-C_8H_{17}	DIBAL	5:95	95	63	<5:>95
n-C_8H_{17}	DIBAL/$ZnCl_2$	>95:<5	92	67	>95:<5

3.8.2.3 Asymmetric Synthesis of Protected α-Hydroxyaldehydes[27]

This procedure utilizes the reduction of optically active β-oxo derivatives of dithioacetal monoxides **28**. It is interesting that when $LiAlH_4$ is used as a reducing agent this reaction is fully stereoselective and gives practically one diastereomer of **29**, which is then transformed into the enantiopure α-methoxyphenylacetaldehyde **30**.

$LiAlH_4$, 76%

(2R,3S)-**28** → (1R,2R,3S)-**29** de >98% ⇒ (-)-(R)-**30** ee ~100%

(3.8.14)

3.8.2.4 Stereoselective Synthesis of (R)-3-Benzoyloxy-2-Butanone[29]

Chiral 2-acyl dithiolane sulfoxide **31** was subjected to reduction using DIBAL and DIBAL/$ZnCl_2$. Interestingly, in both cases the stereochemical outcome was practically the same (no influence of the chelating agent!), the diastereomeric purity of **32** being almost 100%. Benzoylation, followed by a two-step removal of the dithiolane moiety, gave (R)-3-benzoyloxy-butan-2-one **33** with ee > 97%.

DIBAL, -78°C, 91%; 1) benzoylation 2) PBr_3, THF 3) HgO, $BF_3 \cdot Et_2O$, 60%

31 → **32** → **33**

(3.8.15)

3.8.2.5 Synthesis of Chiral 1,3-Diols[30]

(R)-β,δ-Dioxosulfoxides **34** were reduced with DIBAL to give δ-oxo-β-hydroxysulfoxides **35** (note that only the β-oxo group is reduced!) with de > 98%. They were then transformed into optically active diols via a stereoselective reduction of the δ-oxo group and desulfurization. In a similar way, optically active 3-hydroxybutyrates were synthesized.[31]

DIBAL, -78 °C, 70-85%; $Me_4NHB(OAc)_3$, 90%; Raney Ni, 100%

34 → **35** → **36**, R = Me, 88% de → (-)-(R,R)

(3.8.16)

3.8.2.6 Reduction of Fluoroalkyl-Oxosulfoxides

The perfluoro derivatives, e.g., **37**, when reduced with borohydrides give the corresponding β-hydroxysulfoxides with low to moderate stereoselectivity which is explained in terms of their existence in both the keto and hydrated forms.[32-34]

$$p\text{-Tol}S(O)CH_2C(O)CF_3 \ (\mathbf{37}) \xrightarrow{NaBH_4} p\text{-Tol}S(O)CH_2CH(OH)CF_3 + p\text{-Tol}S(O)CH_2CH(OH)CF_3 \quad (3.8.17)$$

However, when the difluoro compounds **38** are reduced with DIBAL, only (R_S, S_C)-β-hydroxy-sulfoxides **39** are produced.[35]

$$p\text{-Tol}S(O)CH_2C(O)CRF_2 \ ((R)\text{-}\mathbf{38}) \longrightarrow p\text{-Tol}S(O)CH_2CH(OH)CRF_2 \ ((R_S,S_C)\text{-}\mathbf{39}) \quad (3.8.18)$$

There are a great number of other applications of the methodology discussed above which cannot be included in this chapter. Some of them are listed below: synthesis of arabinitol,[36] (R,R)-pyrenophorin and (R)-patulolide,[37] 1,6-dioxaspirononanes,[38] functionalized 1,2-diols,[39] enantiopure 3-hydroxyketones,[40] leukotriene B_4,[41] 2- and 3-furylcarbinols,[42] juvenile hormones,[43] and many others.

For other transformations of optically active β-oxosulfoxides see: hydrocyanation, Reference 44; and alkylation at the carbonyl group, References 45, 46. For the application of enantiopure α-sulfinyl N,N-dimethylacetamide see Reference 47.

3.8.3 β-Thioxosulfoxides

The only example of this type of sulfoxide reported thus far is the enantiopure p-tolyl sulfinyl-N,N-dimethylthioacetamide **40**, prepared from (–)-(S) menthyl p-toluenesulfinate and N,N-dimethylthioacetamide. Its aldol condensation with aldehydes gives the corresponding hydroxylakyl adducts **41** as diastereomeric mixtures. Subsequent removal of the sulfinyl moiety affords β-hydroxythioacetamides **42** in good yields and with reasonable optical purity.[48]

$$p\text{-Tol}S(O)CH_2C(S)NMe_2 \ ((+)\text{-}(R)\text{-}\mathbf{40}) \xrightarrow{1)\ t\text{-BuMgBr};\ 2)\ RCHO} p\text{-Tol}S(O)CH(CH(OH)R)C(S)NMe_2 \ (\mathbf{41}) \xrightarrow{Na/Hg} RCH(OH)CH_2C(S)NMe_2 \ ((S)\text{-}\mathbf{42}) \quad (3.8.19)$$

	42	
R	Yield [%]	ee]%]
Me	75	64
i-Pr	75	90
i-Bu	52	40
t-Bu	40	55

3.8.4 β-Iminosulfoxides and Analogs

Among the compounds having a carbon-nitrogen double bond in the β-position with respect to the sulfinyl group, four types, namely α-sulfinyl hydrazones **43**,[49] α-sulfinyl-N-methoxyacetimidates **44**,[50] 2-sulfinyl oxazolines **45**,[51] and 3-sulfinyl-4,5-dihydroisoxazoles **46**[52] were subjected to aldol condensation with aldehydes to give the corresponding adducts which were subsequently transformed into sulfur-free chiral products with various optical purities. The results are briefly illustrated below.

1) base, R^2CHO
2) Na/Hg

Cu^{2+}

43 R^1 = H, Me, Ph, t-Bu, s-Bu

ee up to 88%

1) base
2) excess (±) R^2R^3CH–CHO (kinetic resolution)
3) Na/Hg; Cu^{2+}

(R,R) - major diastereomer

(3.8.20)

R^1	R^2	R^3	dr	ee [%]
Ph	Et	Me	1.2:1	≥80
Ph	$PhCH_2$	Me	1.2:1	≥90
Ph	Ph	Me	≥30:1	100

1) n-BuLi
2) RCHO

Raney Ni

R–CH(OH)–CH_2–CO_2Et

44

R = Ph, i-Pr, n-C_5H_{11}, c-C_6H_{11}

(R) or (S)
ee 75÷94%

(3.8.21)

1) base, RCHO
2) Na-Hg

H_2SO_4

45 R = Me, n-Pr, i-Bu, i-Pr, t-Bu

ee 21÷53%

(3.8.22)

1) t-BuMgBr
2) EtCHO
3) Na-Hg

46

de 50 : 1

(3.8.23)

When optically pure β-iminosulfoxides **47** are reduced with DIBAL/$ZnBr_2$ or L-selectride, the corresponding β-aminosulfoxides are obtained with very high diasteroselection. As may be expected, each reagent leads to a different major epimer.[53]

DIBAL / $ZnBr_2$ or L-selectride

(R)-**47** → ($1R,R_S$)-**48** + ($1S,R_S$)-**48** (3.8.24)

R	ZnBr₂/DIBAL Yield [%]	ZnBr₂/DIBAL (1R,R_S):(1S,R_S)	L-selectride Yield [%]	L-selectride (1R,R_S):(1S,R_S)
Me	80	>97:<3	50	9:91
n-Pr	72	>97:<3	10	<9:>91
i-Pr	82	>97:<3		

In turn, reduction of fluorinated β-iminosulfoxides **49**, which actually exist exclusively in their enamino form **50**, undergo a highly stereoselective reduction by K- or L-selectride to give β-fluoroalkyl-β-aminosulfoxides **51**.[54]

49 ⇌ **50** → [H] → (R_S,2S)-**51** + (R_S,2R)-**51** (3.8.25)

Solvent	(R_S,2S):(R_S,2R)	Yield [%]
CH_2Cl_2	90:10	75
THF	93:7	50

3.8.5 Examples of Experimental Procedures

3.8.5.1 Synthesis of (+)-R t-Butyl p-Tolylsulfinylacetate[1]

Ethylmagnesium bromide is prepared from 2.7 g of magnesium and 12.2 g of EtBr in 159 mL of ether and 10.3 g of i-Pr_2NH are added. The mixture is refluxed for 30 min, then cooled to –35 to –40°C and 10 mL of THF are added. Then a mixture of 8 g of $MeCO_2Bu$-t, 10 g of (–)-(S)menthyl p-toluenesulfinate in 100 mL of ether, and 10 mL of THF are dropped in and the resulting solution is stirred overnight at –35 to –40°C. A saturated solution of NH_4Cl is added and the layers separated. The aqueous layer is extracted with $CHCl_3$ (2 ×), and the combined organic layers are dried (Na_2SO_4) and evaporated. The crude yellow oil is chromatographed on silica gel (ether/petroleum ether 1:1) to give 7.7 g (90%) of the pure product; $[\alpha]^{20}_D$ +149 (c 2.25, EtOH).

3.8.5.2 Hydroxyalkylation of α-Sulfinylcarboxylates: t-Butyl α-(p-Tolylsulfinyl)-β-Hydroxymyristate 5[13]

To a solution of 1.50 g (6 mmol) of (+)-(R)-*tert*-butyl (p-tolylsulfinyl)acetate in 400 mL of THF, cooled at –78°C, is added 40 mL (16 mmol) of *tert*-butylmagnesium bromide (prepared from 3 g of magnesium, 16 mL of *tert*-butyl bromide, and 50 mL of Et_2O) dropwise over a period of 1 h. The mixture is then stirred for 30 min at –78°C, and 1.6 g (12 mmol) of dodecanal in 15 mL of THF are added dropwise. After 1 h at –78°C, the reaction mixture is hydrolyzed with 50 mL of a saturated solution of ammonium chloride and then with 150 mL of water and extracted with chloroform (2 × 100 mL). The extract is dried with sodium sulfate and concentrated. The crude product is then rapidly purified by column chromatography on silica gel with 30/70 ether/hexane as an eluant to remove the excess of dodecanal to give pure **5** in quantitative yield.

3.8.5.3 General Procedure for the Resolution of Methyl Sulfinylcarboxylates via Hydrolyses Catalyzed by *Pseudomonas* sp. K-10 (Table 3.8.1)[5]

To a 1.0-M solution of the racemic methyl sulfinyl carboxylate in toluene is added 8 times the volume of a 0.05-M solution of phosphate buffer (pH 7.5) and 1.0 mass equiv of *Pseudomonas* sp. K-10 (Amano). The heterogeneous mixture is stirred at 25°C for the time indicated. The reaction is filtered through Celite (washing with Et_2O and H_2O) to remove the enzyme and then extracted with several portions of Et_2O. The aqueous fraction is retained and treated as described below. The combined organic fractions are dried, and removal of the volatiles gives the optically active unreacted ester. The aqueous layer is acidified with glacial acetic acid (2 times the original toluene volume), and the sulfinylcarboxylic acid is extracted with $CHCl_3$.The combined organic fractions are dried, and removal of the volatiles gives the optically active acid. Alternatively, the H_2O is removed *in vacuo* from the aqueous layer, and the residue obtained is partially dissolved in $CHCl_3$. Excess diazomethane is then added, and the reaction mixture is stirred for 12 h. The remaining diazomethane is quenched with glacial acetic acid. Removal of the volatiles and purification by flash chromatography give the optically active sulfinylcarboxylate.

3.8.5.4 The "SPAC" Procedure: Methyl γ-Hydroxy-α,β-Unsaturated Esters[15]

A 1.0-M solution of the aldehyde (5.0 equiv) in acetonitrile was added over ~1 h to a solution of piperidine (5.0 equiv) and a 0.5-M solution of the methyl sulfinylacetate (1.0 equiv) in acetonitrile under N_2. The resulting light brown solution was stirred at 25°C for the time specified. Removal of the volatiles *in vacuo* gave the crude product which was purified by flash chromatography.

3.8.5.5 Reduction of β-Oxosulfides; General Procedure[28]

- Reduction with Diisobutylaluminum Hydride (DIBAL): To a solution of β-oxosulfoxide (2 mmol) in THF (20 mL) at –78°C is added dropwise 2.2 mL (2.2 mmol) of a 1-M solution of DIBAL in hexane. After 1 h at –78°C, the reaction mixture is decomposed by adding 20 mL of MeOH. The solvent is then evaporated and the residue diluted with water and extracted with CH_2Cl_2. The organic layer is washed with a 5% NaOH solution, dried, and evaporated. The reaction is quantitative (by TLC). The diastereomeric excesses are determined on the crude product by NMR. Finally the product is purified by chromatography on silica gel (eluent: ether/hexane, 60/40).
- Reduction with DIBAL/$ZnCl_2$: To a solution of β-oxosulfoxide (2 mmol) in THF (20 mL) is added 1.1 equiv of anhydrous zinc chloride (2.2 mmol) in solution in THF (20 mL). After 1 h at room temperature, the reaction mixture is cooled at –78°C and 2.2 mL of a 1-M solution of DIBAL in hexane (2.2 mmol) are added. After 1 h at –78°C the same workup as in the previous part is used.

REFERENCES (3.8)

1. Mioskowski, C., Solladie, G., *Tetrahedron*, **36**, 227, 1980.
2. Abushanab, E., Reed, D., Suzuki, F., Sih, C. J., *Tetrahedron Lett.*, 3415, 1978.
3. Ohta, H., Kato, Y., Tsuchihashi, G., *Chem. Lett.*, 217, 1986.
4. Burgess, K., Henderson, I., *Tetrahedron Lett.*, **30**, 3633, 1989.
5. Burgess, K., Henderson, I., Ho, K.-K., *J. Org. Chem.*, **57**, 1290, 1992.
6. Allenmark, S., Andersson, A. C., *Tetrahedron: Asymmetry*, **4**, 2371, 1993.
7. Cardellicchio, C., Naso, F., Scilimati, A., *Tetrahedron Lett.*, **35**, 4635, 1994.
8. Mikołajczyk, M., Kiełbasiński, P., Żurawiński, R., Wieczorek, M. W., Błaszczyk, J., *Synlett*, 127, 1994.
9. Tamai, S., Miyauch, S., Morizone, C., Miyagi, K., Shimizu, H., Kume, M., Sano, S., Shiro, M., Nagao, Y., *Chem. Lett.*, 2381, 1994.
10. Solladie, G., Matloubi-Moghadam, F., Luttman, G., Mioskowski, C., *Helv. Chim. Acta*, **65**, 1602, 1982.
11. Matloubi, F., Solladie, G., *Tetrahedron Lett.*, 2141, 1979.
12. Mioskowski, C., Solladie, G., *J. Chem. Soc., Chem. Commun.*, 162, 1977.
13. Solladie, G., Matloubi-Moghadam, F., *J. Org. Chem.*, **47**, 91, 1982.
14. Papageorgiou, C., Benezra, C., *Tetrahedron Lett.*, **25**, 1303, 1984.
15. Burgess, K., Cassidy, J., Henderson, I., *J. Org. Chem.*, **56**, 2050, 1991.
16. Burgess, K., Henderson, I., *Tetrahedron Lett.*, **30**, 4325, 1989.
17. Guerrero-de la Rosa, Ordonez, M., Alcudia, F., Llera, J. M., *Tetrahedron Lett.*, **36**, 4889, 1995.
18. Annunziata, R., Cinquini, M., Cozzi, F., Raimondi, L., Restelli, A., *Gazz. Chim. Ital.*, **115**, 637, 1985.
19. Nokami, J., Mandai, T., Nishimuri, A., Takeda, T., Wakabayashi, S., Kunieda, N., *Tetrahedron Lett.*, **27**, 5109, 1986.
20. Burgess, K., Henderson, I., *Tetrahedron*, **47**, 6601, 1991.
21. Trost, B. M., Mallart, S., *Tetrahedron Lett.*, **34**, 8025, 1993.
22. Carreno, M. C., Garcia Ruano, J. L., Martin, A. M., Pedregal, C., Rodriguez, J. H., Rubio, A., Sanchez, J., Solladie, G., *J. Org. Chem.*, **55**, 2120, 1990.
23. Solladie, G., Demaily, G., Greck, C., *Tetrahedron Lett.*, **26**, 435, 1985.
24. Kosugi, H., Konta, H., Ude, H., *J. Chem. Soc., Chem. Commun.*, 211, 1985.
25. Solladie, G., *Pure Appl. Chem.*, **60**, 1699, 1988.
26. Carreno, M. C., Garcia Ruano, J. L. Garrido, M., Ruiz, M. P., Solladie, G., *Tetrahedron Lett.*, **31**, 6653, 1990.
27. Guanti, G., Narisano, E., Pero, F., Banfi, L., Scolastico, C., *J. Chem. Soc., Perkin Trans. 1*, 189, 1984; Guanti, G., Narisano, E., Banfi, L., Scolastico, C., *Tetrahedron Lett.*, **24**, 817, 1983.
28. Solladie, G., Frechou, C., Demailly, G., Greck, C., *J. Org. Chem.*, **51**, 1912, 1986.
29. Barros, M. T., Leitao, A. J., Maycock, C. D., *Tetrahedron Lett.*, **36**, 6537, 1995.
30. Solladie, G., Ghiatou, N., *Tetrahedron Lett.*, **33**, 1605, 1992.
31. Solladie, G., Almario, A., *Tetrahedron Lett.*, **33**, 2477, 1992.
32. Yamazaki, T., Ishikawa, N., *Chem. Lett.*, 889, 1985.
33. Bravo, P., Frigerio, M., Resnati, G., *J. Org. Chem.*, **55**, 4216, 1988.
34. Bravo, P., Frigerio, M., Resnati, G., *Synthesis*, 955, 1988.
35. Bravo, P., Pregnolato, M., Resnati, G., *J. Org. Chem.*, **57**, 2726, 1992.
36. Solladie, G., Frechou, C., Hutt, J., Demailly, G., *Bull. Soc. Chim. France*, 827, 1987.
37. Solladie, G., Gerber, C., *Synlett*, 449, 1992.
38. Solladie, G., Huser, N., Fischer, J., Decian, A., *J. Org. Chem.*, **60**, 4988, 1995.
39. Solladie, G., Almario, A., *Tetrahedron Lett.*, **35**, 1937, 1994.
40. Blase, F. R., Le, H., *Tetrahedron Lett.*, **36**, 4559, 1995.
41. Solladie, G., Stone, G. B., Hamdouchi, C., *Tetrahedron Lett.*, **34**, 1807, 1993.
42. Girodier, L. D., Rouessac, F. P., *Tetrahedron: Asymmetry*, **5**, 1203, 1994.
43. Kosugi, H., Kanno, O., Uda, H., *Tetrahedron: Asymmetry*, **5**, 1139, 1994.
44. Garcia Ruano, J. L., Martin, Castro, A. M., Rodriguez, J. H., *J. Org. Chem.*, **57**, 7235, 1992.
45. Fujisawa, T., Fujimura, A., Ikaji, Y., *Chem. Lett.*, 1541, 1988.
46. Bueno, A. B., Carreno, M. C., Garcia Ruano, J. L., Fischer, J., *Tetrahedron Lett.*, **36**, 3737, 1995.
47. Annunziata, R., Cinquini, M., Cozzi, F., Montanari, F., Restelli, A., *J. Chem. Soc., Chem. Commun.*, 1138, 1983.
48. Cinquini, M., Manfredi, A., Molinari, H., Restelli, A., *Tetrahedron*, **41**, 4929, 1985.
49. Annunziata, R., Cozzi, F., Cinquini, M., Colombo, L., Gennari, C., Poli, G., Scolastico, C., *J. Chem. Soc., Perkin Trans. 1*. 251, 1985; Annunziata, R., Cinquini, M., Cozzi, F., Giraldi, A., Cardani, S., Poli, G., Scolastico, C., *J. Chem. Soc., Perkin Trans. 1*, 255, 1985.
50. Bernardi, A., Colombo, L., Gennari, L., Prati, L., *Tetrahedron*, **40**, 3769, 1984.
51. Annunziata, R., Cinquini, M., Giraldi, A., *Synthesis*, 1016, 1983.
52. Annunziata, R., Cinquini, M., Cozzi, F., Restelli, A., *J. Chem. Soc., Perkin Trans. 1*, 2293, 1985.
53. Garcia Ruano, J. L., Lorente, A., Rodriguez, J. H., *Tetrahedron Lett.*, **33**, 5637, 1992.
54. Bravo, P., Crucianelli, M., Zanda, M., *Tetrahedron Lett.*, **36**, 3043, 1995.

3.9 ACYCLIC α,β-UNSATURATED SULFOXIDES

Enantiomerically pure, acyclic α,β-unsaturated (alkenyl, vinyl) sulfoxides were for the first time applied by Stirling et al.[1] and then by Tsuchihashi et al.[2] as the Michael-type acceptors in the asymmetric C–N and C–C bond formation, respectively. Since then they have been widely used as versatile chiral reagents with their sulfinyl group playing the role of a chiral auxiliary. Three main types of reactions have been the subject of interest: electrophilic addition to the double bond, nucleophilic conjugate addition (both of carbanions and various heteroatomic nucleophiles), and cycloaddition. The general methods of the synthesis of chiral α,β-unsaturated sulfoxides are based on the Andersen reaction and the Horner-Wittig reaction of α-phosphoryl sulfoxides and are described in detail in Chapter 2 and Chapter 3, Section 7, respectively. However, some examples of the synthesis of more sophisticated sulfoxides whose applications are discussed in the following part of the section will be given in the appropriate subsections.

3.9.1 Electrophilic Addition

The electrophilic addition to the ethylenic bond in α,β-unsaturated sulfoxides proceeds with high π-facial stereoselection which is explained in terms of an orbital control.[3]

A model has been put forward in which:

a) the lone electron pair (the "most electron-donating group") occupies the *anti*-position, thus maximizing the σ-π overlap and increasing the HOMO energy level of the alkene;
b) the sulfinyl oxygen atom (the "most electron-withdrawing group") is located in the *inside* position which minimizes the σ*-π overlap and decreases the HOMO energy level.

This model (**A**) has been supported by quantum chemical calculations.[3,4]

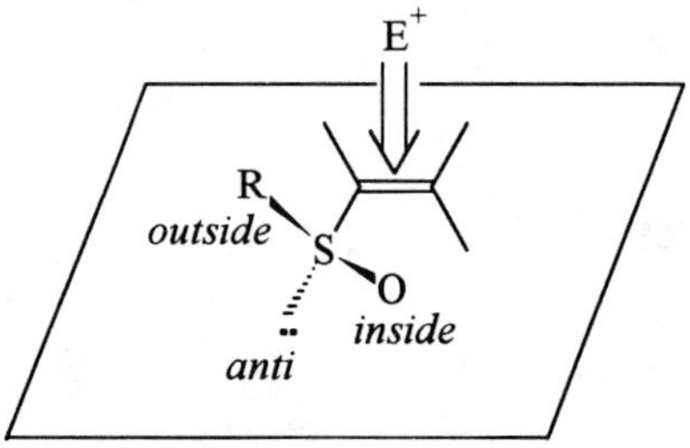

E^+ R *outside* S O *inside* *anti*

A

The scope of this type of reaction is, however, rather limited because only a few electrophiles can be used. These are mainly boranes and halogens. Their reaction with the title compounds leads to hydrogenation of the olefinic bond[3,5] (Equation 3.9.1) and formation of α-halo or α,β-dihalosulfoxides,[3,6] respectively (Equation 3.9.2).

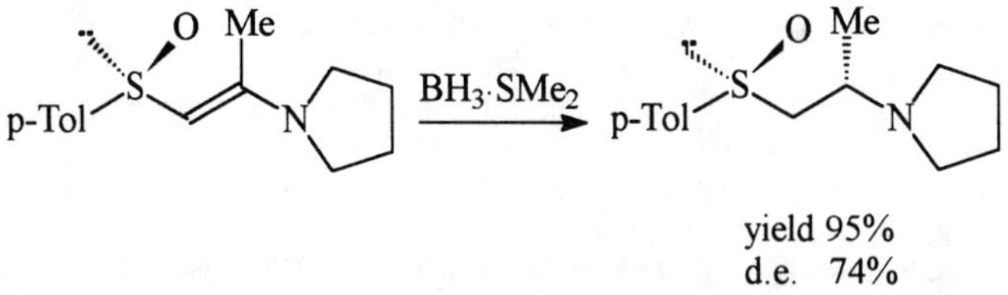

p-Tol S O Me N $\xrightarrow{BH_3 \cdot SMe_2}$ p-Tol S O Me N

yield 95%
d.e. 74% (3.9.1)

$$R^1S(O)C(R^2)=CHR^3 \xrightarrow{E} R^1S(O)CR^2(Br)CH(X)R^3 \qquad (3.9.2)$$

R^1	R^2	R^3	E	X	Yield [%]	Selectivity	Ref.
p-Tol	Me	H	Br_2	Br	68	83:17	3
c-Hex	Me	H	Br_2	Br	79	86:14	3
p-Tol	H	Ph	NBS/H_2O	OH		90:10	6
p-Tol	H	Ph	NBS/MeOH	OMe	72	95:5	6

3.9.2 Michael Addition of Carbon Nucleophiles

a,β-Unsaturated sulfoxides having no additional electron-withdrawing group in the α-position are rather poor Michael acceptors, particularly with respect to carbanionic nucleophiles. Nevertheless, a number of publications appeared which describe successful applications of this reaction with enolate anions[7-9] and alkyl cuprates.[10] However, there are a limited number of examples in which properly β-substituted, optically active vinyl sulfoxides were used and proved to give the addition products with moderate to high asymmetric induction.

The first one is the classical work of Tsuchihashi et al.[2,11] who found that the reaction of (+)-R-*trans*-styryl p-tolyl sulfoxide **1** with diethyl malonate gives a mixture of diastereomers, the ratio of which is strongly dependent on the nature of the counterion and solvent used (Equation 3.9.3).

$$\mathbf{1} \xrightarrow[2)\ H^+]{1)\ M^{\oplus}\ {}^{\ominus}CH(CO_2Et)_2} (R_S, R_C)\text{-}\mathbf{2a} + (R_S, S_C)\text{-}\mathbf{2b} \qquad (3.9.3)$$

Solvent	M^+	2a:2b
EtOH	Na^+	81:19
EtOH	K^+	79:21
THF	K^+	55:45
THF	Na^+	36:64
THF	Li^+	22:78
THF/hexane	Li^+	21:79

The major diastereomer (+)-(R_S,R_C) **2a** was obtained in a pure form by fractional crystallization from ethanol-n-hexane and transformed into (–)-(R)-3-phenylbutyric acid **3**, whose optical purity turned out to be around 95% (Equation 3.9.4).[2]

$$(+)\text{-}(R_S,R_C)\text{-}\mathbf{2a}\ [\alpha]_D+93.1 \xrightarrow[EtOH]{Raney\ Ni} PhCH(Me)CH(CO_2Et)_2 \xrightarrow[2)\ HCl]{1)\ KOH/EtOH\text{-}H_2O} PhCH(Me)\text{-}CH(CO_2H)_2$$

$$\xrightarrow[-CO_2]{149\ ^oC} Me(Ph)CH\text{-}CH_2CO_2H \quad (-)\text{-}(R)\text{-}\mathbf{3}\ [\alpha]_D\text{-}53.9\ (95\%\ op) \qquad (3.9.4)$$

A high asymmetric induction was achieved in the Michael addition of enolates to E-(R)- or Z-(R)-3,3,3-trifluoroprop-1-enyl p-tolyl sulfoxides **4**. Their synthesis has been accomplished in 70% yield via a three-step procedure starting from enantiomerically pure (+)-(R)-methyl p-tolyl sulfoxide as a source of chirality (Equation 3.9.5).[12]

p-Tol–S(O)–CH₃ [α]D+186.6 → (LDA, CF_3CO_2Et, 90%) → p-Tol–S(O)–CH₂–CO–CF_3 [α]D+181.5 →

→ ($NaBH_4$, 87%, (diast. ratio 34:66)) → p-Tol–S(O)–CH₂–CH(OH)–CF_3 [α]D+257.6 and p-Tol–S(O)–CH₂–CH(OH)–CF_3 [α]D+199.1 → (1) separation 2)$MeSO_2Cl$, 30% NaOH TEBA 0 °C → r.t., 89%) → E-(R)-**4** [α]D+504.2; Z-(R)-**4** [α]D-191.1 (3.9.5)

The Michael addition of a variety of enolate anion derived from esters or ketones to the sulfoxides **4** was performed in a THF solution and gave the corresponding products in high yields and with high diastereomeric purity (Equation 3.9.6).[12] It should be emphasized that when the E and Z isomers of **4** are used, the products **5** formed have opposite configurations of the newly created center of chirality (C-CF_3). Some selected examples are collected in Table 3.9.1.

R^1–CO–CH₂–R^2 + **4** → ($LiNEt_2$/THF) → R^1–CO–CH(R^2)–C*H(CF_3)–CH₂–S(O)–Tol-p **5** (3.9.6)

Table 3.9.1 Reaction of **4** with Enolate Anions

	Adduct 5				
4	**R¹**	**R²**	**Yield [%]**	**de [%]**	**Abs. conf. at C-CF_3**
E-4	Ph	H	99	94	R
Z-4	Ph	H	92	>98	S
E-4	t-Bu	H	96	>98	R
E-4	OEt	CO_2Et	95	85	R
E-4	OEt	H	95	>98	R

From Yamazaki, T., Ishikawa, N., Iwatsubo, H., Kitazume, T., *J. Chem. Soc., Chem. Commun.*, 1340, 1987. With permission.

In order to enhance reactivity and to achieve successful additions of various organometallics to vinylic sulfoxides, some attempts have been made to introduce an additional electron-withdrawing group to the sulfoxide molecule. Posner et al.[13] used α-carboxyvinyl and α-methoxycarbonylvinyl sulfoxides for this purpose. As a result of the conjugate addition of dialkylcopper lithiums the adducts were obtained with a moderate diastereomeric purity (up to 65%). They were then transformed into 3-methylcarboxylic acids **6** (e.g., Equation 3.9.7).

Ph S=O n-Hex CO_2Me $\xrightarrow{Me_2CuLi}$ Me SOPh H n-Hex CO_2Me $\xrightarrow{\text{1) Al-Hg, 2) NaOH}}$ Me H C CO_2H n-Hex

(+)-(R)-**6** (ee 65%) (3.9.7)

Another approach is represented by the vinylic sulfoxides **8** and **9**, in which the conjugate addition is enhanced by departure of a good leaving group, namely the mesylate anion. They were prepared by α-lithiation of optically active vinyl sulfoxides **7**, followed by condensation with propanal. Diastereomeric alcohols thus produced in a 45:55 ratio were easily separated and converted into the corresponding mesylates **8** and **9** (Equation 3.9.8), which, due to their instability, were used *in situ* as Michael acceptors in the reaction with organocopper reagents (Equations 3.9.9 and 3.9.10). It should be pointed out that lithium dimethylcuprate which was successfully applied in the example described above, turned out to be insufficiently reactive in this case. Instead, cyanocuprates had to be used. They undergo easy conjugate addition with a concomitant departure of the mesylate anion to give the enantiomerically pure trisubstituted vinyl sulfoxides **10–13** with a high E/Z stereoselectivity. An inspection of the data accompanying Equations 3.9.9 and 3.9.10 reveals that the chiral carbon center rather then sulfinyl moiety controls the asymmetric induction (in fact a chirality transfer) at the β-carbon atom.[14]

O S Tol-p R **7** $\xrightarrow{\text{1) LDA, 2) EtCHO}}$ OH O H Et S Tol-p R / OH O Et H S Tol-p R; separation; $\xrightarrow{MsCl, Et_3N, THF}$ OMs O H Et S Tol-p R **8**; $\xrightarrow{MsCl, Et_3N, THF}$ OMs O Et H S Tol-p R **9**

(3.9.8)

OMs O H Et S Tol-p R **8** $\xrightarrow[\text{THF, -78°→RT}]{[R^2Cu]}$ O Et S Tol-p R2 H R **10** + Et O S Tol-p H R2 R **11**

(3.9.9)

R	$[R^2Cu]$	Product ratio **10:11**	Yield [%]
Ph	MeCuCNLi	6:94	81
Ph	t-BuCuCNLi	9:91	69
n-Bu	MeCuCNLi	9:91	86

[R^2Cu]
THF, -78°→RT

9 → 12 + 13 (3.9.10)

R	[R^2Cu]	Product ratio **12:13**	Yield [%]
Ph	MeCuCNMgBr	6:94	80
Ph	t-BuCuCNMgCl	6:94	71
n-Bu	MeCuCNLi	80:20	86

3.9.3 Michael Addition of Oxygen Nucleophiles

It is known that alcohols add to α,β-unsaturated sulfoxides in the presence of bases in a reversible, thermodynamically controlled process (Equation 3.9.11).[15]

MeONa MeONa

A **A:B** = 39:61 **B** (3.9.11)

However, in some special cases, particularly when conjugate addition proceeds in an intramolecular fashion, it is possible to isolate the kinetically controlled product usually formed with a very high stereoselectivity. This approach has been utilized for the synthesis of several natural or biologically active compounds. Some selected examples of this type of reaction are discussed below.

3.9.3.1 Enantioselective Synthesis of Each Enantiomer of a Sex-Pheromone of an Olive Fly: 1,7-Dioxaspiro [5,5,]Undecane 10[16] and Two Diastereomeric Talaromycines 20[17]

The key intermediate used in the synthesis, namely the cyclic vinylic sulfoxide **14**, was prepared according to Scheme 3.9.1. To accomplish cyclization, the sulfoxide **14** was treated with 5 equiv of sodium hydride in THF at room temperature. The kinetically controlled product **15** with the sulfinyl groups occupying an axial position was obtained in 77% yield as a sole stereoisomer. Its isomerization to the more stable epimer **16** proceeded with inversion of configuration at the spiro-center upon treatment with p-toluenesulfonic acid in methanol. Desulfurization of each diastereomer gave the corresponding enantiomer of **17** (Scheme 3.9.2).[16]

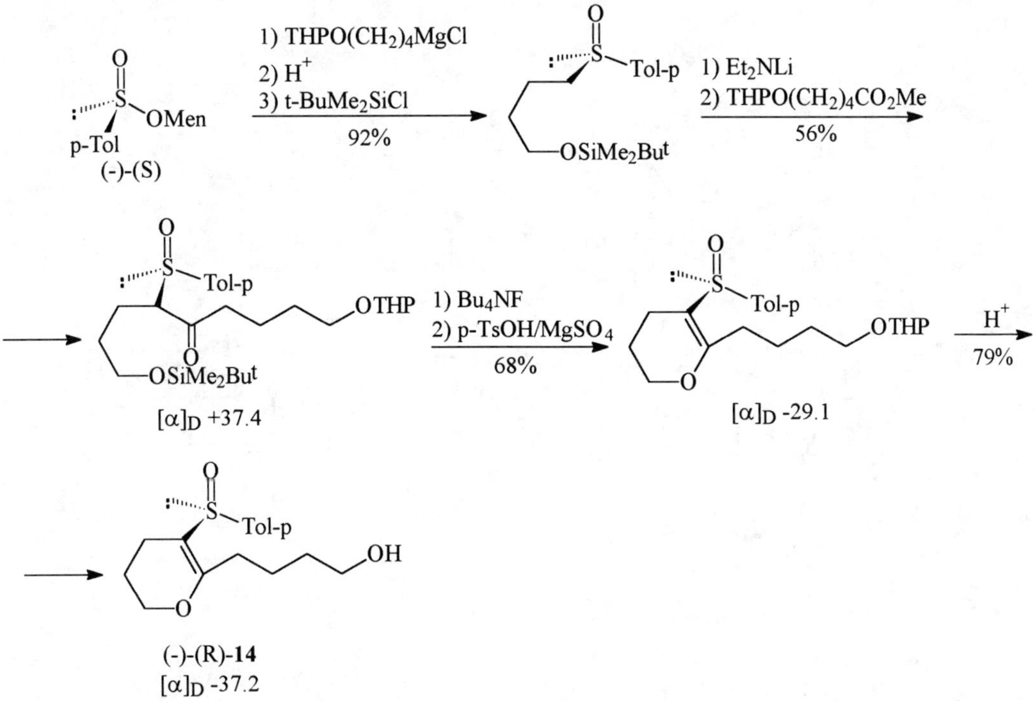

O
S
OMen
p-Tol
(-)-(S)

1) $THPO(CH_2)_4MgCl$
2) H^+
3) $t\text{-}BuMe_2SiCl$
92%

O
S
Tol-p
$OSiMe_2Bu^t$

1) Et_2NLi
2) $THPO(CH_2)_4CO_2Me$
56%

O
S
Tol-p
OTHP
O
$OSiMe_2Bu^t$
$[\alpha]_D$ +37.4

1) Bu_4NF
2) p-TsOH/$MgSO_4$
68%

O
S
Tol-p
OTHP
O
$[\alpha]_D$ -29.1

H^+
79%

O
S
Tol-p
OH
O
(-)-(R)-**14**
$[\alpha]_D$ -37.2

Scheme 3.9.1

(-)-(R)-**14**

NaH/THF
77%

O
S—Tol-p
H
O
O

p-TsOH/MeOH
epimerization

p-Tol
S
O
H
O
O

15
$[\alpha]_D$ +7

16
$[\alpha]_D$ +92.6

78% Raney Ni

83% Raney Ni

O
O

O
O

(-)-(R)-**17**
$[\alpha]_D$ -128

(+)-(S)-**17**
$[\alpha]_D$ +123

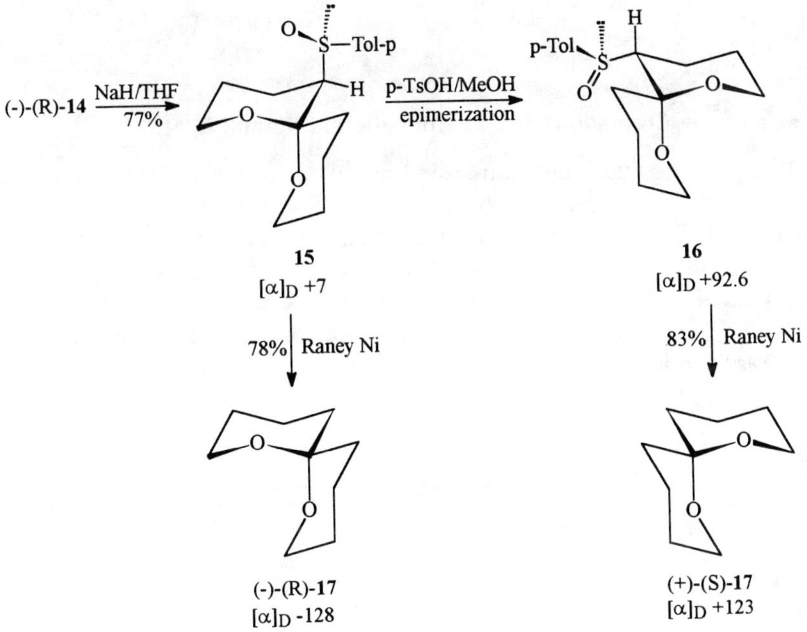

Scheme 3.9.2

In a similar way, two diastereomeric talaromycines were prepared from the cyclic vinylic sulfoxide **18**, which is a properly substituted analog of **14** (Scheme 3.9.3).[17]

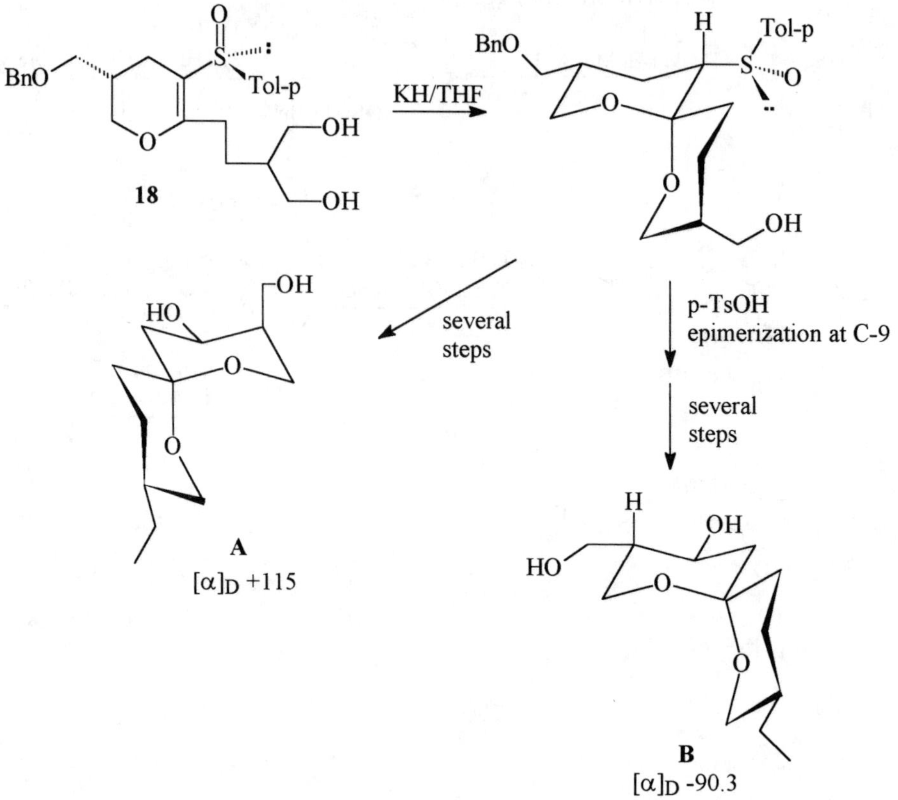

Scheme 3.9.3

3.9.3.2 Synthesis of the Chroman Ring of α-Tocopherol (Vitamin E)[18]

The strategy developed for the title synthesis is based on two crucial steps. The first one is the addition reaction of the enantiomerically pure α-lithiovinyl sulfoxide **19** to the properly protected trimethylhydroquinone aldehyde. The second one involves an intramolecular conjugate addition of the hydroxy group to the vinyl sulfoxide moiety in **21**, followed by elimination of water to produce the chromene **22**. It should be stressed that both reactions proceed with a 100% asymmetric induction giving the corresponding products **20** and **22** as sole diastereomers. The last synthetic steps are straightforward and lead to optically pure (+)-(S)-formylchroman **24** (Scheme 3.9.4).[18]

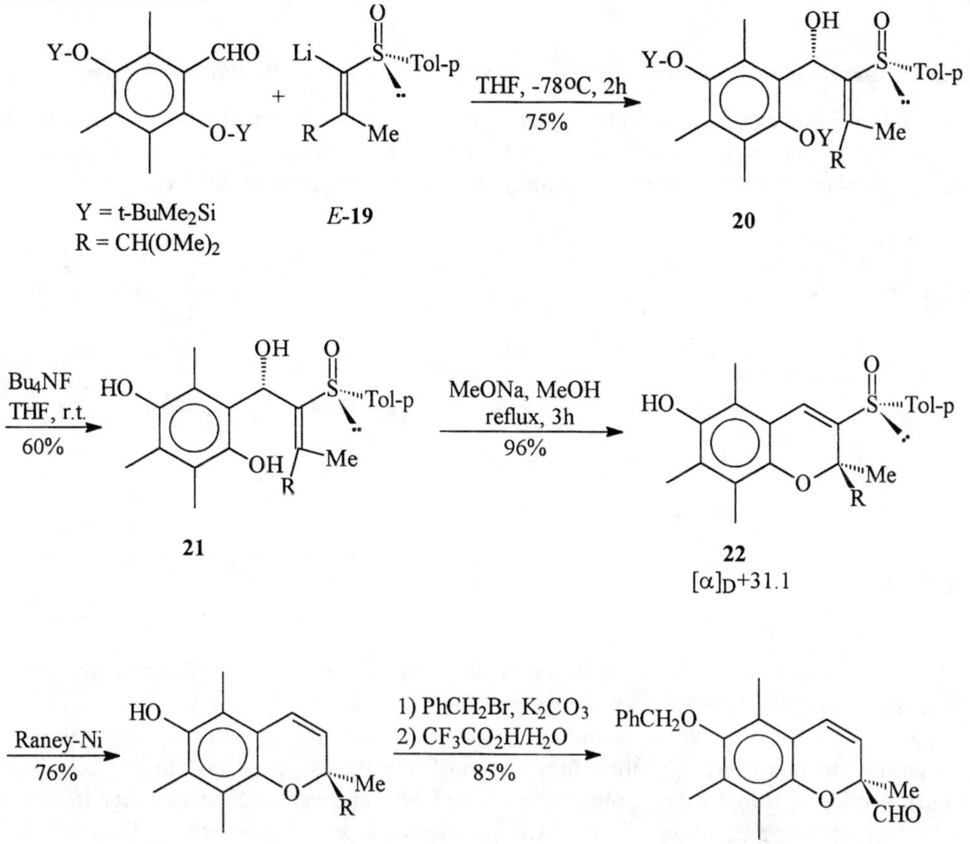

Y = t-BuMe$_2$Si
R = CH(OMe)$_2$

E-19 → THF, -78°C, 2h, 75% → 20

Bu$_4$NF, THF, r.t., 60% → 21 → MeONa, MeOH, reflux, 3h, 96% → 22 $[\alpha]_D$+31.1

Raney-Ni, 76% → 23 → 1) PhCH$_2$Br, K_2CO_3 2) CF_3CO_2H/H_2O, 85% → (S)-24 $[\alpha]_D$+12.5

Scheme 3.9.4

Another interesting point deserving a special explanation is the exclusive formation of the E diastereomer of the lithiosulfoxide **19** from the corresponding sulfoxide **27** (Equation 3.9.12) which is originally obtained as a mixture of both E and Z forms in the Horner-Wittig reaction of α-phosphoryl sulfoxide **25** and the ketone **26** (see Section 3.7). Thus, it is known that α-sulfinylalkenyl carbanions are configurationally unstable[19,20] and usually undergo isomerization to adopt the thermodynamically most stable configuration.[21] A full transformation of the anion of **27** into E-**19** can be explained in terms of such an isomerization, the driving force for it being the chelation of lithium with the two oxygen atoms of the acetal moiety[18] (compare Equation 3.10.3).

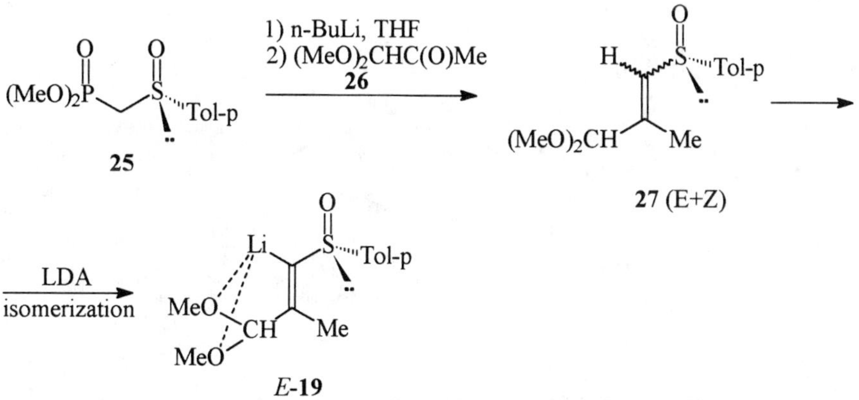

25 → 1) n-BuLi, THF 2) (MeO)$_2$CHC(O)Me **26** → **27** (E+Z) → LDA, isomerization → *E*-**19**

(3.9.12)

3.9.3.3 Synthesis of (+) and (–)-(*cis*-6-Methyltetrahydropyran-2-yl)acetic Acids[22]

The procedure involves a stereoselective formation of *cis* 1,6-disubstituted tetrahydropyrans which takes place in a thermodynamically controlled intramolecular conjugate addition of the hydroxy group to the enantiomerically pure vinylsulfinyl moiety in **28** (Equation 3.9.13).

BnO OH 28 S Tol-p NaH → BnO O S O Tol-p **29a** + BnO O S O Tol-p **30a**

BnO O S O Tol-p **29b** + BnO O S O Tol-p **30b**

29a+b : 30a+b = 38:1

(3.9.13)

The highest *cis/trans* ratio of the products obtained was found to be 38/1 and was achieved by the *trans* → *cis* isomerization involving the elimination-addition sequence. However, the two *cis* diastereomers **29a** and **29b** are formed in almost equal amounts, which proves that there is no asymmetric induction exerted by the sulfinyl group. Nevertheless, they have been easily isolated from the reaction mixture by flash column chromatography, and each of them has been transformed into the corresponding enantiomeric tetrahydropyranylacetic acid (Equations 3.9.14 and 3.9.15).[22]

29a → 1)Ac_2O/AcONa 2)$LiAlH_4$/THF, 76% → **31a** $[\alpha]_D$+23.46 → 1)MsCl/Et_3N 2)$LiAlH_4$/THF 3)H_2/Pd/C, 57% → **32a** $[\alpha]_D$+18.47 → CrO_3/acetone, 71% → HO_2C **33a** $[\alpha]_D$+41.51

(3.9.14)

29b → 76% → **31b** $[\alpha]_D$ -23.62 → 61% → **32b** $[\alpha]_D$ -18.58 → 73% → HO_2C **33b** $[\alpha]_D$ -41.92

(3.9.15)

3.9.4 Conjugate Addition of Nitrogen Nucleophiles

Intermolecular conjugate addition of amines to optically active vinyl sulfoxides was first investigated by Stirling et al.[1,23] and found to proceed with a good β-stereoselection and a poor α-stereoselection. Thus, Z-(–)-(R)-propenyl p-tolyl sulfoxide **34** gave on treatment with piperidine a quantitative yield of the adduct as an 80:20 diastereomeric mixture, from which the pure major component (+)-(R_S, S_C)-**35** was obtained by crystallization (Equation 3.9.16).

Z-(-)-(R)-**34** $[\alpha]_{546}$ -373 —MeOH, piperidine, reflux, 100%→ (+)-(R_S,S_C)-**35** $[\alpha]_{546}$+226 + (R_S,R_C)-**35**

(3.9.16)

On the other hand, addition of piperidine to (+)-(R)-a-methylvinyl p-tolyl sulfoxide **36** resulted in the formation of the adduct in 55% yield only and with de of 29%; also in this case the major diastereomer (+)-(R_S, S_C)-**37** was isolated by crystallization (Equation 3.9.17).

(+)-(R)-**36** $[\alpha]_{546}$ +153 —piperidine, EtOH, reflux, 55%→ (+)-(R_S,S_C)-**37** $[\alpha]_{546}$ +207 + (R_S,R_C)-**37**

(3.9.17)

More recently, Pyne et al.[24,25] have found out that the conjugate addition of benzylamine to isomeric (E) and (Z) chiral sulfoxides is a diastereoconvergent process and gives the same major diastereomeric adduct (Equation 3.9.18).

E-**38** / Z-**38** —$PhCH_2NH_2$, EtOH→ **39** + **40**

(3.9.18)

Substrate	Product Yield [%]	**39**:**40** ratio
(E)-**38**	53	13:87
(Z)-**38**	64	14:86

Since the reaction has been performed under kinetically controlled conditions and all possible equilibrations have been excluded, this unusual feature has been explained in terms of different reactive conformations adopted by the E and Z vinyl sulfoxides. Thus, the E isomer was assumed to adopt the *s-cis* conformation **41** and the Z isomer the *s-trans* conformation **42**, both being attacked by amine from the least-hindered π-face, i.e., *anti* to the tolyl group.

E-41 *s-cis*

Z-42 *s-trans*

In contrast to the intermolecular conjugate addition discussed above, the intramolecular addition of amines to vinyl sulfoxides proceeds in the same diastereofacial sense for E and Z isomers and hence leads to different diastereomers in each case (e.g., Equation 3.9.19).[25] For the intramolecular conjugate addition of allylically situated carbamate nitrogen to vinyl sulfoxides see Reference 26.

$[PhCH_2NEt_3]^+OH^-$, CH_2Cl_2, -40°

(E)-43 → 44 + 45

(Z)-43

Substrate	**44:45** ratio
E-**43**	91:9
Z-**43**	16:84

(3.9.19)

Intramolecular addition of amines to vinyl sulfoxides has been applied for the total synthesis of some alkaloids, and representative examples are presented below.

3.9.4.1 Total Synthesis of (+)-(R)-Carnegine 50[25,27] and (+)- and (–)-Sedamine 54[25]

(+)-(R)-Carnegine has been synthesized according to Scheme 3.9.5. The appropriately substituted vinyl sulfoxides E and Z-**47** were obtained by Horner-Wittig reaction of the aldehyde **46** and α-phosphoryl sulfoxide **25** as a mixture of diastereomers and were readily separated by column chromatography (cf. Section 3.2.4, Equation 3.2.31).

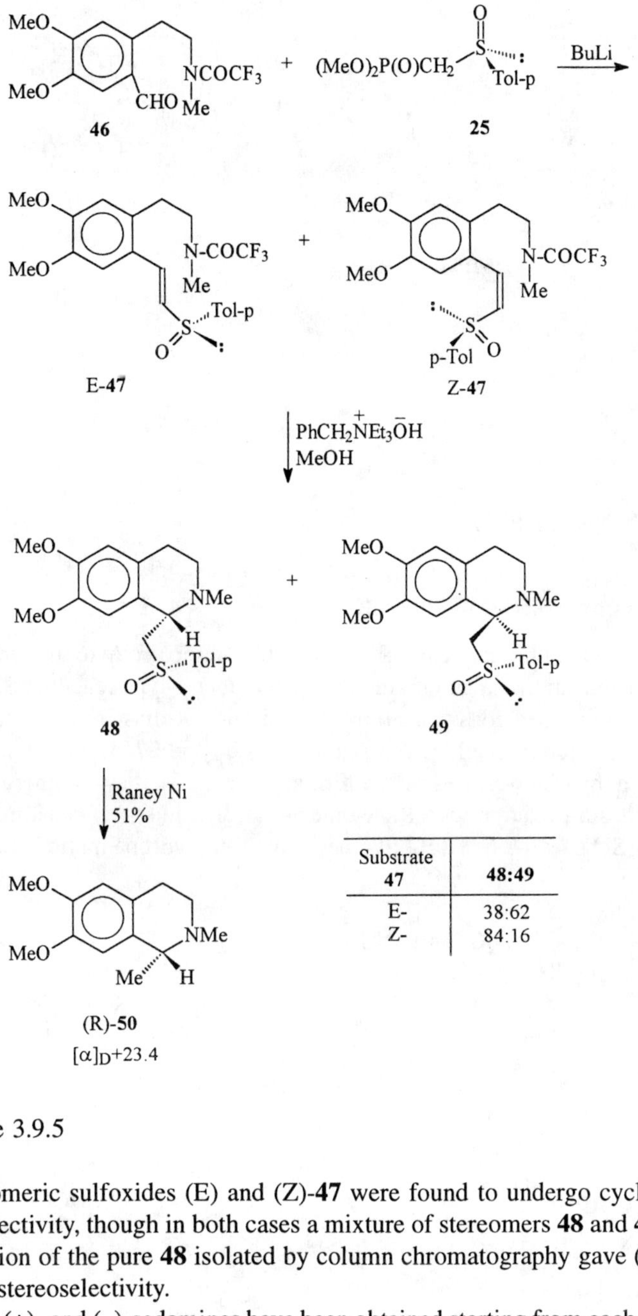

Substrate 47	48:49
E-	38:62
Z-	84:16

Scheme 3.9.5

The isomeric sulfoxides (E) and (Z)-**47** were found to undergo cyclization with the opposite diastereoselectivity, though in both cases a mixture of stereomers **48** and **49** was formed. Reductive desulfurization of the pure **48** isolated by column chromatography gave (+)-(R)-carnegine **50** with almost full stereoselectivity.

In turn, (+)- and (–)-sedamines have been obtained starting from each diastereomeric piperidine derivative **44** and **45**. Deprotonation of **44** or **45** with LDA and then quenching the resulting carbanion with benzaldehyde afforded a mixture of diastereomeric aldol products (four in the case of **44** and three of four possible in the case of **45**) which had to be separated. Reductive removal of the sulfinyl moiety in a two-step procedure (to avoid racemization occurring upon the direct treatment of the sulfoxide with Raney nickel) gave (+)-sedamine **54** and (–)-sedamine **54** (the latter shown in Equation 3.9.20).

45 → 1) LDA 2) PhCHO → **51**-*anti* + **52**-*syn* + **53**-*syn*

51:52:53 = 6:1:1

51-*anti* → 1)Al(Hg) 2)Raney Ni → (-)-**54**

[α]D -87.8 (lit. [α]D -92.5) (3.9.20)

3.9.4.2 Total Synthesis of (+)-(R)-Canadine 59[28]

The title synthesis is based on the intramolecular addition of the amino group to the vinyl sulfoxide moiety in **55** followed by cyclization via an intramolecular Pummerer-type reaction. Starting vinyl sulfoxides (E) and (Z)-**55** were used for cyclization as a mixture of diastereomers to give, after column chromatography, diastereomerically pure isoquinolines **56** and **57**. The Pummerer reaction of **56** with trifluoroacetic anhydride resulted in the formation of a 1:1 diastereomeric mixture of **58** which, after reductive desulfurization with Raney nickel, afforded (+)-(R)-canadine **59** in 81% yield and with 92% ee (Scheme 3.9.6)[28] (cf. Synthesis of Tetrahydropalmatine, Section 3.2.4, Equation 3.2.33).

1)2,3-(MeO)$_2$C$_6$H$_2$CH$_2$NH$_2$
2)LiAlH$_4$
3)(CF$_3$CO)$_2$O/Py
4)K$_2$CO$_3$; 5)CrO$_3$/Py

25, BuLi

E-55 + Z-55

PhCH$_2$NEt$_3$OH
CH$_2$Cl$_2$, -40°C

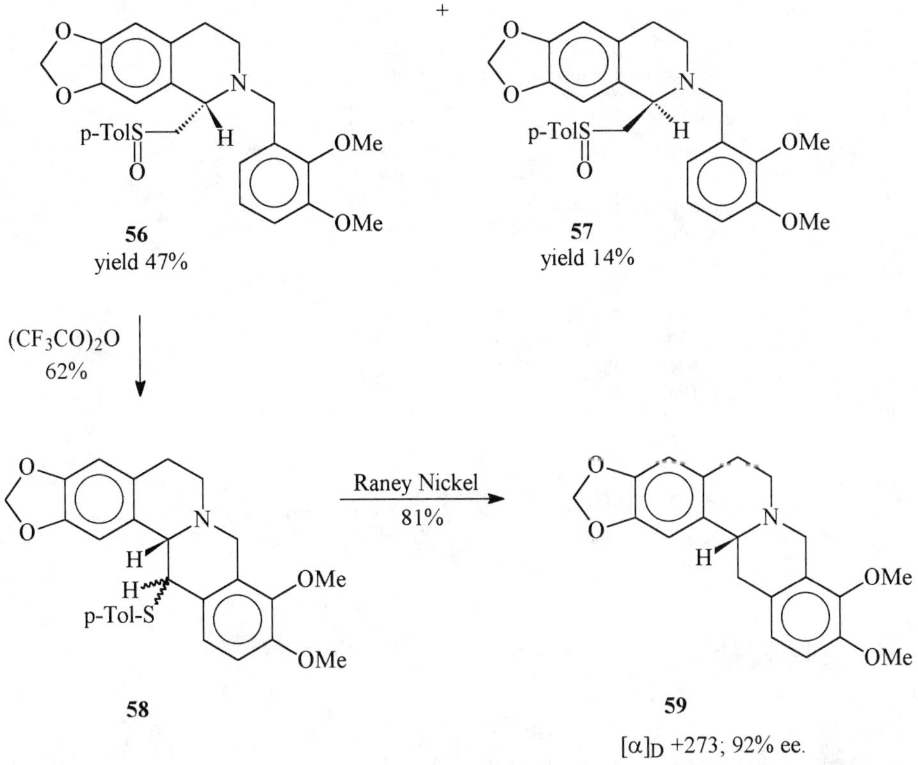

56
yield 47%

57
yield 14%

$(CF_3CO)_2O$
62%

Raney Nickel
81%

58

59

$[\alpha]_D$ +273; 92% ee.

Scheme 3.9.6

3.9.5 Conjugate Addition of Silicon Nucleophiles

Lithium bis(dimethyphenylsilyl)cuprate **61** has been found to react easily with vinyl sulfoxides to give the Michael adducts in good yield and with moderate to good diastereoselection. The procedure allows for asymmetric carbon-silicon bond formation when β-substituted vinyl sulfoxides are applied. The reaction proceeded with the same diastereofacial sense for E and Z isomers of the vinyl sulfoxide and, therefore, opposite ratios of the diastereomeric products were obtained for each isomer (Equation 3.9.21, Table 3.9.2).[29]

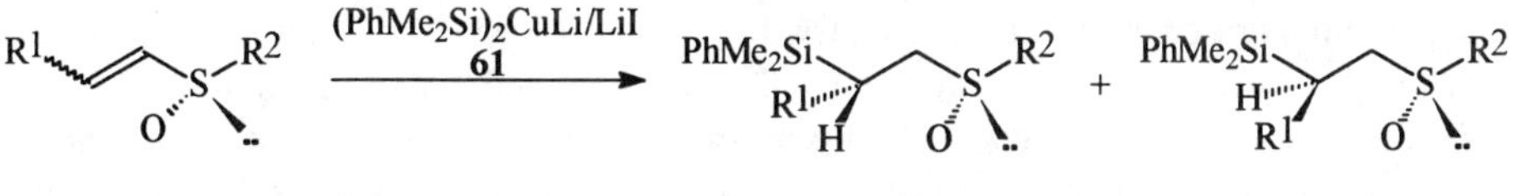

$(PhMe_2Si)_2CuLi/LiI$
61

60 62 63

(3.9.21)

Since the diastereomeric adducts **62** and **63** are easily separated from each other by column chromatography, they can be used as substrates for enantiomerically pure silanes **64** (with the center of asymmetry located on the carbon adjacent to silicon) and methylcarbinols **65** (Equation 3.9.22).

Table 3.9.2 Reaction of Vinyl Sulfoxides **60** with bis(Dimethylphenylsilyl)Cuprate **61**

Sulfoxide	R[1]	R[2]	Total yield [%]	62:63 ratio
E-**60a**	Me	Ph	48	88:12
Z-**60a**	Me	Ph	47	6:94
E-**60b**	n-Pent	Ph	82	80:20
Z-**60b**	n-Pent	Ph	73	25:75
E-**60c**	Ph	Ph	72	79:21
Z-**60c**	Ph	Ph	67	20:80
E-**60d**	n-Pent	p-Tol	67	75:25
Z-**60d**	n-Pent	p-Tol	70	20:80
E-**60e**	Ph	p-Tol	78	85:15
Z-**60e**	Ph	p-Tol	70	23:77

$PhMe_2Si$, $n\text{-}H_{11}C_5$, H, O, S, Tol-p

63d
$[\alpha]_D$ +214

MCPBA
100%

$PhMe_2Si$, $n\text{-}H_{11}C_5$, H, SO_2-Tol-p

$[\alpha]_D$ +13.7

Na-Hg
91%

$PhMe_2Si$, $n\text{-}H_{11}C_5$, H

64
$[\alpha]_D$ -14.2

1)$BF_3 \cdot 2AcOH$
2)KF/MCPBA
62%

HO, $n\text{-}H_{11}C_5$, H

65
$[\alpha]_D$ +12.1 (100% ee)

(3.9.22)

3.9.6 Diels-Alder Cycloaddition

[2+4] Cycloaddition is a reaction in which up to four new chiral centers are formed in one step. Application of optically active vinyl sulfoxides as dienophiles is a fascinating strategy, since the chiral sulfinyl auxiliary is known to exert a high asymmetric induction in the carbon-carbon bond formation. Such a strategy creates a possibility of an easy synthesis of complex products possessing several chiral centers of desired stereochemistry. In recent years this approach has been the subject of a great number of publications. Since several exhaustive and critical reviews devoted to this topic have also come out,[30-33] in this section only those examples will be presented which are of general importance or clearly and unequivocally allow an enantioselective synthesis of natural or biologically active compounds to be carried out.

The Diels-Alder reaction of an optically active vinyl sulfoxide with cyclopentadiene (a characteristic representative of dienes) can lead to four diastereomeric products which are shown in Scheme 3.9.7. To avoid possible confusions concerning the stereochemical nomenclature, the terms "endo" and "exo" used in this section refer to the relative position of the sulfinyl group.

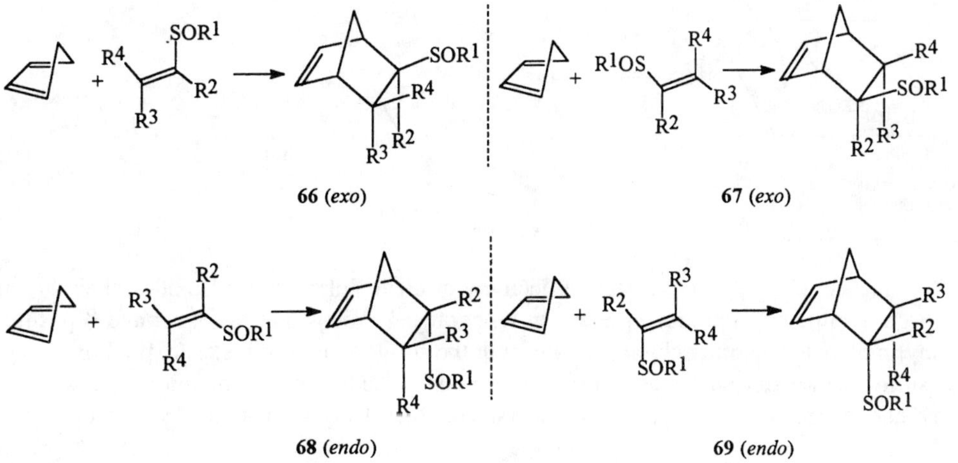

66 (*exo*) 67 (*exo*)

68 (*endo*) 69 (*endo*)

Scheme 3.9.7

It should be emphasized that while the *endo/exo* stereoselection (**66** and **67**, vs. **68** and **69**), consisting of a different orientation of the diene with respect to the sulfoxide substituents in a dienophile, is rather a consequence of steric and/or electronic interaction, the π-facial stereoselectivity (**66** vs. **67** and **68** vs. **69**) arising from the approach of the diene to a different (*'re'* or *'si'*) face of the dienophile, is only a result of the influence of the sulfinyl group chirality. Thus, the π-facial stereoselectivity can be considered a measure of the asymmetric induction exerted by the sulfinyl group.

The most simple representative of sulfinyl dienophiles, namely the unsubstituted tolyl vinyl sulfoxide **70**, proved to require drastic reaction conditions and to afford products with a low stereoselectivity (Equation 3.9.23).[33,34]

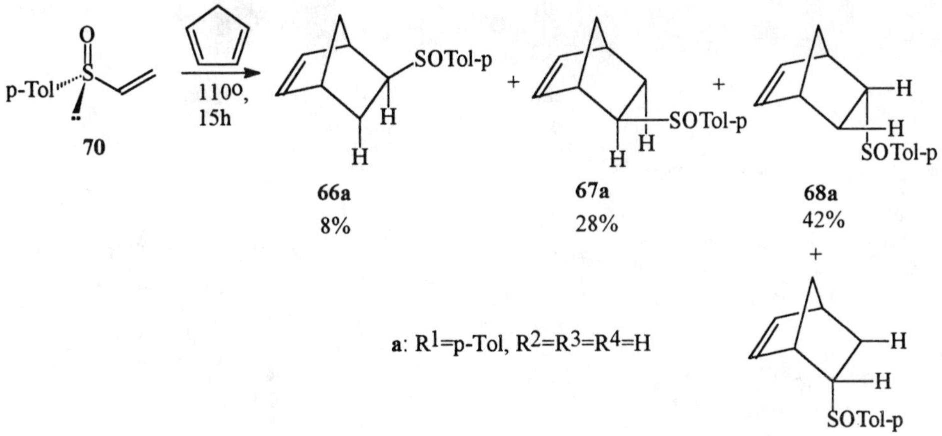

70 → (110°, 15h) **66a** 8% + **67a** 28% + **68a** 42% +

a: R^1=p-Tol, R^2=R^3=R^4=H

(3.9.23)

A poor diastereoselectivity of this reaction, which results in the formation of all possible products is explained in terms of a small energy difference between the two ground state conformations adopted by the dienophile **70** attacked by cyclopentadiene.

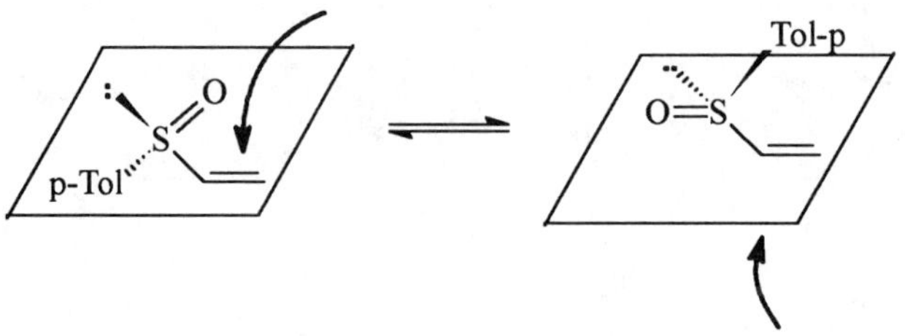
Tol-p
O
p-Tol
O=S

s-cis **70** *s-trans* **70**

To enhance reactivity and diastereoselectivity of dienophilic vinyl sulfoxides, their structures have been modified by introducing electron-withdrawing groups into the α- or/and β-position of the vinyl moiety and another chiral substituent at the sulfur atom. For instance, β-ethoxycarbonylvinyl sulfoxides **71** not only react with cyclopentadiene under milder conditions, but also give a much higher diastereoselection, both *exo/endo* and diastereofacial (Equations 3.9.24 and 3.9.25).[35,36]

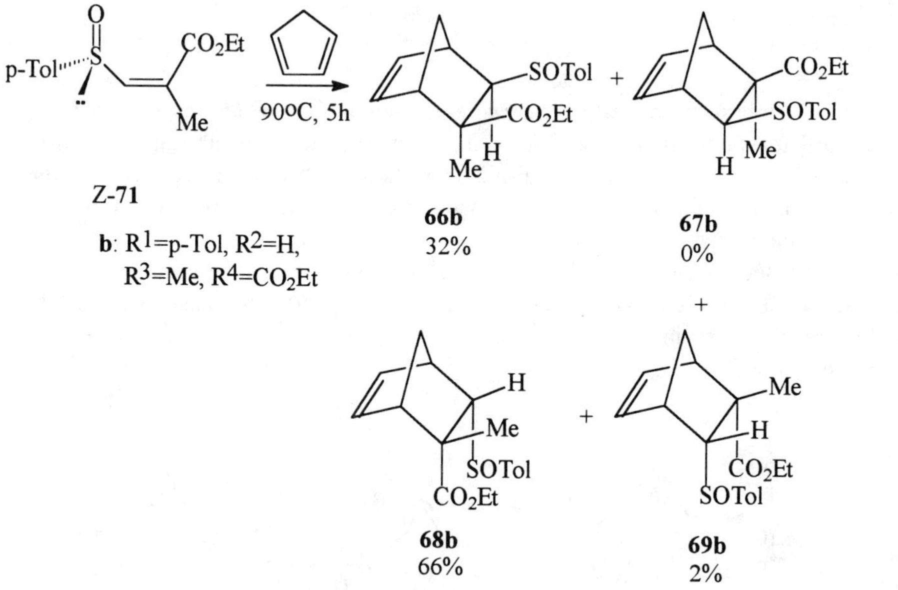
p-Tol
S
O
CO_2Et
Me
90°C, 5h
Z-71
b: R^1=p-Tol, R^2=H, R^3=Me, R^4=CO_2Et
SOTol
CO_2Et
Me
H
66b
32%
67b
0%
68b
66%
69b
2%

(3.9.24)

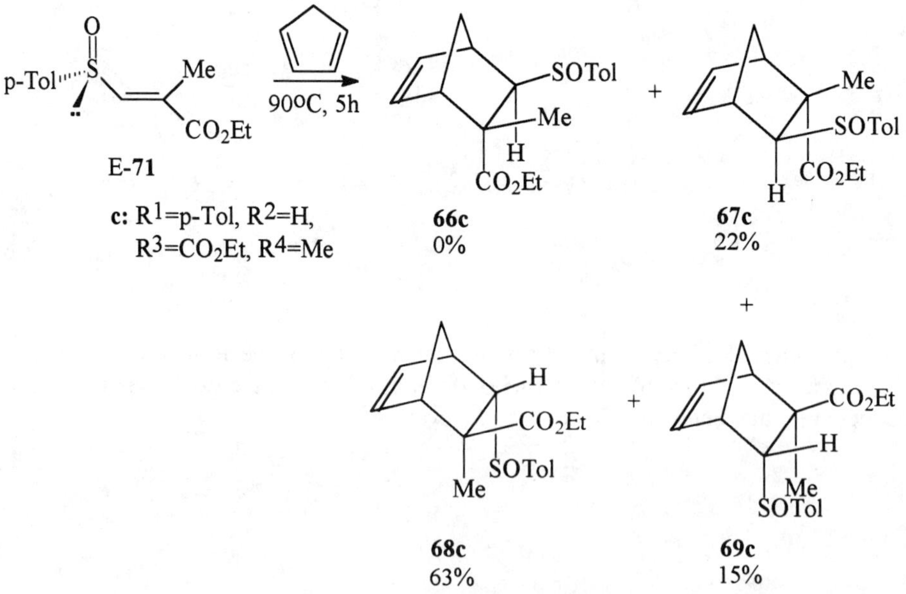
p-Tol
S
O
Me
CO_2Et
90°C, 5h
E-71
c: R^1=p-Tol, R^2=H, R^3=CO_2Et, R^4=Me
SOTol
H
66c
0%
67c
22%
68c
63%
69c
15%

(3.9.25)

A higher diastereoselection observed in this case is most reasonably explained by assuming that of the two conformers of **Z-71 A** and **B**, the latter with the sulfinyl group remote from the ethoxycarbonyl moiety is more stable because of smaller dipole-dipole repulsion. Therefore, the approach of cyclopentadiene takes place preferentially from the less hindered lone-pair side of the conformer **B**.

Z-71-A
s cis

Z-71-B
s-trans

Of much greater importance are the electron-withdrawing substituents at the α-carbon atom, since they allow for a stereodivergent course of the cycloaddition. Thus, when α-ethoxycarbonylvinyl tolyl sulfoxide **72** is reacted with cyclopentadiene in the presence of a Lewis acid as a catalyst, the ratio of diastereomers (due to the π-facial stereoselection) changes dramatically in comparison with the reaction performed without a catalyst (Equation 3.9.26).

72
d: R^1=p-Tol, R^2=CO_2Et, R^3=R^4=H

66d + **67d** + **68d** + **69d**

(3.9.26)

Catalyst	Product ratio 66d	67d	68d	69d
No	64%	11%	2%	23%
$ZnCl_2$	2%	77%	19%	2%

This result is rationalized in terms of different conformations adopted preferentially by the sulfoxide **72**. On basis of the CD spectra it has been proven that the more stable conformation of **72** is *s-cis*, with the S=O bond oriented anti to the carbonyl C=O bond (**A**). On the other hand, addition of a Lewis acid results in the formation of the chelated species (**B**) in which the molecule adopts the *s-trans* conformation with the S=O bond oriented syn to the C=O bond. In each case, cyclopentadiene approaches the dienophile from the less hindered lone pair side, leading to a different diastereomeric product[33] (cf. Section 3.10.3).

72-A
s-cis

Lewis acid

72-B
s-trans

This simple model has turned out, however, to be insufficient to explain all the experimental results obtained. Thus, a surprising effect has been observed when the cycloaddition of benzyl methyl (S)-2-(p-toluenesulfinyl)maleate **73** with cyclopentadiene was performed in the presence of various types of Lewis acid catalysts (Equation 3.9.27, Table 3.9.3).[37] An inspection of Table 3.9.3 reveals that, unexpectedly, almost the same diastereomer ratio resulted from an uncatalyzed reaction (entry 1) and when $Eu(fod)_3$ and $TiCl_4$ were used as catalysts (entries 2 and 3). This is inconsistent with the model presented above, since Lewis acids should form chelated structures and give the reversed diastereomer ratio, as does zinc bromide (entries 4 and 5).

(+)-(S)-**73** → catalyst → **66e** + **67e** + **69e**

e: R^1=p-Tol, R^2=CO_2Bn, R^3=CO_2Me, R^4=H

(3.9.27)

Table 3.9.3 Diels-Alder Reaction of (+)-(S)-**73** with Cyclopentadiene

Entry	Catalyst (equiv.)	Temp. [°C]	Time [h]	Product ratio: 66e	67e	π-Facial select. 66e/67e	69e	*exo/endo* 66e+76e / 69e	Yield [%]
1	None	r.t.	10	73	8	9.1	19	4.3	93
2	$Eu(fod)_3$ (1.2)	–20	2	66	3	22	31	2.2	100
3	$TiCl_4$ (1.2)	–78	2	83	13	6.4	4	24	84
4	$ZnBr_2$ (1.2)	–20	6	7	88	0.08	3	19	95
5	$ZnBr_2$ (1.8)	–20	5	6	91	0.07	3	32.3	96

Note: r.t. = room temperature.

From Alonso, I., Carretero, J. C., Garcia Ruano, J. L., *J. Org. Chem.*, 59, 1499, 1994. Copyright 1994 American Chemical Society. With permission.

Garcia Ruano et al.[37] accounted for these unusual results as follows. The most stable conformation of **73** in the absence of a catalyst is presumably the *s-cis* one (**A**). The europium catalyst, exhibiting a low chelating ability, is mainly associated with the sulfinyl oxygen which additionally stabilizes the *s-cis* conformer and increases the π-facial selectivity. Thus, in both cases this is the *s-cis* conformer which is approached by cyclopentadiene from the *si* face (Figure 3.9.1).

When $TiCl_4$ or $ZnBr_2$ are used as catalysts the chelated structures must be formed from the conformation **B** and act as reactive intermediates. Although at first sight they seem to be identical for each catalyst, the more accurate analysis of their spatial arrangements reveals profound differences (Figure 3.9.2).

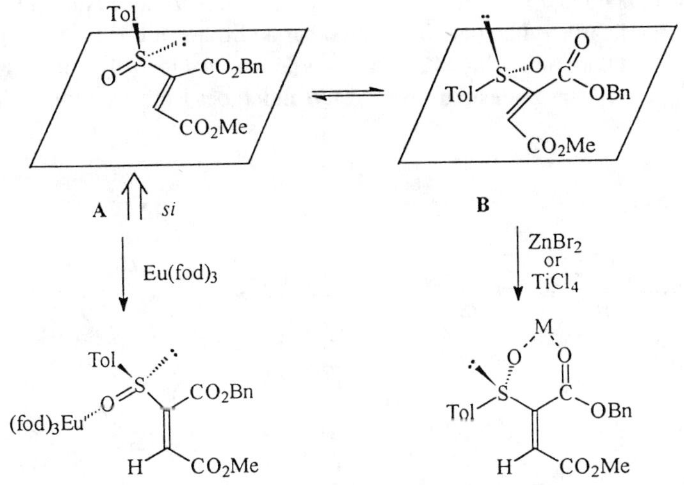

Figure 3.9.1 Reactive conformations of **73**.[37]

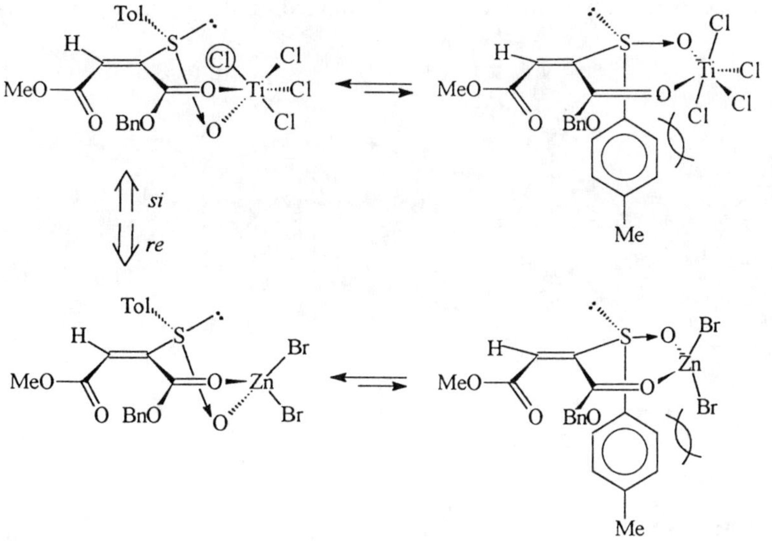

Figure 3.9.2 Spatial arrangements of substituents in chelated species.[37]

Because of the steric interactions between the p-tolyl ring and the halogens of the catalyst molecules, the conformational equilibria of the chelated species must be shifted to the left, particularly in the case of the titanium complex. The *re* face is better accessible in the approach on the $ZnBr_2$ chelate, since the *si* face is sterically and electronically more hindered by the sulfinyl oxygen. On the contrary, in the $TiCl_4$ chelate, where the octahedral substituent arrangement around the titanium atom results in directing one of the chlorine atoms towards the *re* face and hence hindering it, the approach of the diene to the *si* face is easier, which ultimately leads to the inversion of the π-facial selectivity.[37]

To enhance stereoselectivity of the Diels-Alder reactions, several other sulfinyl dienophiles have been prepared which contain additional chiral groups linked to the sulfur atom, such as 2-hydroxy-2-phenylethyl[38] and *exo*-2-hydroxy-10-bornyl. The latter one has been introduced as a substituent into several types of vinyl sulfoxides, among them α-sulfinyl maleates **74**[40] and α-sulfinyl maleimides **77**.[41]

The synthesis of **74** and **77** utilizes 10-mercaptoisoborneol **75** as a chiral reagent. Its addition to dimethyl acetylenedicarboxylate (analogous to other additions of this type)[39] produces a mixture of E and Z **76** (4:1). Oxidation with MCPBA gives almost exclusively and stereoselectively the desired vinyl sulfoxide **74** (the undesired isomer Z-**76** undergoes isomerization during the oxidation) (Equation 3.9.28).[33,40,42]

$MeOC(O)C{\equiv}CC(O)OMe$, Ph_2PMe, 93%; MCPBA

75 E+Z-76 74

(3.9.28)

In turn, addition of **75** to a series of maleimides, followed by chlorination, elimination of hydrogen chloride, and oxidation with MCPBA, affords α-sulfinylmaleimides **77** with almost complete diastereoselectivity, the absolute configuration of the secondary carbinol group of the isoborneol moiety being responsible for the newly formed sulfinyl chiral center (Equation 3.9.29).[41]

Et_3N, 89-98%; NCS, CCl_4, 88-95%; m-CPBA, 98-100%, de ~100%

75 77

I: R=Me, II: R=Ph,
III: R=Bn, IV: R=TBDMS,
V: $R=CH_2CH_2C{\equiv}CH$

(3.9.29)

Both types of substrates give with cyclopentadiene the corresponding cycloadducts in a high yield and with excellent stereoselectivity (Equations 3.9.30 and 3.9.31 and Table 3.9.4).

74 → 66f + 67f + 68f + 69f

f: R^1 = (10-hydroxyisobornyl), $R^2=R^3=CO_2Me$, $R^4=H$ (3.9.30)

Catalyst	Solvent	Temp [°C]	Product ratio **66f:67f:68f:69f**	Yield [%]
None	CH_2Cl_2	25	1:11.7:2.2:0	87
$ZnCl_2$	CH_2Cl_2	–20	15.3:0:0:1	92

77 + cyclopentadiene $\xrightarrow{CH_2Cl_2}$ 66g + 67g

g: R^1 = (10-hydroxyisobornyl), R^2-R^3 = -C(O)-N(R)-C(O)-, R^4=H (3.9.31)

Table 3.9.4 Diels-Alder Reaction of **77** with Cyclopentadiene[33,41,46]

Dienophile Symbol	R	Temp. [°C]	Catalyst	Product ratio 66g:67g	Yield [%]
77-I	Me	0	None	27:73	99
77-I	Me	0	$ZnCl_2$	94:6	95
77-II	Ph	0	None	27:73	98
77-II	Ph	0	$ZnCl_2$	90:10	97
77-III	Bn	r.t.		26:94	98
77-III	Bn	–20	$ZnCl_2$	98:9	100
77-IV	TBDMS	–80	$ZnCl_2$	99:<0.5	93
77-V	CH_2–$CH_2C{\equiv}CH$	–75	$ZnCl_2$	98:2	93

Note: r.t. = room temperature.

It should be added that in contrast to sulfinyl maleates, which are unreactive towards furan (known as an inert diene), maleimides **77** form with furan the corresponding adducts quite smoothly. However, stereoselectivity of this reaction is low, particularly in the absence of a catalyst. In the presence of $ZnCl_2$ the *exo/endo* ratio is strongly dependent on the reaction temperature, which is caused by the isomerization of the *exo*-sulfinyl adducts to the thermodynamically more stable *endo*-sulfinyl isomer. The isomerization proceeds through dissociation to substrates and recombination (Equation 3.9.32).[41]

77-III + furan ⟶ 78 + 79 + 80 + 81

R = (camphor-derived hydroxy group, OH)

Catalyst	Temp. [°C]	Time [h]	Product ratio **78:79:80:81**	Yield [%]
None	0	24	29:22:29:20	56
$ZnCl_2$	0	0.5	71:29:0:0	66
$ZnCl_2$	25	20	55:0:45:0	56

(3.9.32)

To sum up, all the examples discussed above clearly show how flexible can be the cycloaddition reaction of optically active sulfoxides with dienes, and how great, yet unexplored, the possibilities it can create. An appropriate choice of chiral auxiliaries, catalysts, and reaction conditions allows one to obtain practically each desired stereoisomer of a particular product.

Some most representative examples of application of the Diels-Alder reaction of vinyl sulfoxides in the synthesis of natural products or their direct precursors are listed below.

3.9.6.1 Enantiodivergent Synthesis of Fused Bicyclo[2.2.1]Heptane Lactones; Enantioselective Synthesis of (–)-Boschnialactone

The syntheses are based on the application of the Diels-Alder reaction shown in Equation 3.9.30. The major products **66f** and **67f**, obtained from the $ZnCl_2$ catalyzed and noncatalyzed reaction, respectively, are transformed into the title lactones as shown below (Scheme 3.9.8).[40,43]

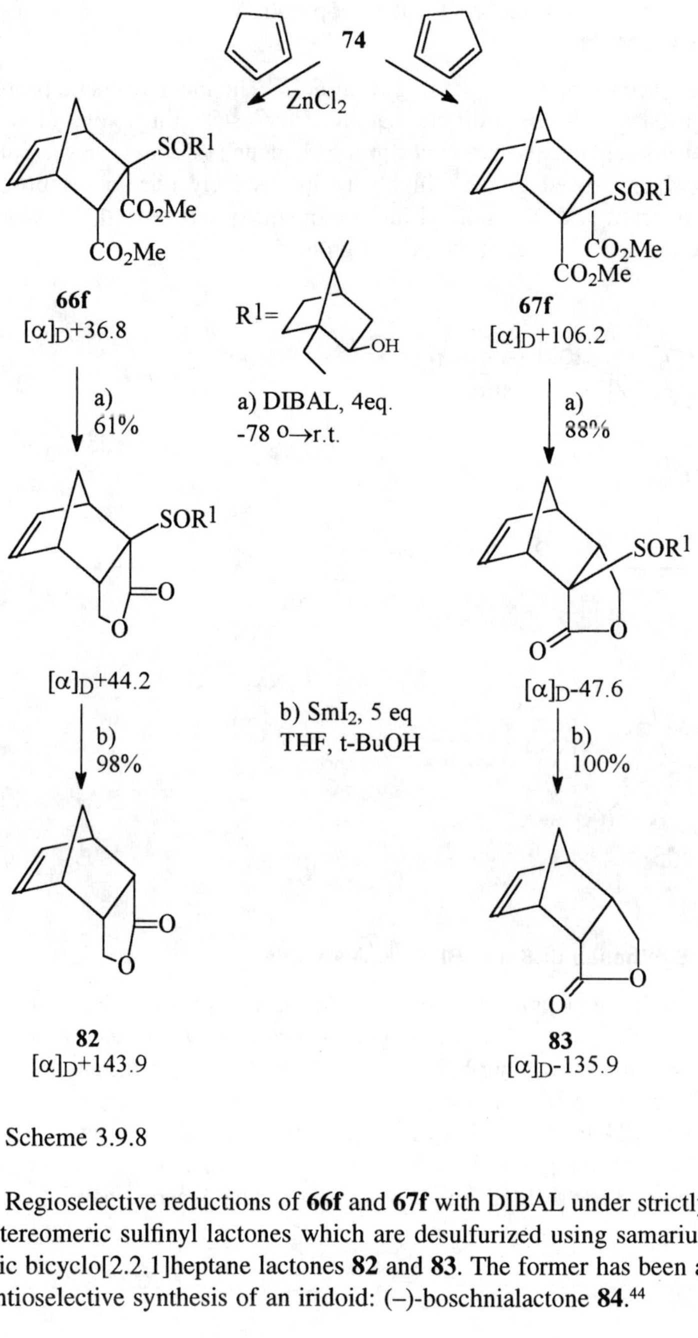

74

$ZnCl_2$

SOR^1, CO_2Me, CO_2Me

66f
$[\alpha]_D$+36.8

R^1= (OH)

67f
$[\alpha]_D$+106.2

a) 61%

a) DIBAL, 4eq.
-78 o→r.t.

a) 88%

$[\alpha]_D$+44.2

$[\alpha]_D$-47.6

b) 98%

b) SmI_2, 5 eq
THF, t-BuOH

b) 100%

82
$[\alpha]_D$+143.9

83
$[\alpha]_D$-135.9

Scheme 3.9.8

Regioselective reductions of **66f** and **67f** with DIBAL under strictly controlled conditions give diastereomeric sulfinyl lactones which are desulfurized using samarium iodide to afford enantiomeric bicyclo[2.2.1]heptane lactones **82** and **83**. The former has been applied as a substrate in the enantioselective synthesis of an iridoid: (–)-boschnialactone **84**.[44]

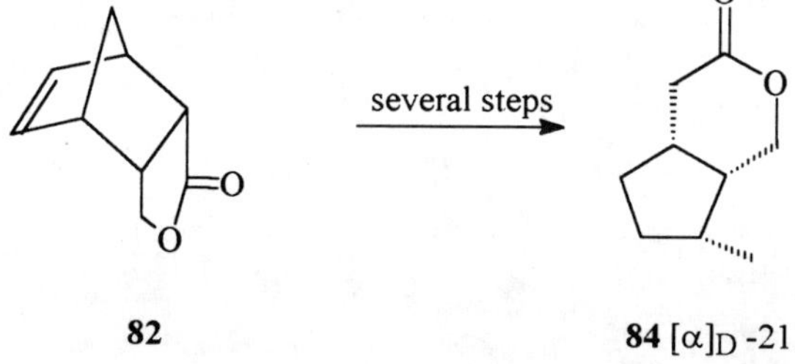

several steps

82

84 $[\alpha]_D$ -21.3

(3.9.33)

3.9.6.2 Synthesis of the Ohno's Intermediate for the Preparation of Carbocyclic Nucleosides

The Ohno's half ester **85**[45] has been synthesized starting from **66f.**[40] The main synthetic problem in this case was connected with the necessary differentiation of the ester groups, one of which should be regioselectively transformed into the carboxylic group. Although this question was solved using $AlBr_3{\cdot}Me_2S$ reagent (which proved to react highly regioselectively due to the possible coordination of the aluminum atom with the sulfinyl and the methoxy oxygen), the next steps required subsequent introduction and removal of the benzyl group.

66f —($AlBr_3$, Me_2S_2, CH_2Cl_2; 71%)→ [SOR^1, CO_2H, CO_2Me] —(BnBr, NaH, 18-crown-6; 86%)→ [SOR^1, CO_2Bn, CO_2Me] →

—(DBU; 90%)→ [CO_2Bn, CO_2Me] —(1)OsO_4, 2) p-TsOH, Me_2CO)→ [CO_2Bn, CO_2Me] →

—(Pd-C, cyclohexadiene; 80%)→ **85** [CO_2H, CO_2Me] → aristeromycin; → neplanocin A

85 $[\alpha]_D$ -29.5

(3.9.34)

3.9.6.3 Stereoselective Syntheses of Some Bicyclic Alkaloids

These syntheses are based on the application of the cycloadducts of cyclopentadiene with the properly N-substituted sulfinyl maleimides **77**. For example, the adduct of **77-V** (see Table 3.9.4) and cyclopentadiene, having the structure of **66g-V**, has been reduced to give the lactam **86** which, in turn, served as a substrate for two other amides: **87** and **88**. The latter ones have been further elaborated into tetracyclic amides **89** and **90** by N-acyliminocyclization. Finally, the retro-cycloaddition performed under the flash vacuum pyrolysis conditions led to two alkaloids: (+)-δ-coniceine **91** and (+)-laburnine **92** (Scheme 3.9.9).[46] In a similar way, elaeokanine A and C have also been synthesized.[47]

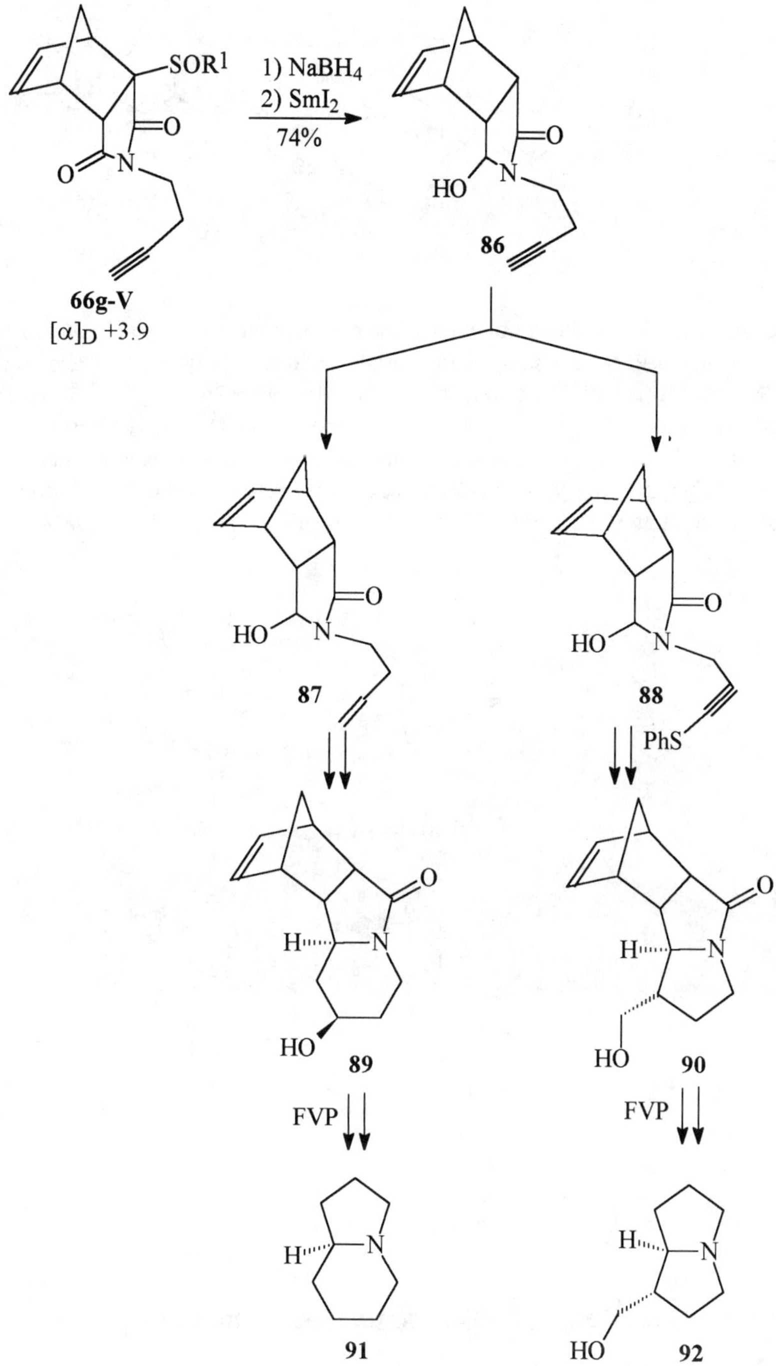

Scheme 3.9.9

3.9.6.4 Reactions of Sulfinyl Dienophiles with Dane's Diene; Synthesis of Steroid Precursors

Dane's diene, 3,4-dihydro-6-methoxy-1-vinylnaphthalene **93**, has been found to react with the sulfinyl maleate (+)-(S)-**73** in the presence of $TiCl_4$ to give only one adduct **94**. This means that the reaction proceeds with practically full *endo* selectivity and π-facial selectivity. The adduct **94** undergoes sulfenic acid elimination to give **95**, which is isolated in 75% yield.[37]

(3.9.35)

Even more interesting results have been obtained when cyclopentenone sulfoxide **96** has been used as a dienophile. In this case, from among eight possible adducts, only the two *exo*-sulfinyl compounds **97** and **98** were obtained. Moreover, it has turned out that the reaction is strongly dependent on the kind and amount of the catalyst used and conditions applied. Therefore, it can lead to the desired stereomer. The adducts obtained have been desulfurized to give both enantiomers of perhydro-cyclopenta[α]phenanthrene **99** [the (–)-one in the optically pure state]. Unfortunately, the stereochemistry of the adducts (*cis,cis* at C_8-C_{14}-C_{13}) is opposite to the usual *trans,trans* in the steroid skeleton.[48]

(3.9.36)

Catalyst (equiv.)	Solvent	T [°C]	t[h]	**97:98** ratio	Yield [%]
$AlCl_3$	Toluene	–20	26	95:5	47
$EtAlCl_2$ (1)	CH_2Cl_2	–25	4	>98:<2	93
$EtAlCl_2$ (2)	CH_2Cl_2	–25	1	29:71	90

3.9.7 Miscellaneous Reactions of α,β-Unsaturated Sulfoxides

3.9.7.1 Additive Pummerer Rearrangement

Cyclization of alkenyl sulfoxides with dichloroketene, generated *in situ* from trichloroacetyl chloride, leads to the β-substituted α,α-dichloro-γ-arylthio-γ-butyrolactones **100**. This reaction is considered as an additive Pummerer rearrangement due to the formation of a typical Pummerer intermediate in its first stage.[49]

100 (3.9.37)

When enantiomerically pure p-tolyl alkenyl sulfoxides are applied in this reaction, the products **100** are formed with almost complete enantioselectivity, which proves that this is one of the most efficient protocols for chirality transfer from sulfur to carbon atoms.[50-52] Some selected, most representative examples are shown in Equations 3.9.38 and 3.9.39.

$Cl_3CC(O)Cl$, Zn/Cu

100 (3.9.38)

R^1	R^2	R^3	**100**, Yield [%]	Ref.
3,4-$(MeO)_2C_6H_3$	Me	H	60	51
H	n-Bu	Me	70	51
n-Bu	Me	H	75	51
H	n-Pr	H	76	52
H	H	n-Pr	50	52
H	n-Bu	H	84	52
H	H	n-Bu	68	52
H	n-C_5H_{11}	H	84	52
H	H	n-C_5H_{11}	49	52
H	n-C_6H_{13}	H	82	52
H	H	n-C_6H_{13}	62	52

$Cl_3CC(O)Cl$, $Zn/Cu/Et_2O$

101

Absolute configuration:

A: STol-p, H

B: STol-p, H

(3.9.39)

Sulfoxide				101			
R^1	R^2	n	Absol. conf.	Yield [%]	$[\alpha]_D$	Absol. conf.	Ref.
H	H	1	(+)-R	70	+68.5	**A**	50
H	H	1	(–)-S	68	–68.3	**B**	50
O		1	(+)-S	60	–91.1	**A**	50
O	O	1	(+)-S	25	+13.9	**A**	50
O		0	(+)-S				53

Marino et al.[54a] applied this reaction in the synthesis of the optically active lactones **106**, direct precursors to porosin. Starting from each enantiomer of the substrate sulfoxide, they succeeded in the preparation of either enantiomer of the lactone. Moreover, on the basis of the mechanism of this reaction it was possible to predict the absolute configuration of the sulfoxide to be used in order to obtain the desired enantiomer of the product. Thus, optically active sulfoxide (E)-(S)-**102** was treated with dichloroketene to give (3R, 4S)-dichlorolactone **103** as a single stereoisomer. The latter was subsequently dechlorinated using aluminum amalgam and desulfurized with Raney nickel. It was found that the use of freshly prepared Raney nickel and the temperature not higher than 0°C were the crucial conditions of a successful outcome of this reaction and allowed obtaining the *cis*-product **105** with only ca. 3% admixture of the *trans*-isomer.* Two consecutive reactions gave the final lactone **106**.[54a]

* Tributyltin hydride was also used for dechlorination and desulfurization of the lactones **104**.[51,52] However, in some instances the latter reaction was found to proceed in a nonstereoselective manner.[51]

(E)-(S)-**102** → ($Cl_3CCOCl/Zn/Cu$, THF, 0 °C, 76%) → **103** → (Ag/Hg, THF, MeOH, H_2O, rt, 73%) → **104** → (Raney Ni, EtOH, 0 °C, 60%) → **105** → (1) LDA, THF, -78 °C; 2) $(MeO)_2CO$, 70%) → (methyl vinyl ketone, Et_3N, MeOH, 0 °C, 72%) → **106**

Ar = 3,4-$(MeO)_2C_6H_3$

(3.9.40)

The use of properly substituted sulfoxides **102** made it possible to synthesize optically pure podorhizon[52] and ring-fused γ-butyrolactones.[54b] Similarly, naturally occurring (–)-physostigmine has been synthesized via the transformation of enantiopure 2-(alkylsulfinyl)indoles into indoline butyrolactones.[54c]

3.9.7.2 Asymmetric Cyclopropanation

Reaction of the vinylic sulfoxide **107** with dimethyl sulfoxonium methylide leads to the corresponding cyclopropane **108**, formed as a 5:1 mixture of diastereomers. The reaction proceeds via an initial formation of the Michael-type cycloadduct. The cyclopropane **108a** has been isolated in a diastereomerically pure form using flash chromatography.[55]

107 → ($H_2C^{\ominus}$–$S^{\oplus}(O)Me_2$) → [intermediate] → (R_C,S_S)-**108a** + (S_C,S_S)-**108b**

(3.9.41)

Similarly, the 2-chloromethylcyclohexenyl sulfoxide **109** produces on treatment with excess allymagnesium bromide the bicyclic sulfoxide **110** as a single diastereomer.[56]

MgBr, 3 equiv., -78 °C

66%

109 **110**

(3.9.42)

3.9.7.3 Asymmetric Radical Cyclizations

Cyclization of enantiopure β-alkoxyvinyl sulfoxides **111** performed under free-radical conditions gives various tetrahydrofuran derivatives in high yields and with good to almost complete stereoselectivity. The sense of chiral induction has been found to be dependent on the configuration around the double bond in the starting sulfoxide.[57]

Bu_3SnH or TTMSS

AIBN, benzene, Δ

111 **a** : R = H
b : R = Me

112 **113**

(3.9.43)

111	Yield [%]	112:113
a (E)	81–89	70:30
a (Z)	87–94	12:88
b (E)	80–93	82:18
b (Z)	79–80	2:98

3.9.8 Cycloadditions of Sulfinyl Dienes

The Diels-Alder reaction of the enantiopure diene **114** (for its synthesis see Section 3.2, Equation 3.2.24) with maleimide gives only one diastereomer (out of four possible) of the adduct **115**.[58]

ether, 25 °C

10 h

(E)-(+)-(R)-**114** **115**

(3.9.44)

On the other hand, the reaction of enantiopure sulfinyl dienes **116**, having an additional center of asymmetry,[59-61] with methyl acrylate gives mixtures of diastereomeric products, the ratio of which depends on the conditions applied.

116a 116b 116c 117abc 118abc 119abc 120abc

(3.9.45)

Diene	Catalyst	Product isomer ratio **117:118:119:120**	Ref.
116a	None	53:32:8:7	60
	$MgBr_2$	73:24:3:0	60
	$LiClO_4$	96:4:0:0	60
116b	None	56:22:16:6	61
	$ZnCl_2$	81:13:0:0	61
	$LiClO_4$	90:6:0:0	61
116c	None	44:26:19:11	61
	$ZnCl_2$	92:2:0:0	61
	$LiClO_4$	93:2:0:0	61

3.9.9 Examples of Experimental Procedures

3.9.9.1 Synthesis of Sulfinyl Precursors of Carnegine 48 and 49 (Scheme 3.9.5)[25]

a) Synthesis of Vinyl Sulfoxides 47

To a solution of the phosphoryl sulfoxide **25** (0.015 mol) in dry THF (25 mL) is added a solution of n-BuLi in hexane (0.016 mol) at –78°C under nitrogen. After 1 h a pale yellow solution is obtained. A solution of the aldehyde **46** (0.015 mol) in dry THF (20 mL) is then added dropwise at –78°C and stirring continued for 30 min. The mixture is then warmed slowly to room temperature and stirred for 2 h. The reaction is quenched by the addition of saturated NH_4Cl solution (50 mL). The organic solvents are evaporated and the aqueous layer extracted with $CHCl_3$ (3 × 25 mL). The combined chloroform extracts are washed with water (25 mL), dried, and evaporated to give the crude product (62%). Pure diastereomers E-**47** and Z-**47** are separated by column chromatography (silica gel, 50% EtOAc/hexane):

(E)-**47**: ^{1}H NMR, δ = 2.42 (s, 3H); 3.0–3.1 (m, 2H); 3.04, 3.05, 3.10 (3xs, 3H, N(CH_3)COCF); 3.61 (m, 2H); 3.84 (s, 3H); 3.87 (s, 3H); 6.68 (s,1H); 6.71 (d, J = 15 Hz, 1H); 6.94 (s, 1H); 7.34 (d, J = 8 Hz, 1H); 7.61 (d, J = 8 Hz, 1H); 7.62 (d, J = 15 Hz, 1H).

(Z)-**47**: ^{1}H NMR, δ = 2.41 (s, 3H); 2.88 (m, 2H); 3.00, 3.02 (s); 3.50 (m, 2H); 3.91 (s,6H); 6.54 (d, J = 10 Hz, 1H); 6.73 (s, 1H); 7.18 (s, 1H); 7.32 (d, J = 8 Hz); 7.35 (d, J = 10 Hz, 1H); 7.54 (d, J = 8 Hz, 1H).

b) (+)-(1R, R_S)-N-Methyl-3,4-Dihydro-6,7-Dimethoxy-1-[(p-Tolylsulfinyl)Methyl] Isoquinoline 48

Benzyltrimethylammonium hydroxide solution (2 M in MeOH, 1.6 mL) is added to a flask under nitrogen and the methanol is evaporated under reduced pressure. CH_2Cl_2 (1 mL) is added and then evaporated to remove traces of methanol. Dry CH_2Cl_2 (5 mL) is added and the solution cooled to –40°C. A solution of (Z)-**47** (0.75 mmol) in CH_2Cl_2 (5 mL) is added and the mixture stirred for 40 h at –40°C until TLC shows no starting vinyl sulfoxide. The mixture is then extracted with AcOEt and water; the AcOEt layer is dried and evaporated. Preparative TLC (50% MeOH/AcOEt) gives **48** (88%), mp 70–72°C, $[\alpha]_D^{23}$ +206 (c 1.1, $CHCl_3$) and **49**(14%), oil.

3.9.9.2 General Procedure for the Reaction of α,β-Unsaturated Sulfoxides 60 with Lithium Bis(Dimethylphenylsilyl)Cuprate 61 (Equation 3.9.21)[29]

Dimethylphenylsilyllithium (0.8 M in THF, 30 mL) is added under nitrogen to a suspension of copper (I) iodide (2.28 g, 12 mmol) in THF (10 mL) at 0°C and stirred for 20 min. Then the solution is cooled to –78°C and HMPT (2.15 g, 12 mmol) is added and stirred for 15 min. The sulfoxide **60** (10 mmol) in THF (20 mL) is added to the solution and quenched at –78°C immediately for **60a** and **60b**, after 5 min for **60d**, or after 15 min for **60c** with aqueous NH_4Cl solution. The solution is warmed to room temperature and poured into saturated NH_4Cl solution. The reaction mixture is extracted with ether, washed with NH_4Cl solution and brine, dried ($MgSO_4$) and concentrated *in vacuo*. The residue is chromatographed on silica gel (hexane/AcOEt, 3:1) to give pure adducts **62** and **63**.

3.9.9.3 (S)-1-Benzyl 4-Methyl 2-p-Tolylsulfinylmaleate 73[37]

To a solution of (+)-(R)-benzyl p-tolylsulfinylacetate (3.0 g, 10.4 mmol, 1 equiv) in DMF (52 mL) cooled to 0°C are added, sequentially, glyoxalic acid monohydrate (2.87 g, 31.2 mmol, 3 equiv), Et_3N (3.15 g, 31.2 mmol, 3 equiv), and pyrrolidine (0.26 g, 3.64 mmol, 0.35 equiv). The mixture is stirred at 0°C for 8 h, and then 1% HCl is added to pH = 1. The solution is extracted with Et_2O (3 × 50 mL). The combined ether phases are washed with water (25 mL), dried ($MgSO_4$), and concentrated. The residue is dissolved in dry DMF (52 mL), and $NaHCO_3$ (2.62 g, 31.2 mmol, 3 equiv), iodomethane (14.76 g, 104 mmol, 10 equiv), and 4-A molecular sieves (1 g) are added. The reaction is kept, under argon, at room temperature for 4 h. The mixture is treated with 29% NH_4Cl (40 mL) and extracted with Et_2O (3 × 50 mL). The combined organic layers are washed with water (25 mL), dried ($MgSO_4$), and concentrated. The residue is purified by chromatography (CH_2Cl_2 and CH_2Cl_2-Et_2O [50:1]. Yield: 2.47 g (67%). mp 66–69°C, $[\alpha]_D^{20}$ = 158.6 (c 1, $CHCl_3$), ee ≥96% [by using Yb(hfc)$_3$ as a chiral shift reagent].

3.9.9.4 Diels-Alder Reaction of 73 with Cyclopentadiene Catalyzed by $ZnBr_2$ (Equation 3.9.27)[37]

A solution of dienophile **73** (41.6 mg, 0.12 mmol, 1.0 equiv) in 0.4 mL of CH_2Cl_2 is added, under argon atmosphere, to a suspension of $ZnBr_2$ (31.4 mg, 0.14 mmol, 1.2 equiv) in 0.2 mL of CH_2Cl_2 at –20°C. The mixture is stirred for 10 min and 60 μL (0.72 mmol, 6 equiv) of cyclopentadiene are added. Stirring is continued for 6 h. Then, 10% $NaHCO_3$ (5 mL) is added. The organic layer is separated and the aqueous layer is extracted with CH_2Cl_2 (2 × 10 mL). The combined organic layers are washed with water (2 mL), dried ($MgSO_4$), and concentrated *in vacuo*. The mixture of adducts is analyzed by ^1H NMR and purified by flash chromatography (hexane/ethyl acetate 4:1). A 41.5 mg (84%) portion of *exo* adducts **66e** and **67e** and 4.3 mg (9%) of *endo* adduct **69e** are obtained (93% overall yield).

3.9.9.5 Synthesis of (R_S)-2{[(1S,2R,4R)-2-Hydroxy-7,7-dimethylbicyclo-[2.2.1]heptan-1-yl]methylsulfinyl}maleate 74 (Equation 3.9.28)[40]

To a solution of a mixture of (E + Z)-**76** (442 mg, 1.3 mmol) in dry CH_2Cl_2 (10 mL) at 0°C is added dropwise a solution of MCPBA (80% purity, 305 mg, 1.4 mmol) in dry CH_2Cl_2 (10 mL). The mixture is stirred for 3 h at 0°C and then washed successively with diluted sodium thiosulfate, 20% aqueous NaOH, and brine. After being dried, the organic layer is concentrated under reduced pressure. The residual oil is chromatographed on silica gel (hexane/AcOEt 3:1) to give compound **74** in 89% yield: oil, $[\alpha]_D^{25}$ +32.8 (c 3.1, $CHCl_3$).

3.9.9.6 Diels-Alder Reaction of Compound 74 with Cyclopentadiene in the Presence of $ZnCl_2$ (Equation 3.9.30)[40]

To a suspension of compound **74** (2.3 g, 6.7 mmol) and $ZnCl_2$ (1.36 g, 10 mmol) in dry CH_2Cl_2 (50 mL) at –20°C is added freshly distilled cyclopentadiene (5.5. mL, 67 mmol). After being stirred at this temperature for 2 h, the mixture is poured onto cold 1M HCl. The aq. layer is extracted with CH_2Cl_2 (3 × 20 mL). The combined organic phase is washed with brine, dried, and evaporated under reduced pressure. The residual oil is chromatographed on silica gel. Elution with hexane gives cyclopentadiene dimer. Elution with hexane/AcOEt (2:1 → 1:2) gives the adduct **66f** (2.359 g, 86%) and **69f** (0.153 g, 5.6%). Compound **66f**: mp 126–127°C; $[\alpha]_D^{25}$ +36.8 (c 1.2, $CHCl_3$). Compound **69f**: mp 159–160°C; $[\alpha]_D^{25}$ –54.7 (c 0.13, $CHCl_3$).

3.9.9.7 Asymmetric Synthesis of the Lactone 105 (Equation 3.9.40)[54a]

a) 3R,4S)-2,2-Dichloro-4-(3,4-Dimethoxyphenyl)-3-Methyl-4-p-Tolylthiobutyrolactone 103

A solution of trichloroacetyl chloride (10.4 g, 57.0 mmol) in anhydrous THF (10 mL) is rapidly added to a suspension of vinyl sulfoxide **102** (3.6 g, 11.4 mmol) and zinc-copper couple (14.9 g, 228.0 mmol) in dry THF (390 mL) at 0°C. The mixture is stirred at 0°C for 10 min, and then filtered through a pad of Celite into ice-cold saturated $NaHCO_3$ solution (100 mL). The zinc is washed with ether (3 × 20 mL) and the combined filtrate and washings are placed in a separatory funnel. The organic layer is decanted and the aqueous layer extracted with ether (3 × 20 mL). The combined organic extract is washed with saturated brine (50 mL) and then dried ($MgSO_4$). The solvent is removed on a rotary evaporator and the residue chromatographed, eluting with 10% EtOAc/hexane to afford dichlorolactone **103** as a viscous orange oil; yield: 3.7 g (76%); $[\alpha]_D^{25}$ –73.6°C (c 3.79, $CHCl_3$).

b) (3S,4S)-(3,4-Dimethoxyphenyl)-3-Methyl-4-p-Tolyl-Thio-γ-Butyrolactone 104

A solution of dichlorolactone **103** (3.6 g, 8.4 mmol) in THF (50 mL) is added to a stirred suspension of aluminum amalgam (prepared from 5.0 g of 20-mesh aluminum pellets according to the procedure given in Reference 62) in THF (60 mL), followed by a 1:1 mixture of $MeOH/H_2O$ (50 mL). The suspension is stirred at room temperature for 24 h, and then filtered through a pad of Celite. The amalgam is rinsed with ether (3 × 20 mL) and the combined filtrate and washing are dried ($MgSO_4$). The solvent is removed on a rotary evaporator and the residue chromatographed, eluting with 20% EtOAc/hexane to afford lactone **104** as a colorless oil; yield: 2.2 g (73%); $[\alpha]_D^{25}$ –75.0°C (c 2.0, $CHCl_3$).

c) (3S,4R)-4-(3,4-Dimethoxyphenyl)-3-Methyl-γ-Butyrolactone 105

A solution of lactone **104** (2.1 g, 5.8 mmol) in absolute EtOH (25 mL) is added to a stirred suspension of Raney nickel (ca. 30.0 g, activated as W-5 according to the procedure given in Reference 63) in absolute EtOH (50 mL) at 0°C. The mixture is stirred at 0°C until TLC (2% ether/CH_2Cl_2) indicates complete consumption of the starting material (ca. 2-4 h). The ethanolic solution is then decanted and the Raney nickel washed with CH_2Cl_2 (3 × 20 mL). The combined organic extract is filtered through a Celite pad and concentrated *in vacuo*. The residue is chromatographed eluting with 2% ether/CH_2Cl_2 to afford **105** as a colorless oil; 822 mg (60%); $[\alpha]_D^{25}$ +31.1°C (c 3.21, $CHCl_3$).

REFERENCES (3.9)

1. Abbott, D. J., Colonna, S., Stirling, C. J. M., J. *Chem. Soc. D, Chem. Commun.*, 471, 1971.
2. Tsuchihashi, G., Mitamura, S., Inoue, S., Ogura, K., *Tetrahedron Lett.*, 323, 1973.
3. Fujiata, M., Ishida, M., Manako, K., Sato, K., Ogura, K., *Tetrahedron Lett.*, **34**, 645, 1993.
4. Kahn, S., Hehre, W. J., *J. Am. Chem. Soc.*, **108**, 7399, 1986.
5. Ogura, K., Tomori, H., Fujita, M., *Chem. Lett.*, 1407, 1991.
6. Tsuchihashi, G., Mitamura, S., Ogura, K., *Tetrahedron Lett.*, 455, 1974.
7. Tanikaga, R., Sugihara, H., Tanaka, K., Kaji, A., *Synthesis*, 299, 1977.
8. Ono, N., Miyake, H., Kamimura, A., Tsukui, N., Kaji, A., *Tetrahedron Lett.*, **23**, 2957, 1982.
9. Bruhn, J., Heimgartner, H., Schmidt, H., *Helv. Chim. Acta*, **62**, 2630, 1979.
10. Sugihara, H., Tanikaga, T., Tanaka, K., Kaji, A., *Bull. Chem. Soc. Jpn.*, **51**, 655, 1978.
11. Tsuchihashi, G., Mitamura, S., Ogura, K., *Tetrahedron Lett.*, 855, 1976.
12. Yamazaki, T., Ishikawa, N., Iwatsubo, H., Kitazume, T., *J. Chem. Soc., Chem. Commun.*, 1340, 1987.
13. Posner, G. H., Mallamo, J. P., Miura, K., *J. Am. Chem. Soc.*, **103**, 2886, 1981.
14. Marino, G. H., Viso, A., Fernandez de la Pradilla, R., Fernandez, P., *J. Org. Chem.*, **56**, 1349, 1991.
15. Tsuchihashi, G., Mitamura, S., Ogura, K., *Tetrahedron Lett.*, 2469, 1973.
16. Iwata, C., Fujita, M., Hattori, K., Uchide, S., Imanishi, T., *Tetrahedron Lett.*, **26**, 2221, 1985.
17. Iwata, C., Fujita, M., Moritani, Y., Hattori, K., Imanishi, T., *Tetrahedron Lett.*, **28**, 3135, 1987.
18. Solladie, G., Moine, G., *J. Am. Chem. Soc.*, **106**, 6097, 1984.
19. Posner, G. H., Tang, P. W., Mallamo, J. P., *Tetrahedron Lett.*, 3995, 1978; Posner, G. H., in *Asymmetric Synthesis*, vol. 2, Morrison, J. D., ed., Academic Press, New York, 1983, 225.
20. Okamura, H., Mitsushira, Y., Miura, M., Takei, H., *Chem. Lett.*, 517, 1978.
21. Fawcett, J., House, S., Jenkins, P. R., Lawrence, N. J., Russell, D. R., *J. Chem. Soc., Perkin Trans. 1*, 67, 1993.
22. Mandai, T., Ueda, M., Kashiwagi, K., Kawada, M., Tsuji, J., *Tetrahedron Lett.*, **34**, 111, 1993.
23. Abbott, D. J., Colonna, S., Stirling, C. J. M., *J. Chem. Soc., Perkin Trans. 1*, 492, 1976.
24. Pyne, S. G., Griffith, R., Edwards, M., *Tetrahedron Lett.*, **29**, 2089, 1988.
25. Pyne, S. G., Bloem, P., Chapman, S. L., Dixon, C. E., Griffith, R., *J. Org. Chem.*, **55**, 1086, 1990.
26. Hirama, M., Hioki, H., Ito, S., Kabuto, C., *Tetrahedron Lett.*, **29**, 3121, 1988.
27. Pyne, S. G., Chapman, S. L., *J. Chem. Soc., Chem. Commun.*, 1688, 1986.
28. Pyne, S. G., *Tetrahedron Lett.*, **28**, 4737, 1987.
29. Takaki, K., Maeda, T., Ishikawa, M., *J. Org. Chem.*, **54**, 58, 1989.
30. De Lucchi, O., Pasquato, L., *Tetrahedron*, **44**, 6755, 1988.
31. Koizumi, T., *Phosphorus, Sulfur Silicon*, **58/59**, 111, 1991.
32. Arai, Y., Koizumi, T., *Rev. Heteroatom Chem.*, **6**, 202, 1992.

33. Arai, Y., Koizumi, T., *Sulfur Reports*, **15**, 41, 1993.
34. Maignan, C., Raphael, R. A., *Tetrahedron*, **39**, 3245, 1983.
35. Koizumi, T., Hakamada, I., Yoshii, E., *Tetrahedron Lett.*, **25**, 87, 1984.
36. Arai, Y., Kuwayama, S., Takeuchi, Y., Koizumi, T., *Tetrahedron Lett.*, **26**, 6205, 1985.
37. Alonso, I., Carretero, J. C., Garcia Ruano, J. L., *J. Org. Chem.*, **59**, 1499, 1994.
38. De Lucchi, O., Buso, M., Modena, G., *Tetrahedron Lett.*, **28**, 107, 1987.
39. De Lucchi, O., Lucchini, V., Marchioro, C., Valle, G., Modena, G., *J. Org. Chem.*, **51**, 1457, 1986; corr. *J. Org. Chem.*, **54**, 3245, 1989.
40. Arai, Y., Hayashi, K., Matsui, M., Koizumi, T., Shiro, M., Kuriyama, K., *J. Chem. Soc., Perkin Trans. 1*, 1709, 1991.
41. Arai, Y., Matsui, M., Koizumi, T., Shiro. M., *J. Org. Chem.*, **56**, 1983, 1991.
42. Arai, Y., Hayashi, K., Koizumi, T., Shiro, M., Kuriyama, K., *Tetrahedron Lett.*, **29**, 6143, 1988.
43. Arai, Y., Matsui, M., Koizumi, T., *J. Chem. Soc., Perkin Trans. 1*, 1233, 1990.
44. Arai, Y., Kawanami, S., Koizumi, T., *Chem. Lett.*, 1585, 1990.
45. Arita, M., Adachi, K., Ito, Y., Sawai, H., Ohno, M., *J. Am. Chem. Soc.*, **105**, 4049, 1983.
46. Arai, Y., Kontani, T., Koizumi, T., *Chem. Lett.*, 2135, 1991.
47. Arai, Y., Kontani, T., Koizumi, T., *Tetrahedron: Asymmetry*, **3**, 535, 1992.
48. Alonso, I., Carretero, J. C., Garcia Ruano, J. L., Martin Cabrejas, L. M., Lopez-Solera, I., Raithby, P. R., *Tetrahedron Lett.*, **35**, 9461, 1994.
49. Marino, J. P., Neiser, M., *J. Am. Chem. Soc.*, **103**, 7687, 1981.
50. Marino, J. P., Perez, A. D., *J. Am. Chem. Soc.*, **106**, 7643, 1984.
51. Marino, J. P., de la Pradilla, R. F., *Tetrahedron Lett.*, **26**, 5481, 1985.
52. Kosugi, H., Tagami, K., Takahashi, A., Kanna, H., Uda, H., *J. Chem. Soc., Perkin Trans. 1*, 935, 1989.
53. Posner, G. H., Asirvatham, E., Ali, S., *J. Chem. Soc., Chem. Commun.*, 542, 1985.
54. a) Marino, J. P., de la Pradilla, R. F., Laborde, E., *Synthesis*, 1088, 1987;
b) Marino, J. P., Laborde, E., Paley, R. S., *J. Am. Chem. Soc.*, **110**, 966, 1988;
c) Marino, J. P., Bogdan, S., Kimura, K., *J. Am. Chem. Soc.*, **114**, 5566, 1992.
55. Hamdouchi, C., *Tetrahedron Lett.*, **33**, 1701, 1992.
56. Imanishi, T., Ohra, T., Sugiyama, K., Ueda, Y., Takemoto, Y., Ywata, C., *J. Chem. Soc., Chem. Commun.*, 269, 1992.
57. Zahouily, M., Journet, M., Malacria, M., *Synlett*, 366, 1994.
58. Gosselin, P., Bonfand, E., Hayes, P., Retoux, R., Maignan, C., *Tetrahedron: Asymmetry*, **5**, 781, 1994.
59. Aversa, M. C., Bonaccorsi, P., Gianetto, Jafari, S. M. A., Jones, D. N., *Tetrahedron: Asymmetry*, **3**, 701, 1992.
60. Adams, H., Jones, D. N., Aversa, M. C., Bonaccorsi, P., Giannetto, P., *Tetrahedron Lett.*, **34**, 6481, 1993.
61. Aversa, M. C., Bonaccorsi, P., Gianetto, P., Jones, D. N., *Tetrahedron: Asymmetry*, **5**, 805, 1994.
62. Fieser, L. F., Fieser, M., *Reagents for Organic Synthesis*, vol. 1, J. Wiley & Sons, New York, 1967, p. 20.
63. Adkins, H., Billica, H. R., *J. Am. Chem. Soc.*, **70**, 695, 1948.

3.10 CYCLOALKENONE SULFOXIDES

Simple vinyl sulfoxides were for some time considered as being insufficiently electrophilic to undergo successful conjugate nucleophilic addition.[1,8,9] Although this opinion has ultimately turned out to be not fully true, since a number of papers devoted to this reaction have come out,[2] it resulted in the development of new types of the sulfinyl group containing Michael acceptors in which electrophilicity is enhanced by the presence of other electron-withdrawing groups (cf. Section 3.9.2).

Posner et al. were the first to develop new classes of these reagents, namely enantiomerically pure 2-sulfinyl-2-cycloalkenones **4** and 2-sulfinyl-2-alkenolides **8**. These reagents are successfully applied as Michael acceptors in a highly stereocontrolled conjugate addition in which the sulfinyl group plays the role of a chiral auxiliary. Recently, they have also been used as substrates in asymmetric radical reactions.

3.10.1 Synthesis of 2-Sulfinyl-2-Cycloalkenones

2-Sulfinyl-2-cycloalkenones **4** are synthesized in a stereospecific reaction of 2-metallated 2-cycloalkenone ketals **1** with diastereomerically pure menthyl arenesulfinates **2** which serve here as a source of chirality (Equation 3.10.1). It should be noted that the nucleophilic substitution by **1** at the sulfur atom in **2** proceeds with complete inversion of configuration at sulfur. 2-Sulfinyl-2-cycloalkenones thus obtained are listed in Table 3.10.1, and the appropriate experimental procedure is given at the end of this section.

Br– (ketal) $\xrightarrow{\text{BuLi}}$ Li– (ketal) **1** $\xrightarrow{\text{Ar–S(O)–O-Men, (-)-(S)-2}}$ **3** $\xrightarrow[\text{acetone}]{\text{CuSO}_4\text{ anh.}}$ **4**

(3.10.1)

The cycloalkenone sulfoxides **4** are crystalline stable compounds which can be stored in a dessicator at 0°C for more than 1 year without evidence of decomposition.

3.10.2 Synthesis of 2-Sulfinyl-2-Alkenolides

Two general methods of the preparation of enantiomerically pure 2-sulfinyl-2-alkenolides **8** have been developed so far.

The first one (method A), described by Posner et al.[5,6] and Posner and Hamill,[12] consists in the lithiation of the open-chain substrates **5** followed by their reaction with diastereomerically pure menthyl arenesulfinates **2**, which also in this case serve as a source of the chiral sulfinyl grouping. It should be pointed out that the use of **5** and not the proper alkenolides is necessary, since the direct lithiation at the α-position in alkenolides cannot be achieved by any method. The ω-hydroxyalkenyl sulfoxides **6** obtained are treated with methyllithium to generate an α-carbanion and subsequently with carbon dioxide to give the α-carboxy-ω-hydroxyalkenyl sulfoxides **7**. The latter undergo lactonization to give the desired products **8** (Equation 3.10.2, Table 3.10.2).

Table 3.10.1 Synthesis of 2-Sulfinyl-2-Cycloalkenones **4**

	2			4				
Entry	**Ar**	**Absol. config. at sulfur**	**n**	**Yield [%]**	**Absol. config. at sulfur**	**[α]$_D$ (solvent)**	**m.p. [°C]**	**Ref.**
a	p-Tol	(–)-S	1	54	S	+142 ($CHCl_3$)	125–126	3b
b	p-$MeOC_6H_4$	(–)-S	1	76	S	+141 (acetone)	120.5–121.5	4
c	1-Naph	(–)-S	1	65	S	+292 (acetone)	96.5–97.0	3a,b
d	p-Tol	(–)-S	2	66	S	+210 ($CHCl_3$)	101–102	3b
e	p-$MeOC_6H_4$	(–)-S	2	~50	S	+187 (acetone)	106.5–107	4
f	3,5-di-t-Bu-4-$MeOC_6H_2$	n.r.[a]	1	n.r.	R	–73.5 ($CHCl_3$)	n.r.	20
g	2,4,6,-tri-i-PrC_6H_2	—[b]	1	n.r.	S	+229.2 ($CHCl_3$)	n.r.	20

[a] n.r. = not reported.
[b] Prepared from the predominant diastereomer of the DAG sulfinates[21] (cf. Section 2.1.1, Scheme 2.1.5).

1) BuLi
2) (-)-(S)-**2**
3) Bu_4NF

1) MeLi
2) CO_2

H^+
$CHCl_3$

(3.10.2)

Table 3.10.2 Synthesis of 2-Sulfinyl-2-Alkenolides **8**

					2-Sulfinyl-2-alkenolide 8				
Method	**Symbol**	**n**	**Ar**	**R**	**Overall yield [%] 5 → 8 or 9 → 8**	**mp [°C]**	**[α]$_D$ ($CHCl_3$)**	**Abs. conf.**	**Ref.**
A	**8a**	1	p-Tol	H	20	121–125	+244	S	5
B	**8a**	1	p-Tol	H	49	115–118	+242	S	16
A	**8b**	2	p-Tol	H	22	93–94	+212.8	S	6
A	**8c**	2	p-An	H	—	—	+242.9	S	12
B	**8d**	1	p-Tol		61.5	56–58	+150	S	16
B	**8e**	1	p-Tol	Me	56	79–82	+169	S	16
B	**8f**	1	p-Tol	n-Bu	64	61–63	+156	S	16

The second method (B), described by Holton and Kim,[16] is based on the application of optically active α-phosphoryl sulfoxide **9** as a source of the chiral sulfinyl moiety. The Horner-Wittig reaction of the lithio derivatives of **9** with ω-silyloxycarbonyl compounds leads to ω-silyloxyalkenylsulfoxides **10** formed as mixtures of **E** and **Z** diastereomers. Treatment with LDA causes isomerization of the latter and provides predominantly the **E** lithio-derivatives, which are quenched with carbon dioxide and subsequently treated with p-toluenesulfonic acid to give 2-sulfinyl-2-alkenolides **8** (Equation 3.10.3, Table 3.10.2).

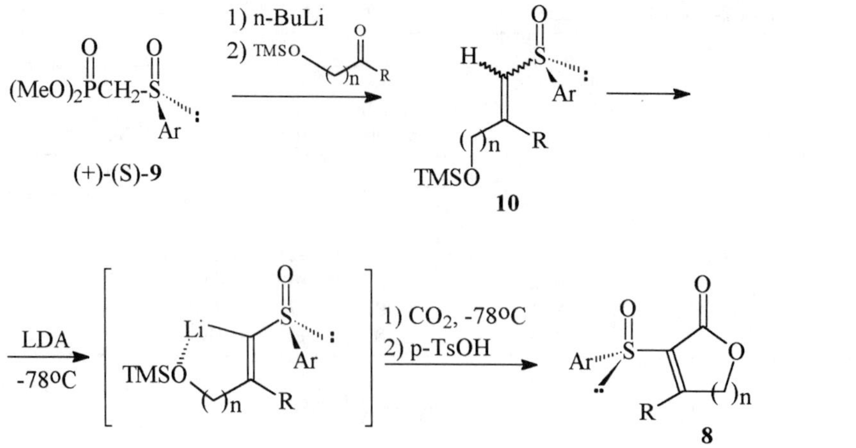

(3.10.3)

3.10.3 Conjugate Addition of Nucleophiles to Sulfinyl Cycloalkenones

Conjugate addition of organometallic reagents to enantiomerically pure 2-sulfinyl-2-cycloalkenones **4** proceeds very smoothly and gives, as anticipated, products in excellent diastereomeric purity. Moreover, this reaction is a stereodivergent process, i.e., depending on the conditions applied it enables one to obtain either enantiomer of the final product from the same stereoisomer of the substrate. This feature is explained in terms of the conformations adopted preferentially by the substrates. Thus, X-ray analysis revealed that the ground-state conformation of the sulfoxides **4** has the sulfinyl S=O bond oriented *anti* to the carbonyl C=O bond (model A). On the other hand, however, addition of chelating metal salts like $ZnBr_2$ results in a complete change of the preferred conformation. In this case a zinc-chelated conformer B becomes predominant, which locks the β-oxosulfoxide moiety into a *syn* conformation. In each case the chirality at the sulfinyl sulfur atom might control the approach of an organometallic reagent to the *si* or *re* face of the prochiral β-carbon atom of the substrate. This is believed to be achieved by the presence of the bulky aryl group attached to sulfur which shields one diastereotopic face of the enone system (different in each conformation **A** and **B**), thus determining the absolute configuration of the newly formed carbon chiral center.

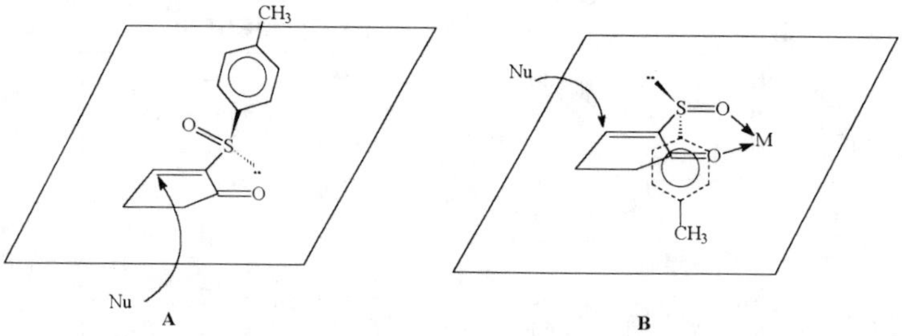

As a result of the asymmetric Michael reaction discussed above, followed by removal of the sulfoxide moiety, a broad series of optically active 3-substituted cycloalkanones have been obtained from (+)-(S)-2-sulfinyl-2-cycloalkenones **4*** (Equations 3.10.4 and 3.10.5).

* Since both enantiomeric menthyl p-toluenesulfinates **2** (Ar=p-Tol), namely (–)-(S) and (+)-(R), are now commercially available, either enantiomer of the sulfoxide **4** can easily be obtained. This enables one to choose the proper procedure (with or without addition of a chelating agent) depending on its stereoselectivity and the required absolute configuration of the product.

(+)-(S)-4 —R-Met, non-chelate model→ —Al-Hg→ (S)

(3.10.4)

(+)-(S)-4 —R-Met, chelate model→ —Al-Hg→ (R)

(3.10.5)

It should be emphasized that a large variety of organometallic reagents can be applied, such as Grignard reagents, organomagnesium and organotitanium compounds, and enolate nucleophiles as well. The most representative examples of the conjugate additions to **4** are collected in Tables 3.10.3 and 3.10.4.

Table 3.10.3 Conjugate Addition of Organometallic Reagents to 2-Sulfinyl-2-Cycloalkenones **4** According a Nonchelate Model

Ar	R-Met	Solvent	n	3-Substituted cycloalkanone R	Yield [%]	ee [%]	Abs. conf.	Ref.
p-Tol	Me_2Mg	THF	1	Me	69[a]	97	S	7,8
p-Tol	Et_2Mg	DME	1	Et	88	81	S	7,8
p-Tol	$(neo\text{-}C_5H_{11})_2Mg$	DME	1	neo-C_5H_{11}	77[a]	91	S	7
p-Tol	$(CH_2{=}CH)_2Mg$	DME	1	$CH_2{=}CH$	74[a]	57	S	7
p-Tol	Ph_2Mg	THF	1	Ph	72	>98	S	7,8
p-Tol	Me_2Mg	DME	2	Me	67[a]	79	S	7,8
p-Tol	i-Pr_2Mg	DME	2	i-Pr	53[a]	50	S	7
p-Tol	s-Bu_2Mg	DME	2	s-Bu	67[a]	62	S	7
1-Naph	MeMgI	THF	1	Me	76	72	S	3a
1-Naph	Me_2CuLi	THF	1	Me	42	57	S	3a
p-Tol	$Me_3SiCH(Li)CO_2Me$	DME	1	MeO_2CCH_2[b]	95	70	S	6
p-Tol	$Me_3SiCH(Li)CO_2Me$	DME	2	MeO_2CCH_2[b]	78	95	S	6
p-Tol	$PhCH_2OCH_2OCH_2Li$	Hexane/DMTHF	1	$HOCH_2$[c]	76	78	S	15
p-Tol	$PhCH_2OCH_2OCH_2Li$	Hexane/DMTHF	2	$HOCH_2$[c]	84	93	S	15

Note: DMTHF= *cis* and *trans* 2,5-dimethyltetrahydrofuran.

[a] One equivalent of 18-crown-6 was also used.

[b] The primary adduct was subsequently treated with Al-Hg and KF (to remove the Me_3Si group).

[c] After hydrogenolysis of the primary adduct.

Some interesting points should be highlighted in Tables 3.10.3 and 3.10.4. Firstly, for the reactions proceeding according to the nonchelate model the p-tolyl group attached to sulfur seems to be the substituent of choice. Thus, p-toluenesulfinyl cycloalkenones **4a** and **4d** are not only easy to synthesize (because of the commercial availability of menthyl p-toluenesulfinates), but they also give products of conjugate addition in better enantiomeric excess than their 1-naphthyl analog **4c**. The latter, which was expected to direct nucleophilic addition more effectively due to its larger size, gives less satisfactory stereoselection, most probably because of its preferred conformation in which the naphthyl group is spatially distant from the enone unit. Secondly, for the reactions performed in the presence of chelating agents, the degree of asymmetric inductions is higher when the aromatic substituent bonded to the sulfinyl sulfur atom is more electron releasing and the solvent less coordinating. Thus, replacing the p-tolyl by the p-anisyl group causes an increase in diastereoselectivity and so does

Table 3.10.4 Conjugate Addition of Organometallic Reagents to 2-Sulfinyl-2-Cycloalkenones **4** According to a Chelate Model

				3-Substituted cycloalkanone				
Ar	**R-Met**	**Solvent**	**n**	**R**	**Yield [%]**	**ee [%]**	**Abs. conf.**	**Ref.**
p-Tol	MeMgCl	THF	1	Me	93	<98	R	3a,8,9
p-Tol	$EtMgCl/ZnBr_2$	THF	1	Et	84–90	80	R	4,9
p-Tol	$MeMgI/ZnBr_2$	THF	1	Me	71–89	87	R	3a,8
p-Tol	neo-$C_5H_{11}MgCl/ZnBr_2$	THF	1	neo-C_5H_{11}	32	18	R	4
p-An[a]	neo-$C_5H_{11}MgCl/ZnBr_2$	THF	1	neo-C_5H_{11}	69	87	R	4
p-Tol	$MeTi(Oi\text{-}Pr)_3$	THF/CH_2Cl_2	1	Me	90	90	R	3a,8
p-Tol	$EtTi(Oi\text{-}Pr)_3$	THF	1	Et	67	98	R	9
p-Tol	$t\text{-}BuMgCl/ZnBr_2$	THF	1	t-Bu	98	86	R	9
p-Tol	$CH_2{=}CHMgBr/ZnBr_2$	THF	1	$CH_2{=}CH$	75	98	R	9
p-Tol	$PhMgCl/ZnBr_2$	THF	1	Ph	70	92	R	9
p-Tol	p-$TolMgBr/ZnBr_2$	THF	1	p-Tol	89	58	R	4,9
p-An	p-$TolMgBr/ZnBr_2$	THF	1	p-Tol	86	69	R	4,9
p-Tol	p-$TolMgBr/ZnBr_2$	DMTHF	1	p-Tol		86	R	9,10
p-Tol	$MeMgBr/ZnBr_2$	THF	2	Me	95	62	R	8
p-Tol	$MeTi(Oi\text{-}Pr)_3$	THF	2	Me	85	87	R	4,9
p-Tol	MeTi(Oi-Pr)3	DMTHF	2	Me		96	R	9,10
p-An	$MeTi(Oi\text{-}Pr)_3$	THF	2	Me	85	96	R	4
p-Tol	$(Me_3Si)_2C(Li)CO_2Me$	THF	1	MeO_2CCH_2[b]	95	98	S[c]	13

Note: DMTHF = *cis* and *trans* 2,5-dimethyltetrahydrofuran.

[a] p-An = p-$MeOC_6H_4$.

[b] The primary adduct was subsequently treated with Al-Hg and KF (to cleave the Me_3Si groups).

[c] Starting from (–)-(R)-**4a**.

replacing THF by dimethyltetrahydrofuran (DMTHF). Unfortunately, these two benefits cannot be combined since the p-anisylsulfinyl cycloalkenones **4b** and **4e** are insoluble in DMTHF.

The above strategy of a stereoselective conjugate addition to 2-sulfinyl-2-cycloalkenones has been found to work well also for the reaction of 3-substituted 2-sulfinyl-2-cyclopentenones **11** leading, after reductive cleavage of the sulfinyl group, to 3,3-disubstituted cyclopentanones **12**.[11] This procedure constitutes an efficient method for a stereocontrolled formation of chiral quaternary carbon centers (Equation 3.10.6, Table 3.10.5).

p-Tol–S(O)– **11** (R) → 1) R^1-Met, THF, -78° → 0°; 2) Al-Hg → **12** (R, R^1) (3.10.6)

Table 3.10.5 Asymmetric Synthesis of 3,3-Disubstituted Cyclopentanones

			12		
11, R		**R^1-Met**	**R^1**	**Yield [%]**	**ee [%]**
a	p-Tol	Me_2CuLi	Me	58	78
a	p-Tol	Me(PhS)CuMgBr	Me	77	73
a	p-Tol	n-Bu(PhS)CuMgCl	n-Bu	69	81
b	Me	p-Tol_2CuLi	p-Tol	53	90–93
b	Me	n-Bu(PhS)CuMgCl	n-Bu	79	53
b	Me	n-Bu(PhO)CuMgCl	n-Bu	61	88

From Posner, G. H., Kogan, T. P., Hulce, M., *Tetrahedron Lett.*, 25, 383, 1984. With permission.

It is interesting that reversing the substituent R in **11** and R^1 in the organometallic reagent enables one to obtain either enantiomer of the product **12** whose configuration always fits the chelate model. An unexpected feature of this reaction is the fact that the use of the organomagnesium reagents/zinc bromide system produces exclusively 1,2- and not 1,4-adducts. For this reason the appropriate organocopper reagents must always be used instead.

3.10.4 Conjugate Addition of Nucleophiles to Sulfinyl Alkenolides

2-Sulfinyl-2-alkenolides **8** behave in the same way as 2-sulfinyl-2-cycloalkenones toward nucleophilic reagents, i.e., they act as Michael acceptors. Thus, treatment of **8** with a $ZnBr_2$/Grignard reagent system or an organotitanium compound, followed by Raney nickel cleavage of the sulfoxide moiety, results in the formation of 3-substituted alkenolides **13**, the stereochemistry of which is consistent with the chelate model (Equation 3.10.7, Table 3.10.6).[5,12]

1) R-Met
2) Raney-Ni

(+)-(S)-**8** → **13** (3.10.7)

Table 3.10.6 Conjugate Addition of Nucleophiles to 2-Sulfinyl-2-Alkenolides **8**

Substrate						13			
	n	Ar	Solvent	R-Met	R	Abs. conf.	Yield [%]	ee [%]	Refs.
8a	1	p-Tol	DMTHF	(3,4-methylenedioxyphenyl)MgBr/ZnBr$_2$	3,4-methylenedioxyphenyl	S[a]	70	98	5
8b	2	p-Tol	THF	MeTi(OiPr)$_3$	Me	R[a]	44	55	12
8c	2	p-An	THF	MeTi(OiPr)$_3$	Me	R[a]	51	93	12
8a	1	p-Tol	Hexane/DMTHF	$PhCH_2(OCH_2)_2Li$	$HOCH_2$[c]	S[b]	48	87.3	15

[a] Chelate mode of addition.
[b] Nonchelate mode of addition.
[c] After hydrogenolysis over Pd-C of the primary adduct.

On the other hand, asymmetric Michael addition of various ester enolate anions to alkenolide sulfoxides **8** proceeds in a nonchelate mode to produce, after reductive elimination of the sulfinyl group and in some cases also of the arylthio group, 1,5-dicarbonyl adducts **14** (Equation 3.10.8, Table 3.10.7).[6]

(+)-(S)-**8** + RO(C=O)CH(Li)X —Raney-Ni→ **14** (3.10.8)

Table 3.10.7 Conjugate Addition of Ester Enolate Anions to 2-Cycloalkenolides **8**

Substrate				14		
Symbol	n	R	X	Yield [%]	Abs. conf.	ee [%]
8a	1	Me	H	65	(+)-S	80
8a	1	t-Bu	H	82	(+)-S	43
8a	1	Me	PhS	100	(+)-S	91
8a	1	$MeOCH_2$	PhS	79	(+)-S	78
8b	2	$MeOCH_2$	H	62	(–)[a]	>96
8b	2	Me	PhS	92	(–)[a]	91
8b	2	$MeOCH_2$	PhS	94	(–)[a]	>96
8b	2	t-Bu	p-TolS	29	(–)[a]	88

[a] Not established, presumably *S.*

From Posner, G. H., Weitzberg, M., Hamill, T. G., Asirvatham, E., Cun-Heng, H., Clardy, J., *Tetrahedron,* 42, 2919, 1986. With permission from Elsevier Science Ltd., Kidlington, U.K..

3.10.5 Applications of the Conjugate Addition of Nucleophiles to 2-Sulfinyl-2-Cycloalkenones and 2-Sulfinyl-2-Alkenolides in the Synthesis of Natural Products: Selected Examples

3.10.5.1 Formal Total Synthesis of 11-Oxoequilenin 15[3a]

In the preceding examples, the 3-substituted 2-sulfinyl cycloalkanones formed in the highly efficient asymmetric conjugate addition step have been treated with a desulfurating agent to remove the sulfoxide moiety. It has turned out, however, that the sulfinyl group is not only an effective chiral auxiliary, but also an excellent means of controlling the regiochemistry of enolate ion formation. This allows one to introduce other substituents in the position 2 of the new cycloalkanone formed. Such an approach has been used in the synthesis of a 9, 11-seco-steroid and ultimately 11-oxoequilenin **15** (Equation 3.10.9).

p-Tol S O O (+)-**8a** 1) MeO MgBr 2) MeI/ N Me O p-TolS O Me O H MeO *cis/trans* mixture 78%

1) Me_2CuLi 2) $BrCH_2CO_2Me$/HMPA MeOCCH$_2$ O Me O MeO 90% ee~100% several steps O O H HO **15**

(3.10.9)

3.10.5.2 Synthesis of (+)-α-Cuparenone 16

In this synthesis the approach based on a stereocontrolled formation of a chiral quaternary carbon center in the position 3 has been applied. The product was obtained in 6.5% overall yield and ee = 71%.[11]

1) p-Tol_2CuLi
2) m-CPBA
3) t-BuOK,MeI
4) Me_2CuLi, MeI

11b **16** (3.10.10)

3.10.5.3 Synthesis of Natural (–)-Methyl Jasmonate 19[13]

It has been found that the conjugate addition of α-monosubstituted lithium acetate esters to 2-sulfinyl-2-cycloalkenones proceeds in a nonchelate mode[6] (cf. Table 3.10.3). On the contrary, an analogous reaction of α,α disubstituted lithium enolates proceeds via the chelated forms of the sulfoxide (cf. Table 3.10.4). This feature has been used in the asymmetric synthesis of (–)-methyl jasmonate.[13] Note that in this case the other enantiomer of the starting sulfoxide, namely (–)-(R)-**4a**, has been used as a substrate.

1) $(Me_3Si)_2C(Li)$-CO_2Me
2) P_2I_4
3) KF
54%

NaH, Br–(pentenyl)
69%

(-)-(R)-**4a** **17**

Raney-Ni
45%

18 **19** ee >98% (3.10.11)

It is interesting that attempts to pentenylate the β-oxo sulfoxide formed after initial conjugate addition have led to predominant O-alkylation. For this reason the sulfoxide must be reduced to the sulfide **17** prior to alkylation.

3.10.5.4 Synthesis of (+)-Estrone Methyl Ether 22[14]

A similar difference in the stereochemical course of the conjugate addition of α-mono- and α,α-disubstituted ketone lithium enolates has been used in the first step of the total synthesis of (+)-estrone methyl ether.[14] By reacting monosubstituted enolate **20** with the sulfoxide (–)-(R)-**4a** and separately by reacting disubstituted enolate **21** with the opposite enantiomer of the sulfoxide (+)-(S)-**4a**, two Michael adducts have been prepared, each having the same desired natural absolute configuration at the carbon 3 of the cyclopentanone ring. Several further transformations have led to (+)-estrone methyl ether **22** in 6.3% overall yield and 97.3% ee.[14]

LiO, X, MeO, (-)-(R)-**4a**, (+)-(S)-**4a**, m-CPBA, p-TolSO$_2$, O, X, H, MeO

20: X=H
21: X=Br

several steps → OH, H, H, H, MeO

22

(3.10.12)

3.10.5.5 Synthesis of (+)-A Factor 25[15]

In this case the synthetic strategy is based on a conjugate addition of a synthetic equivalent of the hydroxymethyl carbanion, HO–CH_2^-, to 2-sulfinyl-2-butenolide **8a**. As expected, the addition proceeds in a nonchelate mode. The relatively low ee value of the final product **25** is due to the racemization during the hydrogenolysis of **24**.

p-Tol, S, O, (+)-(S)-**8a**; 1) Ph–O–O–Li 2) Raney-Ni, 48%; Ph, O, O, **23** ee=87.3%

1) LHMDS 2) $Me_2CH(CH_2)_4COCl$, 79%; Ph, O, O, **24**; H_2/Pd-C, 91%; HO, **25**, ee~50%

(3.10.13)

3.10.5.6 Synthesis of Aphidicolin 31[17]

A key reaction in the total synthesis of the title compound involves the Michael addition of lithium dienolate **26** to 3-substituted 2-sulfinyl butenolide **8d**, which results in the stereoselective formation of the quaternary center at C-3. The adduct is formed as a 7.4:1 mixture of diastereomers, from which **27** is isolated by recrystallization. (The minor diastereomer has opposite absolute configuration at C-6 in the cyclohexenone moiety.) Addition of vinyllithium, followed by treatment with HF, gives the conjugated dienone **28** which undergoes smooth cyclization in the presence of sodium methoxide to **29**. It should be stressed that the whole sequence (**8d** → **29**) has also been carried out in a single synthetic operation to produce the tricyclic product **29** in 45% yield. After removal of the sulfinyl moiety and several further transformations, aphidicolin **31** was obtained in the enantiomerically pure form.

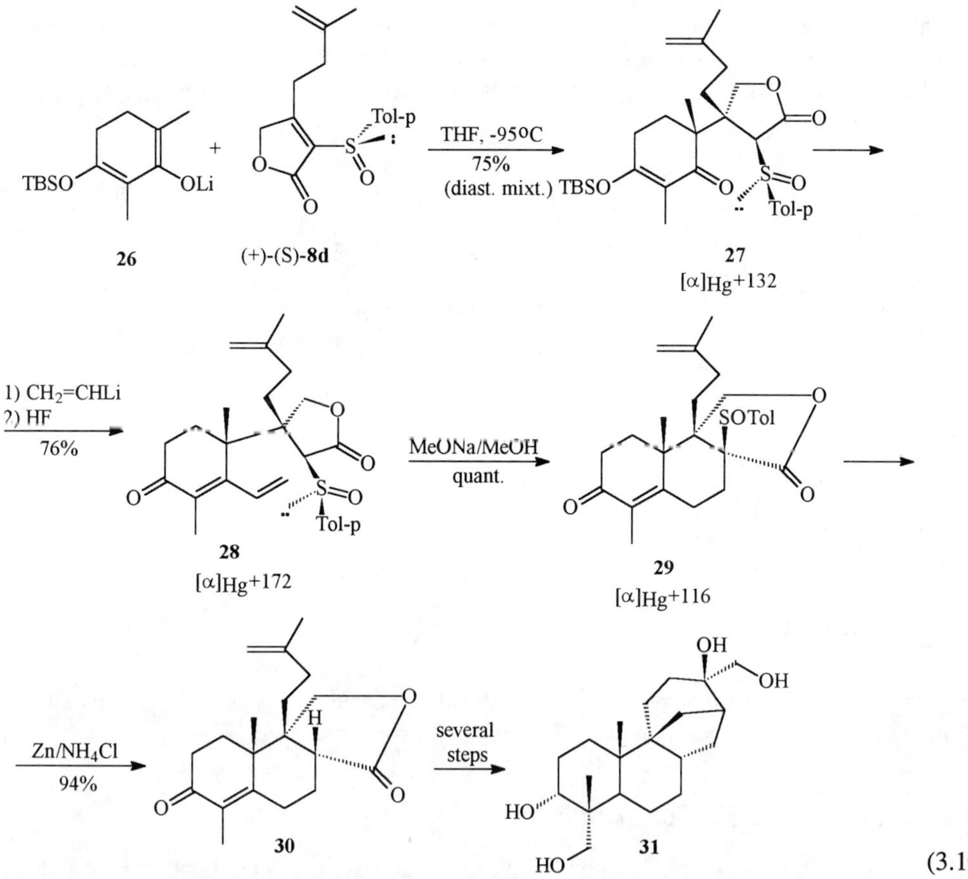

(3.10.14)

3.10.5.7 Preparation of Chiral 2-Substituted Chroman-4-Ones 37[18]

The asymmetric Michael addition to the chiral vinyl sulfoxide system also has an interesting application in the synthesis of optically active 2-substituted chromanones. The appropriate substrate, namely 3-(p-toluenesulfinyl)chromone **35**, has been synthesized by the reaction sequence shown below, involving acylation of the anion of (+)-*(R)*-methyl p-tolyl sulfoxide **32** with the protected salicylate **33**, deprotection of the hydroxy group, formylation of the β-oxosulfoxide **34** (R=H), and final cyclization.

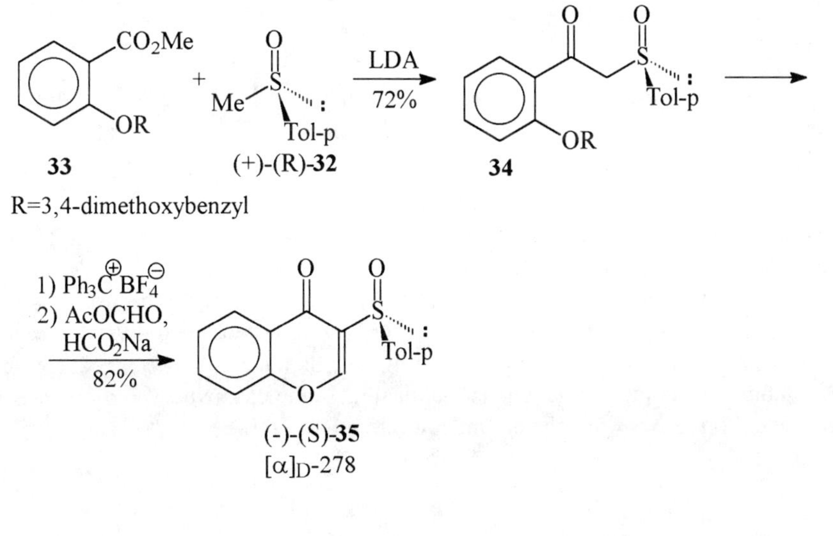

(3.10.15)

Reaction of **35** with lithium dimethylcuprate proceeds in a chelate mode and gives a mixture of chromanones from which the *cis* isomer **36** can be isolated in 24% yield. The reductive removal of the sulfinyl moiety and subsequent oxidation of the chromanol formed gives the chromanone **37** in at least 90% ee.

Me_2CuLi 90%; separation 24%

(-)-(S)-**35** **36**

1) Al-Hg
2) PDC
59-65%

(-)-(S)-**37**
$[\alpha]_D$-50

(3.10.16)

For other, so far described, applications of chiral sulfinyl cycloalkenones **4** and sulfinyl alkenolides **8** in asymmetric synthesis see Reference 12 [(–)-β-vetivone] and Reference 5 [(–)-podorhizon].

3.10.6 Asymmetric Radical Addition

Addition of alkyl radicals to chiral 2-sulfinyl-2-cyclopentenones **4** has been found to proceed with a high β-stereoselection and to give exclusively *trans* adducts. However, for this purpose, sulfinyl cyclopentenones **4f** and **4g** with sterically demanding substituents at the sulfinyl sulfur atom must be used.[20] Thus, *tert*-butyl radical, generated from *tert*-butyl iodide and triethylborane, adds to **4f** in the presence of a Lewis acid to give the product **38** with diastereomeric excess up to 90% (dr 95:5). In this case effective coordination of the sulfinyl and carbonyl oxygen atoms by the Lewis acid seems to be crucial for fixing the reactive conformer, since no diastereoselection is observed in the absence of the catalyst.

1) 1.1 equiv $TiCl_2(O\text{-}i\text{-}Pr)_2$
2) 5 equiv t-BuI
3) 10 equiv Et_3B
86%

(-)-(R)-**4f** **38a** + **38b** 95:5

(3.10.17)

However, when the more hindered 2-sulfinylcyclopentenone **4g**[20] is applied and trialkylboranes are used as a sole source of alkyl radicals, the chelation turns out to be too weak to dominate the influence of the two bulky isopropyl groups of the phenyl ring. Therefore, the use of strongly chelating agents, like $TiCl_2(Oi\text{-}Pr)_2$ or Et_2AlCl, gives almost no diastereoselectivity, while application of $MgCl_2$, $ZnBr_2$, and $ZrCl_4$ leads to the product **39** with a high stereoselection. Moreover, the stereochemical outcome of the latter reactions is identical as in the case when no catalyst is used and opposite to that expected from the chelate mode of addition. (Table 3.10.8).[20]

(+)-(S)-4g → 1) R_3B 2) catal. → 39a + 39b (3.10.18)

Table 3.10.8 Reaction of Sulfinylcyclopentenone 4g with Trialkylboranes

Entry	R	Catalyst	Yield [%]	Ratio 39a:39b
1	Et	$TiCl_2(Oi\text{-}Pr)_2$	60	42:58
2	Et	Et_2AlCl	92	45:55
3	Et	$ZnBr_2$	52	>98:<2
4	Et	$ZrCl_4$	82	>98:<2
5	Et	—	95	>98:<2
6	*i*-Pr	—	94	>98:<2
7	*c*-C_6H_{11}	—	71	>98:<2
8	*t*-Bu	—	66	>98:<2

From Toru, T., Watanabe, Y., Tsusaka, M., Ueno, Y., *J. Am. Chem. Soc.*, 115, 10464, 1993.

3.10.7 Examples of Experimental Procedures

3.10.7.1 (+)-(S)-2-(p-Toluenesulfinyl)-2-Cyclopentenone Ethylene Ketal 3a[3b]

A 250-mL, three-necked, round-bottomed flask equipped with two rubber septa, a nitrogen inlet, 125-mL pressure-equalizing dropping funnel, and a magnetic stirring bar is flame dried under nitrogen. The flask is charged with 70 mL of anhydrous THF and cooled in an isopropyl alcohol-dry ice bath. Stirring is begun as 42 mL (60.8 mmol) of 1.45 M butyllithium in hexane is added slowly through the dropping funnel over 10–30 min. After another 10 min a solution of 11.3 g (55.1 mol) of 2-bromo-2-cyclopentenone ketal[19] is added from the dropping funnel over 30 min. The colorless or pale yellow solution is stirred and cooled at –78°C for 1.5 h. A 1-L, three-necked, round-bottomed flask equipped with a magnetic stirring bar, two rubber septa, and a stopcock connected to a bubbler gas exit is flushed with nitrogen and charged with 24.4 g (82.9 mmol) of (–)-(S)-menthyl p-toluenesulfinate **2** and 460 mL of anhydrous THF. The sulfinate suspension is stirred vigorously and cooled at –78°C as the vinyllithium reagent in the first flask is then transferred into the second flask through a cooled cannula by means of nitrogen pressure. (Note: lower yields are obtained when the vinyllithium reagent is allowed to warm above –78°C during the transfer.) As the 50 min transfer proceeds, the sulfinate suspension becomes yellow. The mixture is stirred for another 15 min at –78°C, the cooling bath is removed, and 125 mL of saturated aqueous sodium dihydrogen phosphate is added. When the contents have warmed to room temperature, the THF is removed by rotary evaporation. The residue is partitioned between 300 mL of water and 200 mL of chloroform. The aqueous layer is extracted with three 100-mL portions of chloroform. The chloroform extracts are combined and dried over anhydrous potassium carbonate. Evaporation of the chloroform gives 40–55 g of a viscous brown oil consisting of the sulfinyl ketal, menthol, menthyl sulfinate, minor by-products, and residual chloroform. The sulfinyl ketal is isolated by modified flash chromatography on 500 g of Woelm silica gel (32–64 μ) packed in dry diethyl ether in a 6.5- × 45-cm column. The crude product is applied to the column in 25 mL of chloroform and the column is eluted with ether under sufficient compressed air pressure to achieve a flow rate of 60 mL per min. After thirty 60-mL fractions are collected, the solvent is changed to ethyl acetate, and another forty 60-mL fractions are collected and analyzed by thin-layer chromatography.

Combination and evaporation of fractions 40–60 provide 9.05–9.75 g (62–67%) of crude (+)-(S)-2-(p-toluenesulfinyl)-2-cyclopentenone ethylene ketal **3a** as a pale, yellow oil, $[\alpha]_D^{25}$ +78, ($CHCl_3$, *c* 0.25) ^{1}H-NMR ($CDCl_3$) δ: 2.0–2.2 (m, 2H, CH_2), 2.3–2.6 (m, 2H, C=CCH_2), 2.37 (s, 3H, CH_3), 3.7–3.9 (m, 4H, OCH_2CH_2O), 6.67 (t, 1H, J = 2, C=CH), 7.24 and 7.57 (2d, 4H, J = 8, aryl H).

3.10.7.2 (+)-(S)-2-(p-Toluenesulfinyl)-2-Cyclopentenone 4a[3b]

A magnetic stirring bar, 100 g of anhydrous copper (II) sulfate, and a solution of 0.05–9.75 g of the sulfinyl ketal in 300 mL of acetone are placed in a 500-mL Erlenmeyer flask. The flask is flushed with nitrogen and stoppered. The suspension is stirred vigorously overnight, the copper sulfate is separated by filtration, and washed thoroughly with 500–700 mL of acetone. Concentration of the combined filtrates gives 7.36–7.58 g of tan crystals. Recrystallization is carried out by dissolving the product in a minimum volume of ethyl acetate (ca. 80 mL) at room temperature, treating with Norite, diluting with an equal volume of diethyl ether, and cooling to –20°C. After the resulting crystals are collected, the mother liquor is evaporated under reduced pressure at room temperature, and the procedure is repeated twice. The mother liquor is again evaporated and the residue (1.4–1.8 g) is purified by flash chromatography on 110 g of Woelm silica gel using ethyl acetate as eluent. Combination of appropriate fractions, evaporation, and recrystallization affords two additional crops of crystalline product (0.4–0.7 g). The yield of (+)-(S)-2-(p-toluenesulfinyl)-2-cyclopentenone **4a**, mp 125–126°C, $[\alpha]_D^{25}$ +148 ($CHCl_3$, *c* 0.11), is 6.02–6.60 g (50–54% based on bromo ketal). IR (CCl_4) cm^{-1}: 924 (m), 1715 (s), 1287 (m), 1152 (s), 1083 (s), 1054 (s), 728 (m); ^{1}H-NMR ($CDCl_3$) δ: 2.2–2.5 (m, 2H, CH_3), 2.30 (s, 3H, CH_3), 2.6–2.8 (m, 2H, C = CCH_2), 7.19 and 7.58 (2d, 4H, J = 8, aryl H), 8.03 (t, 1H, J = 2, C = CH); mass spectrum (70 eV), *m/z* (rel intensity): 220 (M^+, 30), 172 (100), 139 (48), 129 (72).

3.10.7.3 Conjugate Addition to 2-Sulfinyl-2-Cyclopentenone (+)-(S)-4a — a Nonchelate Mode; Synthesis of (–)-(S)-3-Methoxycarbonylmethylcyclopentanone[6]

A flame-dried, 10-mL, round-bottomed flask was charged with 1 mL of dry THF and 266 μL (1.26 mmol) of hexamethyldisilazane and cooled to –78°C. After 10 min, 800 μL of 1.5 M n-BuLi (1.20 mmol) was added and the mixture was stirred for 40 min at –78°C under N_2. Methyl trimethylsilylacetate (Aldrich, 197 μL 1.20 mmol) was added and, after 2 h, a –78°C solution of 132 mg (0.60 mmol) of **4a** in 5 mL of THF was added dropwise over 10 min via a precooled cannula. After the addition, the pale yellow solution was stirred for 30 min at –78°C. The mixture was then quenched by adding a saturated solution of sodium hydrogen phosphate and warmed to room temperature. The contents in the flask were extracted with Et_2O (3 × 10 mL) and the combined organic layers were washed with H_2O and dried over $MgSO_4$. Filtration and solvent evaporation gave a white solid which was used directly in the next step without further purification.

The crude conjugate adduct from the previous reaction was dissolved in 10 mL of aqueous THF solution (THF-H_2O, 9:1) and cooled to –15°C. Al amalgam (6.0 mmol) was added and the mixture was warmed slowly to room temperature and stirred overnight. Anhydrous $MgSO_4$ was added to the gray slurry and the organic layer was filtered off. The slurry was washed with Et_2O (2 × 10 mL). Evaporation of the solvent under reduced pressure gave a pale yellow liquid which was purified by column chromatography (eluting solvent: Et_2O-hexane, 1:9) to give 133 mg (97% from **4a**) of the desulfurized keto ester: ^{1}H-NMR ($CDCl_3$), δ 0.06 (S, 9H), 2.10–3.15 (m, 8H), 3.61 (s, 3H).

Protodesilylation was carried out in a 20% MeOH solution (8 mL) of the above trimethylsilyl ester (130 mg, 0.54 mmol) with KF (63 mg, 1.14 mmol) stirred at room temperature for 6 h. The mixture was concentrated under reduced pressure and the residue was extracted with CH_2Cl_2 (3 × 10 mL). The combined extracts were washed with H_2O and dried over $MgSO_4$. Filtration and solvent evaporation gave 88 mg of a pale yellow liquid which was purified by column chromatography (eluting solvent: Et_2O-hexane, 2:8) to give 80 mg (95%) of (–)-(S)-3-methoxycarbonylmethylcyclopentanone IR ($CHCl_3$) 1740 cm^{-1}; ^{1}H-NMR ($CDCl_3$) δ 1.4–2.8 (m, 9H), 3.70 (s, 3H); mass spectrum *m/z* 156 (M^+). Kugelrohr distillation gave 70 mg of a colorless liquid: $[\alpha]_D^{18}$ –82.2 (*c*, 1.46, $CHCl_3$). Lit. value: $[\alpha]_D^{18}$ –121.0 (*c*, 1.47, $CHCl_3$).

3.10.7.4 Preparation of (+)-(R)-3-Methylcyclohexanone via Methyltitanium Triisopropoxide

Conjugate Addition — a Chelate Mode[4]

A flame-dried, 1-neck, 10-mL, round-bottom flask equipped with a serum cap, a gas needle inlet, and a magnetic stirring bar was flushed with argon and charged with triisopropoxytitanium chloride (0.75 mL, 1.4 M in THF, 1.1 mmol), then cooled to –78°C. This solution then was treated with MeLi (0.75 mL, 1.4 M, 1.1 mmol) dropwise via a gastight syringe. The resultant solution was stirred at –78°C for 15 min, and 1.0 mL of anhydrous THF was added slowly. To this bright yellow solution, sulfoxide **4e** (0.088 g, 0.35 mmol) in 1 mL of anhydrous THF was added dropwise, the solution was stirred at -78°C for 1 h, then was warmed to 0°C and stirred for 2 h; subsequent quenching with 0°C 10% HCl aqueous and rotary evaporation at 0°C gave an aqueous concentrate which was extracted with ether. The ethereal solution was dried over $MgSO_4$ and the diethyl ether was evaporated at –10°C. The flask was then equipped with a magnetic stirring bar and placed in a –10°C bath. 15 mL of 0°C THF-H_2O (9:1) was added and the solution treated with 10 equivalents of freshly prepared aluminum amalgam. The resultant mixture was stirred with slow warming to room temperature overnight. Anhydrous $MgSO_4$ was added, the slurry filtered through a fine sintered glass filter funnel, and the cake washed well with dimethyl ether. The combined filtrates were rotary evaporated at –10°C and the residue was dissolved in CH_2Cl_2 for chromatography. Preparative TLC (SiO_2, 10:0.6 benzene:diethyl ether) gave (+)-(R)-3-methylcyclohexanone (32.2 mg, 82%). The sample for specific rotation was purified by Kugelrohr distillation (40°C, 10 mm Hg, –78°C receiving bulb): $[\alpha]_{365}^{25}$ = +171.0 (*c*, 0.56, $CHCl_3$) with 97% ee. The product was pure by TLC (SiO_2, 10:0.6 benzene: diethyl ether, R_f 0.32) and GLPC (retention time 22.2 min, 120°C). The diastereomeric ketal prepared from this –3-methylcyclohexanone had an optical purity of 96 ± 4% by ^{13}C NMR.

REFERENCES (3.10)

1. Posner, G.H., Asymmetric synthesis using α-sulfinyl carbanions and β-unsaturated sulfoxides, in *The Chemistry of Sulphones and Sulphoxides*, Patai, S., Rappoport, Z., Stirling, C. J. M., eds., John Wiley & Sons, Chichester, 1988, 823.
2. For recent reviews see: Drabowicz, J., Kiełbasiński, P., Mikołajczyk, M., Synthesis of Sulphoxides, In Reference 1, 233; Drabowicz, J., Kiełbasiński, P., Mikołajczyk, M., Appendix to: "Synthesis of Sulphoxides", in *Syntheses of Sulphones, Sulphoxides and Cyclic Sulphides*, Patai, S., Rappoport, Z., eds., John Wiley & Sons, Chichester, 1994, 255.

3a. Posner, G. H., Mallamo, J. P., Hulce, M., Frye, L. L., . *J. Am. Chem. Soc.*, **104**, 4180, 1982.

3b. Frye, L. L., Kogan, T. P., Mallamo, J. P., Posner, G. H., *Org. Synth.*, **64**, 196, 1985.

4. Posner, G. H., Frye, L. L., Hulce, M., *Tetrahedron*, **40**, 1401, 1984.
5. Posner, G. H., Kogan, T. P., Haines, S. R., Frye, L. L., *Tetrahedron Lett.*, **25**, 2627, 1984.
6. Posner, G. H., Weitzberg, M., Hamill, T. G., Asirvatham, E., Cun-Heng, H., Clardy, J., *Tetrahedron*, **42**, 2919, 1986.
7. Posner, G. H., Hulce, M., *Tetrahedron Lett.*, **25**, 379, 1984.
8. Posner, G. H., Addition of organometallic reagents to chiral vinyl sulfoxides, in *Asymmetric Synthesis*, vol. 2, Morrison, J. D., ed., Academic Press, New York, 1983, 225.
9. Posner, G. H., *Acc. Chem. Res.*, **20**, 72, 1987.
10. Posner, G. H., Frye, L. L., *Isr. J. Chem.*, **24**, 88, 1984.
11. Posner, G. H., Kogan, T. P., Hulce, M., *Tetrahedron Lett.*, **25**, 383, 1984.
12. Posner, G. H., Hamill, T. G., *J. Org. Chem.*, **53**, 6031, 1988.
13. Posner, G. H., Asirvatham, *J. Org. Chem.*, **50**, 2589, 1985.
14. Posner, G. H. Switzer, C., *J. Am. Chem. Soc.*, **108**, 1239, 1986.
15. Posner, G. H. Weitzberg, M., Jew, S., *Synth. Commun.*, **17**, 611, 1987.

16. Holton, R. A., Kim, H. -B., *Tetrahedron Lett.,* **27**, 2191, 1986.
17. Holton, R. A., Kennedy, R. M., Kim, H. -B., Krafft, M. E., *J. Am. Chem. Soc.,* **109**, 1597, 1987.
18. Saengchantara, S. T., Wallace, T. W., *J. Chem. Soc., Chem. Commun.,* 1592, 1986.
19. Smith, A. B., III, Branca, S. J., Guaciaro, M. A., Wovkulich, P. M., Korn, A., *Org. Synth.,* **61**, 65, 1983.
20. Toru, T., Watanabe, Y., Tsusaka, M., Ueno, Y., *J. Am. Chem. Soc.,* **115**, 10464, 1993.
21. Llera, J. M., Fernandez, I., Alcudia, F., *Tetrahedron Lett.,* **32,** 7299, 1991.

Chapter 4

Chiral Sulfilimines and Sulfoximines

Sulfilimines **1** and sulfoximines **2** can be considered as the nitrogen analogs of sulfoxides and sulfones, respectively.

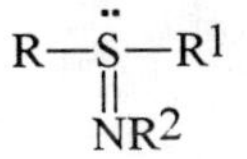

R—S(:)(=NR2)—R1

1

R—S(=O)(=NR2)—R1

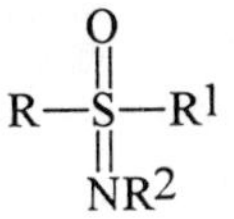

2

Considering the chemistry of sulfilimines, it should be noted that, although the first optically active member of this family of chiral organosulfur compounds was prepared as early as 1927,[1] studies of this class of compounds are much less abundant as compared with chiral sulfoxides. On the other hand, in a sharp contrast to sulfones, which can be chiral only in a special case of the O^{16}/O^{18} isotopic substitution, a rich family of chiral sulfoximines can be easily prepared. A few of them have found very useful applications as chiral reagents in asymmetric synthesis. Below, the most common routes for the preparation of synthetically useful chiral sulfilimines and sulfoximines will be described, and this compilation will be followed by presentation of their synthetic potential as chiral auxiliaries.

4.1 SYNTHESIS OF CHIRAL SULFILIMINES

The first example of a chiral sulfilimine was reported by Clarke and co-workers[1] in 1927, who were able to resolve the racemic sulfilimine **3a** containing a carboxyl group via diastereomeric salts with optically active alkaloid amines (brucine, cynchonidine). Soon thereafter, a similar procedure was reported for the ethyl analog **3b**.[2]

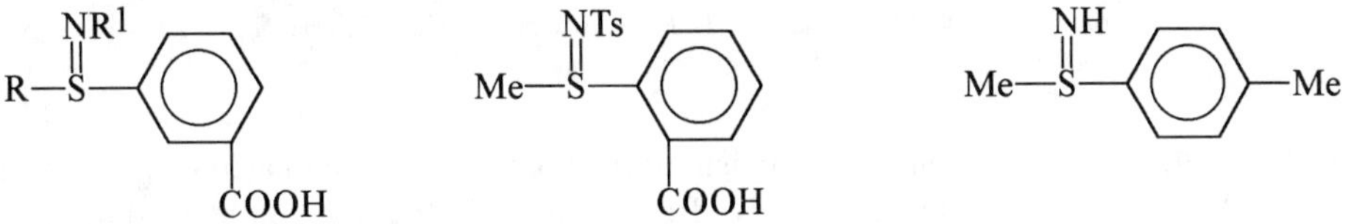

3 (R—S(=NR1)—C6H4—COOH)

4 (Me—S(=NTs)—C6H4—COOH)

5 (Me—S(=NH)—C6H4—Me)

a, R = Me, R[1] = p-TolSO$_2$
b, R = Et, R[1] = p-TolSO$_2$
c, R = Me, R[1] = PhSO$_2$

More recently, Kresze and Wustrow[3] and Bohman and Allenmark[4] separated in the same way the enantiomers of other structurally related sulfilimines **3c** and **4**. The stable, optically active unsubstituted sulfilimine **5** was obtained[5] upon resolution of the racemic mixture with (+)-α-bromo-π-camphorosulfonic acid.

A series of chiral N-substituted o-anisyl aryl sulfilimines **6** were prepared by the reaction of the corresponding sulfides **7** and t-butyl hypochlorite in the presence of (–)-menthol and the appropriate amide anions **8**.[6] This asymmetric synthesis was believed to proceed via an initial formation of diastereomeric menthoxysulfonium chlorides **9** which react further with N-arylsulfonyl amide or N-benzamide anions, affording optically active sulfilimines **6** (Scheme 4.1.1).

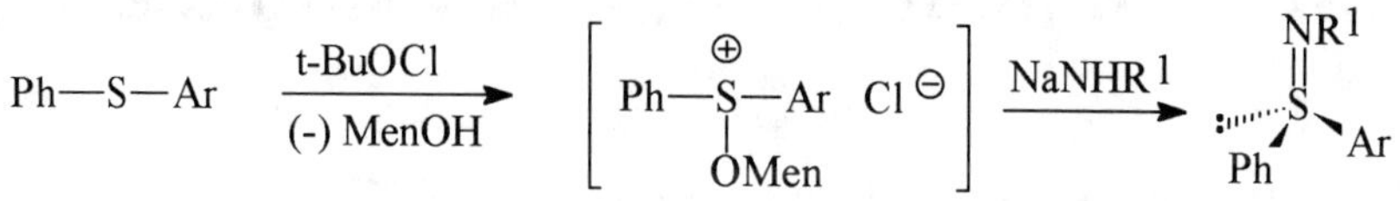

Ph—S—Ar (**7**) —[t-BuOCl, (-) MenOH]→ [Ph—$S^{\oplus}$(OMen)—Ar $Cl^{\ominus}$] (**9**) —[$NaNHR^1$]→ Ph(Ar)S=NR^1 (**6a-g**)

	Ar	R^1
a	o-An	p-$TolSO_2$
b	o-An	p-$ClC_6H_4SO_2$
c	o-An	p-$MeOC_6H_4SO_2$
d	o-An	$PhSO_2$
e	o-An	o-$TolSO_2$
f	o-Tol	p-$TolSO_2$
g	o-An	PhCO

Scheme 4.1.1

Taking advantage of the fact that the specific rotation of the sulfilimines thus obtained increased by repeated recrystallization from acetone-hexane, all the compounds shown above were isolated as pure enantiomers with the values of optical rotation given in Table 4.1.1.

Table 4.1.1 Asymmetric Synthesis of Aryl N-Substituted Sulfilimines **6**

No	Ar	R^1	Yield [%]	$[\alpha]_D$[a]	M.p. [°C]
a	o-An	p-$TolSO_2$	15	–48.0	161.5–162.0
b	o-An	p-ClC_6H	12	–78.2	160.5–161.0
c	o-An	p-MeO-C_6H_4	2	–88.7	149.5–150.5
d	o-An	$PhSO_2$	3	–79.3	143.5–144.0
e	o-An	o-$MeC_6H_4SO_2$	6	–49.1	149.0–152.0
f	o-Tol	p-$TolSO_2$	11	–41.0	122.0–122.5
g	o-An	PhCO	65	–13.0	Oil

[a] Measured in chloroform (c = 1.00).

From Moriyama, M., Yoshimura, T., Furukawa, N., Numata, T., Oae, S., *Tetrahedron,* 32, 3003, 1976. With permission.

When optically active N-p-toluenesulfonylsulfilimines **6a** and **6f** were treated with concentrated sulfuric acid, the corresponding optically active free sulfilimines **10a,b** were obtained in relatively good yields (Scheme 4.1.2).

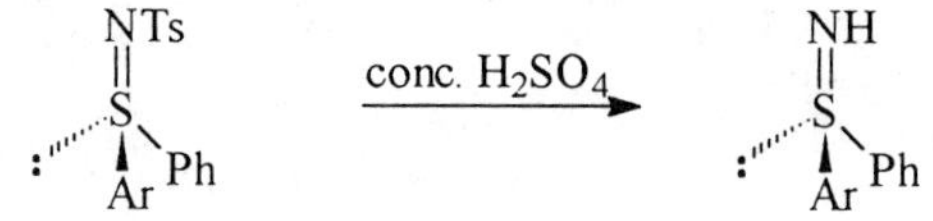

Ar(Ph)S=NTs (**6a,f**) —[conc. H_2SO_4]→ Ar(Ph)S=NH (**10a,b**)

No	Ar	$[\alpha]_D$	No	Yield	$[\alpha]_D$
6a	o-An	–98.0	**10a**	57	–194.0
6f	o-Tol	–41.0	**10b**	70	–85.0

Scheme 4.1.2

Optically active N-acyl-**11a–e** and N-arylsulfonyl sulfilimines **11f–g** were obtained easily by the reaction of the sulfilimine (–)-(S)-**10a** with carboxylic anhydride and arenesulfonyl chloride, respectively. N-β-Cyanoethylsulfilimine **11h** was also prepared by treatment of the free sulfilimine **10a** with acrylonitrile at room temperature (Scheme 4.1.3).

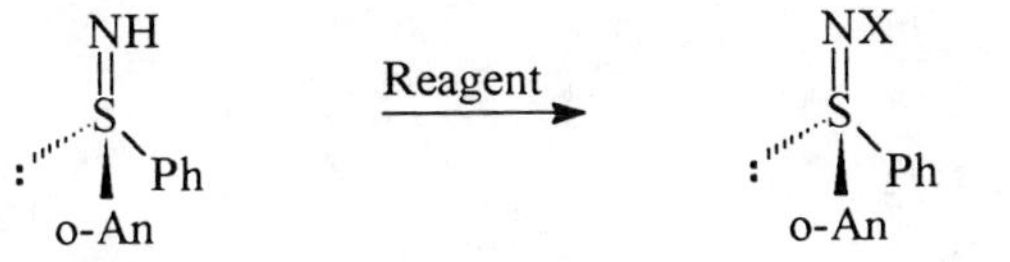

NH, S, Ph, o-An — Reagent → NX, S, Ph, o-An

10a **11a-g**

X	11	Reagent	$[\alpha]_D$ 10a	$[\alpha]_D$ 11	Yield of **11** [%]
MeCO	a	$(MeCO)_2O$	–107.0	86.2	77
PhCO	b	$(PhCO)_2O$	–107.0	–43.6	83
Me_2CHCO	c	$(Me_2CHCO)_2O$	–157.0	–123.0	76
CF_3CO	e	$(CF_3CO)_2O$	–157.0	–66.7	77
p-TolSO_2	f	p-TolSO_2Cl	–194.0	–98.0	95
PhSO_2	g	PhSO_2Cl	–157.0	–77.0	93
CH_2CH_2CN	h	CH_2=CHCN	–138.0	–96.0	65

Scheme 4.1.3

The simplest and most versatile method for the synthesis of chiral sulfilimines of very high optical purity is based on a highly stereoselective reaction of chiral sulfoxides with various iminating reagents such as bis (N-tosylsulfurdiimide),[7,8] N-sulfinyl-p-toluenesulfonamide,[7] p-toluenesulfinyl nitrene,[9] and arylsulfonamide in the presence of P_2O_5 and Et_3N[7a,10] (Equations 4.1.1–4.1.5). With all these reagents the sulfoxide → sulfilimine conversion carried out in C_6H_6 or CH_2Cl_2 occurs with retention of configuration at sulfur. However, when pyridine is used as a solvent, the reaction of sulfoxides with bis(N-tosylsulfurdiimide) and N-sulfinyl p-toluenesulfonamide proceeds with inversion of configuration at sulfur in accordance with the original observation of Day and Cram.[7a]

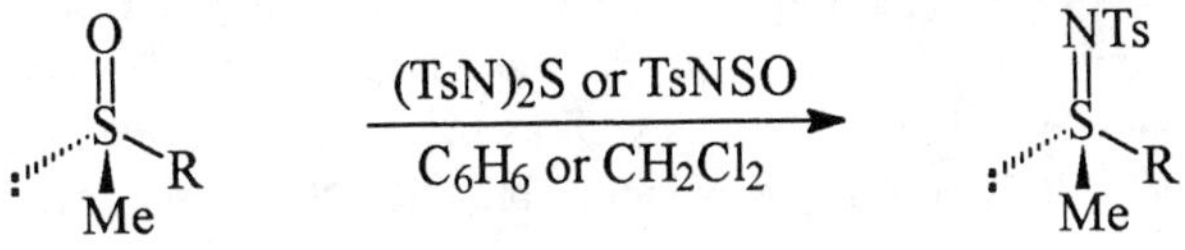

O, S, R, Me — $(TsN)_2S$ or TsNSO / C_6H_6 or CH_2Cl_2 → NTs, S, R, Me

12a,g **a**, R=p-Tol **b**, R=n-Bu **13a,b** (4.1.1)

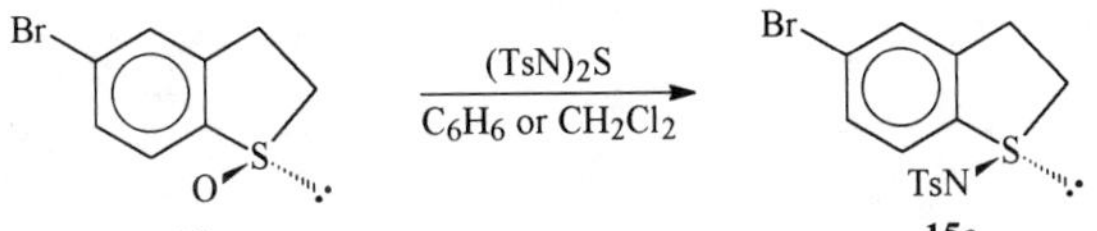

Br, O, S — $(TsN)_2S$ / C_6H_6 or CH_2Cl_2 → Br, TsN, S

14a **15a** (4.1.2)

X, S, O — $(TsN)_2S$ or TsNSO / pyridine → X, TsN, S

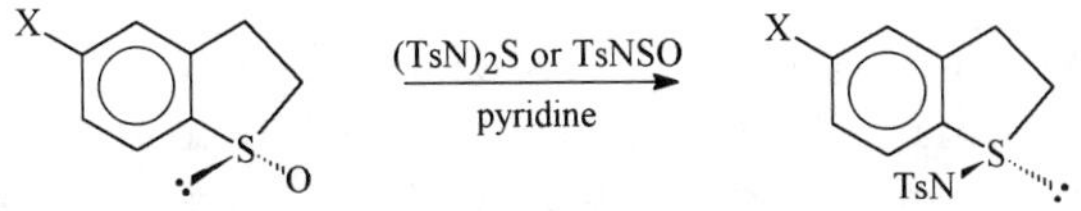

14a,b **15a,b**

a, X=Br
b, X=H (4.1.3)

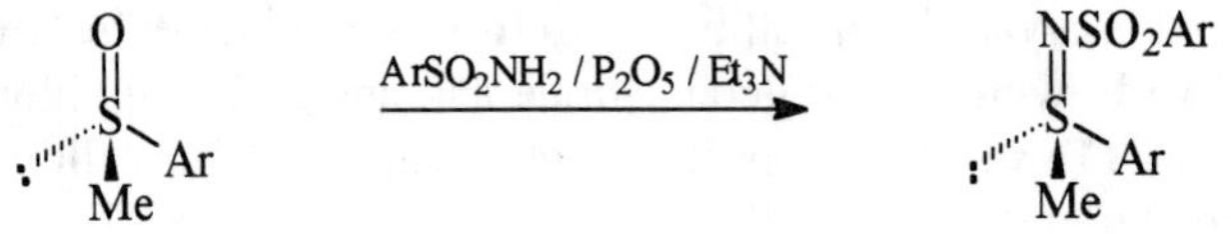

$ArSO_2NH_2 / P_2O_5 / Et_3N$

16 → 17 (4.1.4)

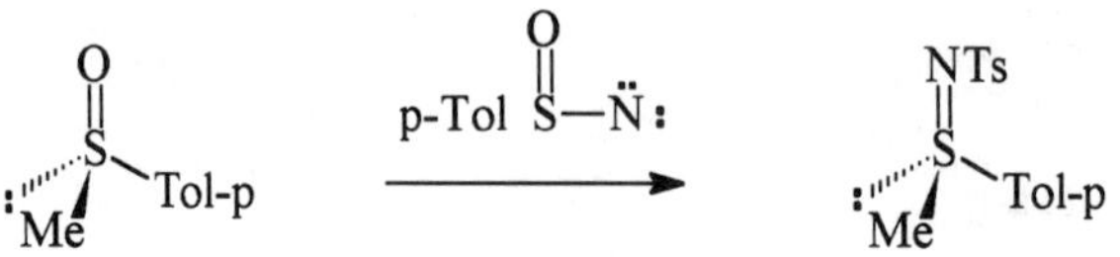

p-Tol S(O)–N:

12a → 13a (4.1.5)

The enantiomerically pure sulfilimine **13a** was obtained in the reaction of the diastereomerically pure O-menthoxy precursor **18** with methylmagnesium bromide (Equation 4.1.6).[11]

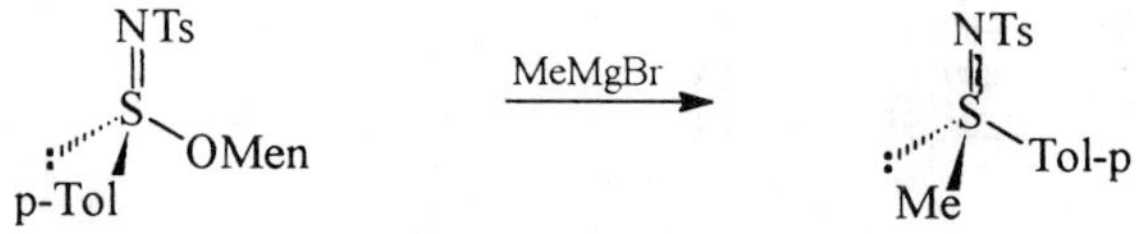

MeMgBr

(-)(S)-**18** → (+)(S)-**13a** (4.1.6)

4.2 SYNTHESIS OF CHIRAL SULFOXIMINES

As a matter of fact, it should be noted that the sulfoximine as a functional group was unknown to organic chemistry until 1950, when Bentley and co-workers[12a] reported the isolation and structure elucidation of the diastereomerically pure methionine sulfoximine **19a**. The (2S, 5S)-isomer of this compounds was found to be a potent inhibitor of the enzyme glutamine synthetase.[13] It is responsible for the mammalian neurotoxicity resulting from the ingestion of nitrogen trichloride-bleached flour. This sulfoximine was later found to occur as a natural product in the form of the tripeptide N-phosphate **19b**,[14] the first naturally occurring sulfoximine to be discovered. These observations and remarkable biological activity of many sulfoximines **19**, as well as the interesting diversity of reaction associated with the functionality itself, have stimulated increasing interest in stereochemical aspects of the sulfoximine chemistry.

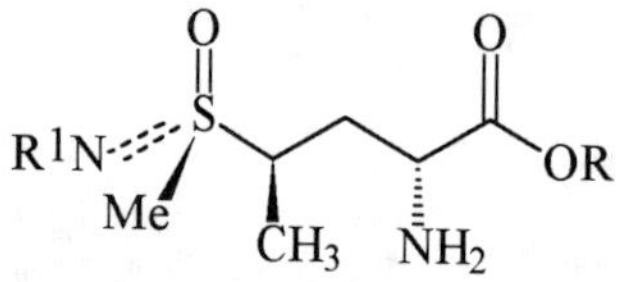

19a R = OH, R^1 = H

19b R = ala-ala, R^1 = PO_3H_2

The first example of a chiral, laboratory-made sulfoximine was reported by Barash[15] in 1960. He succeeded in separation of the diastereomeric sulfoximines **20a,b** obtained by condensation of racemic methyl-p-nitrophenylsulfoximine **21** with the chiral sulfonyl chloride **22** (Equation 4.2.1).

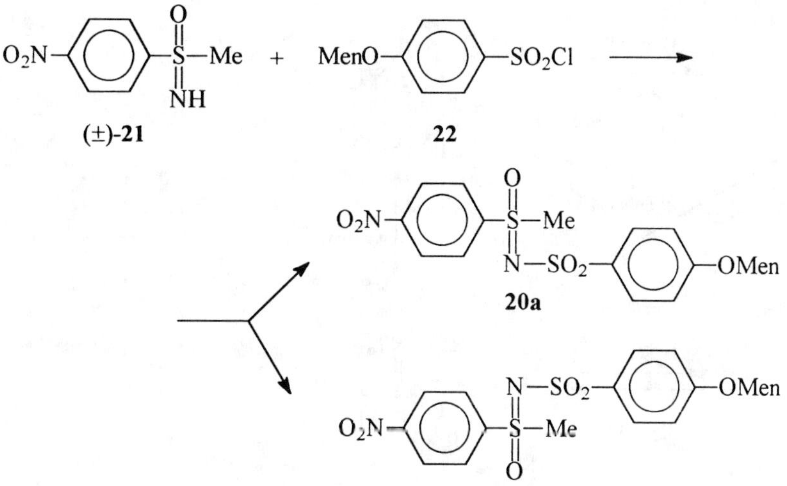

(±)-21 22

20a

20b (4.2.1)

Soon thereafter, the resolution of the racemic sulfoximine **23** containing a carboxylic groups, via diastereomeric salts with α-methylbenzyl amine, was reported by Kresze and Wustrow.[3] Fusco and Tenconi[16] took advantage of the basic nature of the imino group of the free sulfoximines and resolved methylphenylsulfoximine **24** by means of (+) camphoro-10-sulfonic acid. Later on, this procedure was improved by Johnson and Schroeck.[17] They also resolved in a similar way mesityl methyl sulfoximine **25**.[17,18] Another improvement of the original procedure of Fusco and Tenconi[16] has been reported to give both enantiomers of **24** in greater than 99.9% enantiomeric purity.[19] These sulfoximines have also been resolved via chromatography on chiral stationary phase.[20,21]

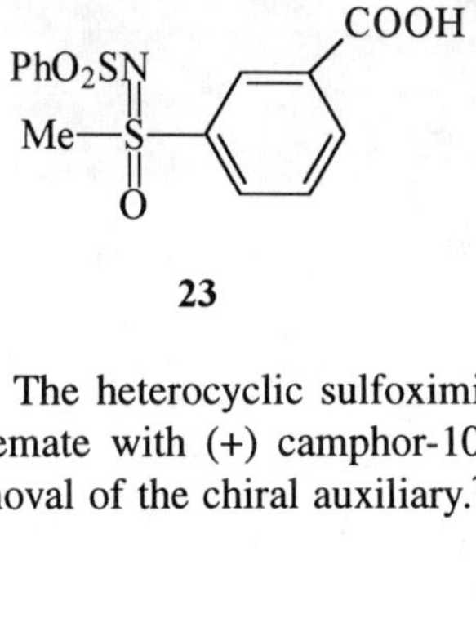

23 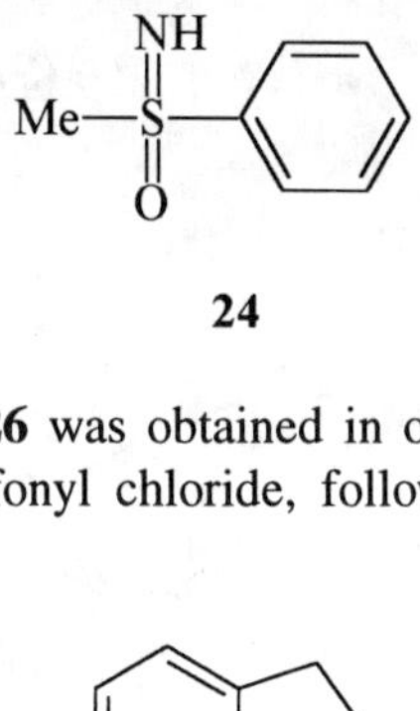 24 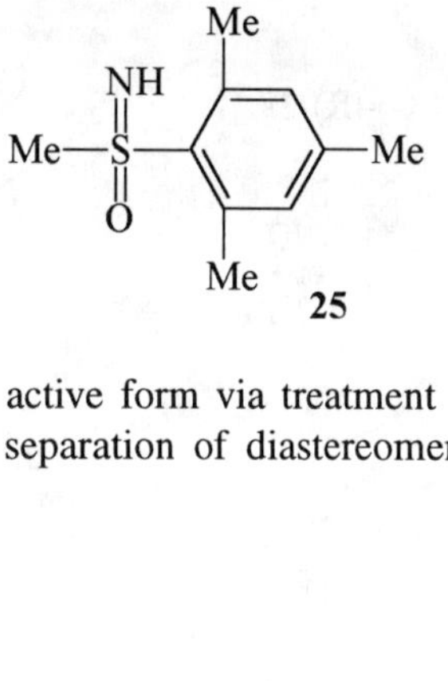 25

The heterocyclic sulfoximine **26** was obtained in optically active form via treatment of the racemate with (+) camphor-10-sulfonyl chloride, followed by separation of diastereomers and removal of the chiral auxiliary.[7a]

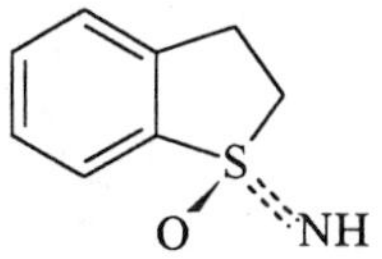

26

Other approaches to the preparation of chiral sulfoximines are based on the reaction of optically active sulfoxides with arenesulfonyl azides in the presence of copper and on the oxidation of optically active sulfilimines. These reactions occur with full retention of configuration at sulfur[7,10,11] and were utilized in the construction of the sulfoxide-sulfilimine-sulfoximine interconversion cycles shown in Schemes 4.2.1–4.2.3.

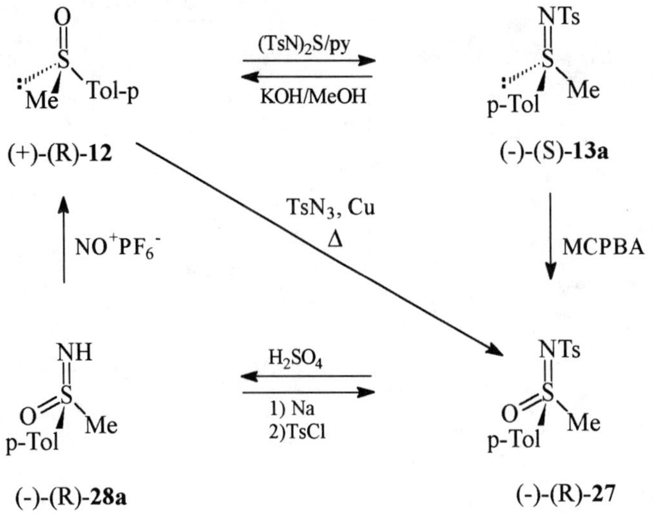

(+)-(R)-**12** $\xrightarrow{(TsN)_2S/py}$ (-)-(S)-**13a**; (-)-(S)-**13a** $\xrightarrow{KOH/MeOH}$ (+)-(R)-**12**

(+)-(R)-**12** $\xrightarrow{TsN_3, Cu, \Delta}$ (-)-(R)-**27**

(-)-(S)-**13a** $\xrightarrow{MCPBA}$ (-)-(R)-**27**

(-)-(R)-**27** $\xrightarrow{H_2SO_4}$ (-)-(R)-**28a**; (-)-(R)-**28a** $\xrightarrow{1) Na, 2) TsCl}$ (-)-(R)-**27**

(-)-(R)-**28a** $\xrightarrow{NO^+PF_6^-}$ (+)-(R)-**12**

Scheme 4.2.1

(-)-(R)-**29** $\xrightarrow{TsNSO/py}$ (-)-(S)-**15b**; (-)-(S)-**15b** $\xrightarrow{KOH/MeOH}$ (-)-(R)-**29**

(-)-(R)-**29** $\xrightarrow{TsN_3, Cu, \Delta}$ (-)-(R)-**30**

(-)-(S)-**15b** $\xrightarrow{KMnO_4}$ (-)-(R)-**30**

(+)-(R)-**26** $\xrightarrow{TsCl/Pyridine}$ (-)-(R)-**30**; (-)-(R)-**30** $\xrightarrow{H_2SO_4}$ (+)-(R)-**26**

(+)-(R)-**26** $\xrightarrow{HNO_2 \text{ or } NO^+PF_6^-}$ (-)-(R)-**29**

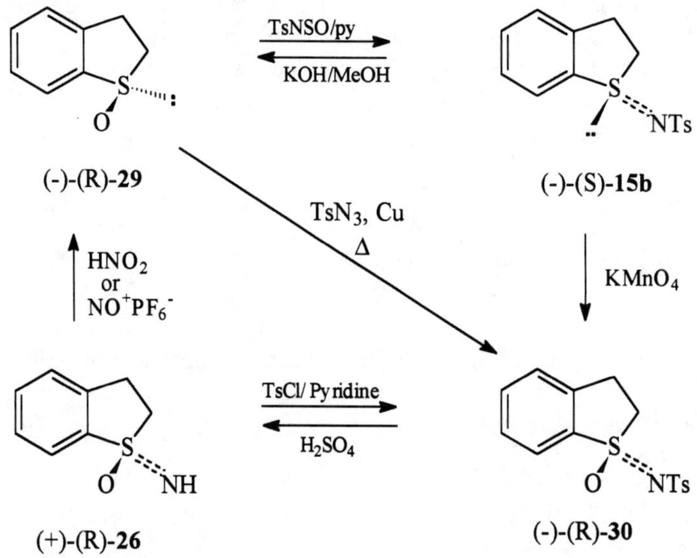

Scheme 4.2.2

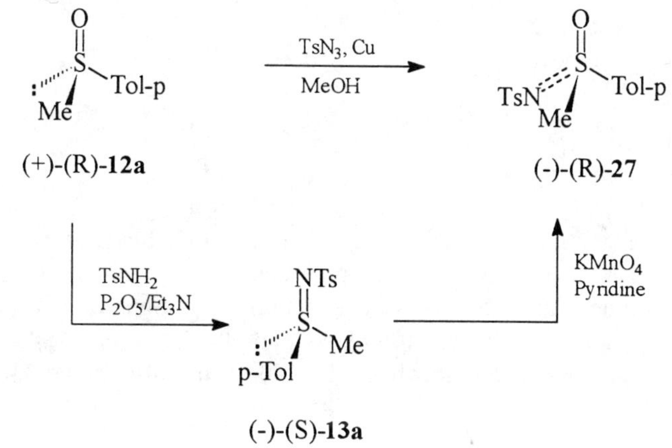

(+)-(R)-**12a** → (TsN_3, Cu; MeOH) → (-)-(R)-**27**

(+)-(R)-**12a** → ($TsNH_2$; P_2O_5/Et_3N) → (-)-(S)-**13a** → ($KMnO_4$; Pyridine) → (-)-(R)-**27**

Scheme 4.2.3

Chiral N-arylsulfonylsulfoximines, produced either in the copper-catalyzed reaction of chiral sulfoxides with arenesulfonyl azide or by the oxidation of arylsulfonyl sulfimines, may be hydrolyzed with strong acids to optically active N-H sulfoximines.[7] However, this conversion often fails and/or results in decomposition. Therefore, it is interesting to note that a series of chiral N-unsubstituted p-tolyl alkyl (aryl)sulfoximines **28** have been easily prepared by treatment of chiral sulfoxides **12** with O-mesitylsulfonyl hydroxylamine (MSH)[22,23] (Equation 4.2.2 and Table 4.2.1). As expected, this imidation process occurs with retention of configuration and results in the formation of sulfoximines **28** of very high optical purity.

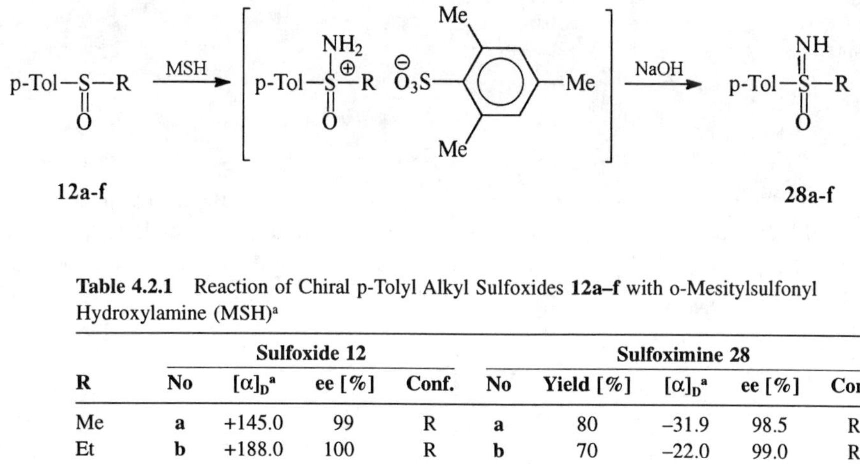

p-Tol—S(=O)—R (**12a-f**) —MSH→ [p-Tol—$S^{\oplus}(NH_2)$(=O)—R $^{\ominus}O_3S$—mesityl] —NaOH→ p-Tol—S(=NH)(=O)—R (**28a-f**) (4.2.2)

Table 4.2.1 Reaction of Chiral p-Tolyl Alkyl Sulfoxides **12a–f** with o-Mesitylsulfonyl Hydroxylamine (MSH)[a]

	Sulfoxide 12				Sulfoximine 28				
R	**No**	**$[\alpha]_D$[a]**	**ee [%]**	**Conf.**	**No**	**Yield [%]**	**$[\alpha]_D$[a]**	**ee [%]**	**Conf.**
Me	**a**	+145.0	99	R	**a**	80	−31.9	98.5	R
Et	**b**	+188.0	100	R	**b**	70	−22.0	99.0	R
i-Pr	**c**	+191.1	100	R	**c**	79	−12.1	99.0	R
n-Bu	**d**	+193.8	100	R	**d**	77	−17.2	99.0	R
$PhCH_2$	**e**	+234.0	93	R	**e**	60	+4.7	92.0	R
Ph	**f**	+21.0	99.5	R	**f**	19	+5.0	99.0	R

[a] All rotations measured in acetone solution.

From Johnson, C. R., Kirchoff, R. A., Corkins, H. G., *J. Org. Chem.*, **39**, 2458, 1974. With permission.

Very recently, this reaction has been applied[24] for the preparation of unsubstituted sulfoximines **31** and **24**.

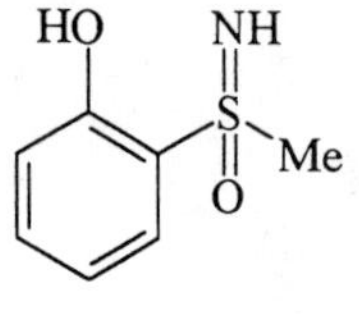

31 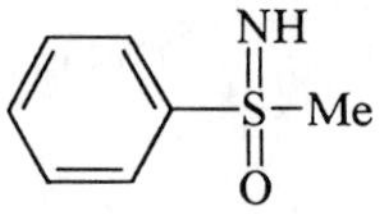 24

A closely related stereoselective amination of chiral sulfoxides reported by Colonna and Stirling[25] is based on reaction with N-aminophthalimide in the presence of lead tetra-acetate. The corresponding N-phthalimidosulfoximines **33a–c** formed with retention of configuration may be reconverted to the starting sulfoxides **12a,e,h** by treatment with sodium ethoxide (Scheme 4.2.4). This reconversion occurs even in the absence of a base when a sample of the sulfoximine **33a** is kept for a longer time in methanol.[26]

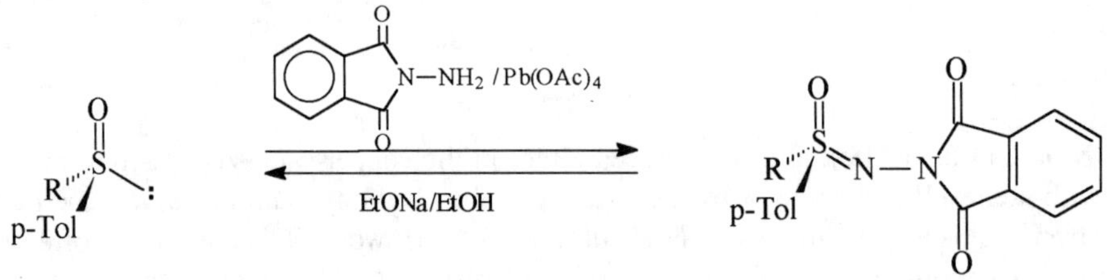

N–NH$_2$ /Pb(OAc)$_4$

EtONa/EtOH

(R)-**12** (R)-**33a-f**

No	R	$[\alpha]_{589}$	No	$[\alpha]_{589}$	Ref.
a	Et	203.0	**a**	–125.8	26
b	n-Pr	–171.6	**b**	+142.7	26
c	t-Bu	+156.7	**c**	–166.6	26
d	CH_2Ph	a	**d**	–19.0	25
e	$CH_2CH=CH_2$	a	**e**	+30.0	25
f	1-Npht	a	**f**	–26.8	25

[a] Not reported.

Scheme 4.2.4

Chiral unsubstituted sulfoximines react with electrophilic reagents at the nitrogen atom, affording a variety of the N-substituted sulfoximines. The monomethylation, dimethylation, and chlorination reactions of (–)-(R)-methyl p-tolyl sulfoximine **28a** illustrate this type of functionalization (Scheme 4.2.5).

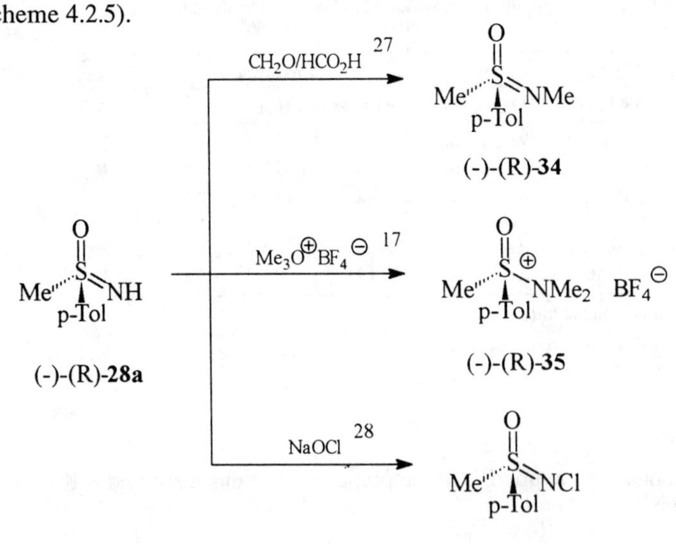

CH_2O/HCO_2H 27

(-)-(R)-**34**

$Me_3O^{\oplus}BF_4^{\ominus}$ 17

(-)-(R)-**28a**

(-)-(R)-**35**

NaOCl 28

(-)-(R)-**36**

Scheme 4.2.5

Similar modifications of (+)-(S) methyl phenyl sulfoximine **24** are collected in Scheme 4.2.6.[29]

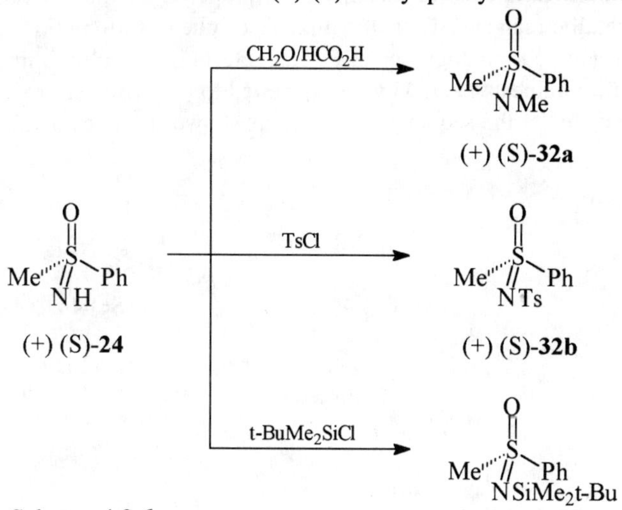

(+) (S)-**24** — CH_2O/HCO_2H → (+) (S)-**32a**

(+) (S)-**24** — TsCl → (+) (S)-**32b**

(+) (S)-**24** — t-$BuMe_2SiCl$ →

Scheme 4.2.6

Unsubstituted sulfoximine **37** was obtained[30] with almost full retention of configuration at sulfur upon alkaline hydrolysis of N-chlorosulfilimine **38** (Equation 4.2.3).

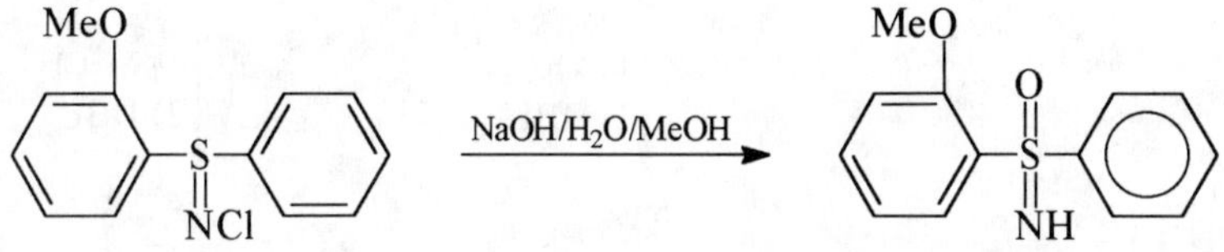

38 — $NaOH/H_2O/MeOH$ → **37** (4.2.3)

Reaction of racemic allylic sulfoximines **33d**, **39**, and **40** with a deficiency of a (–)(1R, 2S)-ephedrine-**41** resulted in kinetic resolution of these sulfoximines (Scheme 4.2.7).[31]

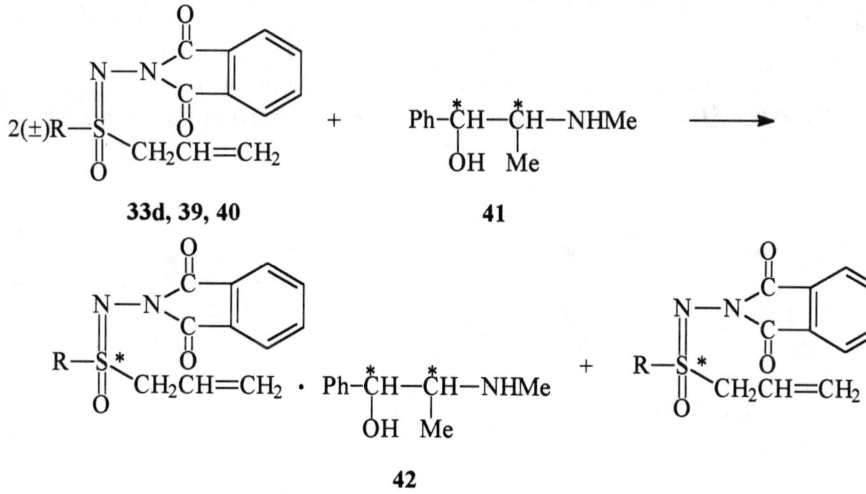

33d, 39, 40 + **41** → **42** +

Sulfoximine No	Sulfoximine R	Amine	T/°C	Time/day	Recovered sulfoximine $[\alpha]_D$	Recovered sulfoximine ee [%]
33d	p-Tol	(–)-**41**	25	2	–11.5	38
33d	p-Tol	(–)-**41**	–30	2	–13.8	46
39	$PhCH_2$	(–)-**41**	–30	2	+14.7	45
40	1-$C_{10}H_7$	(–)-**41**	–30	3	–3.0	12

Scheme 4.2.7

A few diastereomerically pure sulfoximines have recently been prepared by the reaction of chiral sulfoximidates with organometallic reagents. Thus, the mixture of phenylsulfonimidates **43** upon treatment with methyllithium gave chromatographically separable (+)-(S_S)-sulfoximine **44** and (+)(R_S)-diastereomer **45**. The sulfoximines **44** and **45** were converted to (S_S)-vinyl sulfoximine **46** and (R_S)-sulfoximine **47**, respectively, by the sequence of reactions shown in Scheme 4.2.8.[32]

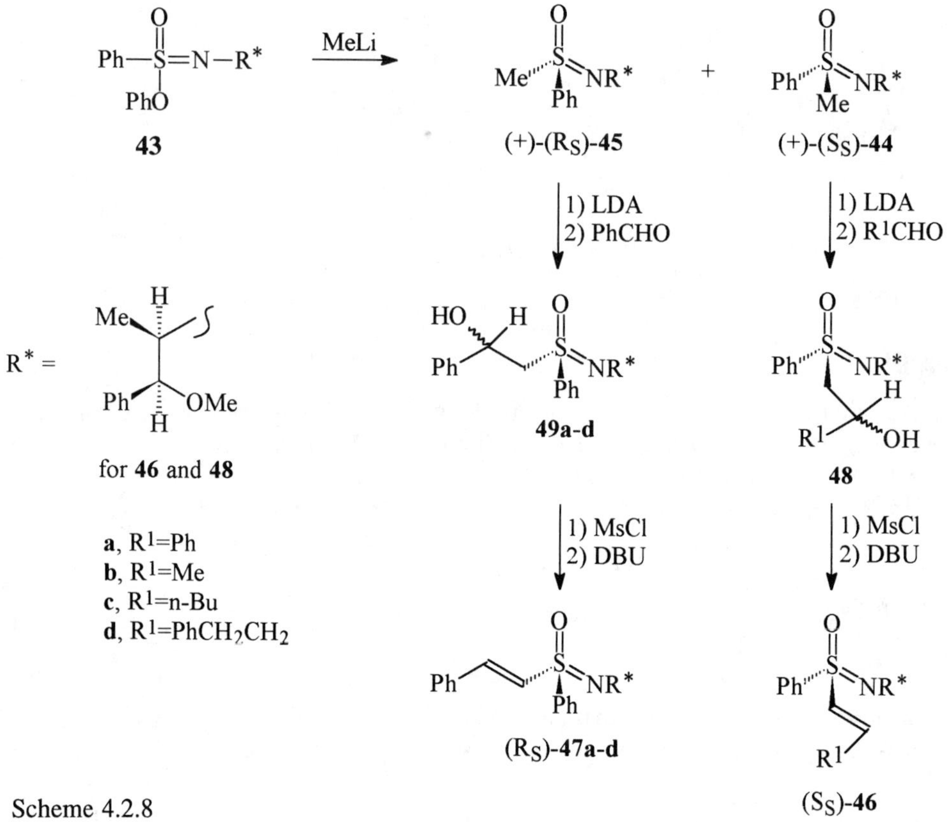

Scheme 4.2.8

A similar sequence of reactions were used[33] for the preparation of diastereomerically pure sulfoximines **50a,b** and **51a,b** starting from the parent methyl phenyl sulfoximines **52** and **53** (Scheme 4.2.9).

(R$_S$)-**53** → 1) LDA; 2) R^{1}CHO; 3) MsCl; 4) DBU → (R$_S$)-**51a,b**

(S$_S$)-**52** → 1) LDA; 2) R^{1}CHO; 3) MsCl; 4) DBU → (S$_S$)-**50a,b**

R* = —CH(Me)—Ph

a, R^1= n-Bu
b, R^1= Me

Scheme 4.2.9

The diastereomerically pure sulfoximines **54** and **55** were obtained[34] in the reaction of the cyclic sulfoximines **56** and **57** derived from valinol with organometallic reagents (Equation 4.2.4 and Scheme 4.2.10).

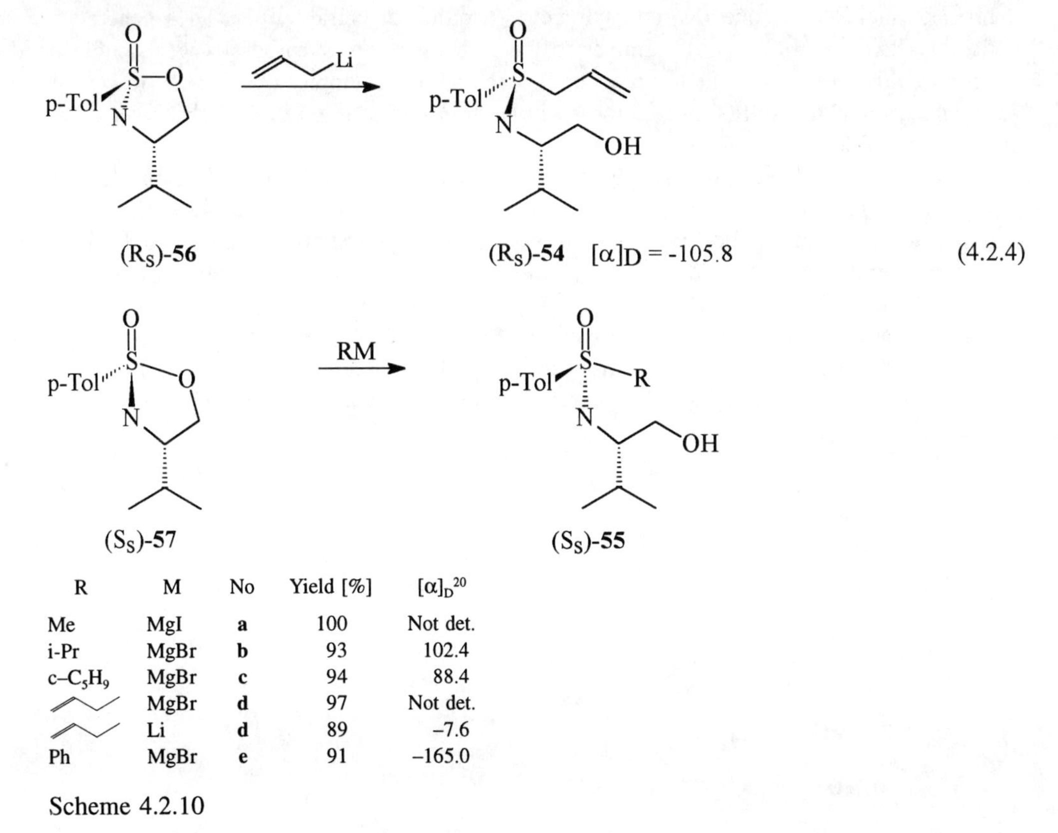

R	M	No	Yield [%]	$[\alpha]_D^{20}$
Me	MgI	a	100	Not det.
i-Pr	MgBr	b	93	102.4
c–C_5H_9	MgBr	c	94	88.4
	MgBr	d	97	Not det.
	Li	d	89	–7.6
Ph	MgBr	e	91	–165.0

Scheme 4.2.10

4.3 CHIRAL SULFILIMINES AND SULFOXIMINES IN ASYMMETRIC SYNTHESIS

The chemical literature contains a single report on the use of sulfilimine as a chiral auxiliary.[35] It describes asymmetric synthesis of the chiral α-acylaziridines **59** in the reaction between optically active o-methoxyphenyl phenyl sulfilimine **10a** and olefines **60** (Scheme 4.3.1).

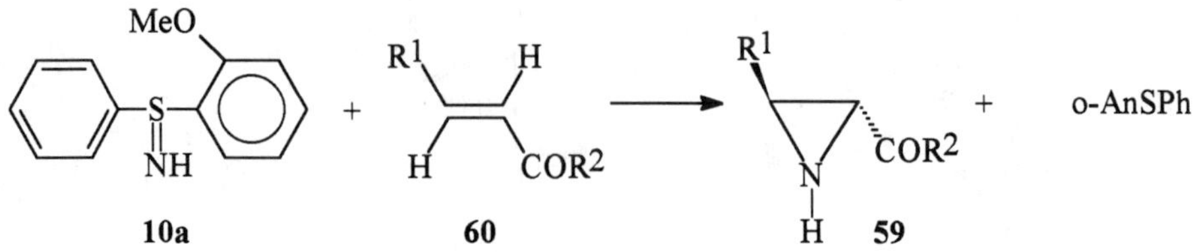

MeO

S NH + R^1 H H COR^2 → R^1 COR^2 N H + o-AnSPh

10a **60** **59**

R^1	R^2	Yield [%]	$[\alpha]_{589}$	ee [%]
PhCO	Ph	30	–32.9	n.r.
Ph	Ph	66	–88.0	28.7
Ph	Me	23	–60.0	n.r.

Scheme 4.3.1

This reaction involves a typical Michael addition of **10a** to the carbon-carbon double bond followed by elimination of the corresponding sulfide and gives the chiral α-acylaziridines **59** with ee about 30%. Unfortunately, this asymmetric synthesis is limited to strongly electrophilic olefines such as α,β-dicarbonyl derivatives and therefore has no general character.

On the other hand, in the last two decades chiral sulfoximines have found wide application in asymmetric synthesis. The first example of asymmetric induction in the transfer of chirality from the chiral sulfoximine sulfur atom to prochiral carbon atom was described by Johnson and Schroeck[36] in 1968. It concerns the asymmetric addition of chiral sulfur ylides generated from optically active methyl aryl sulfoximines to the carbon-carbon or carbon-oxygen double bonds. Therefore, this type of asymmetric reaction, as well as two examples of asymmetric synthesis of aziridines using chiral sulfoximine anions, will be discussed later on in the subchapter describing chiral sulfur ylides (Section 5.2).

A very convenient asymmetric synthesis of alcohols developed also by Johnson and Stark[37] is based on the use of optically active β-hydroxy sulfoximines as the reagents that induce optical activity. Generally, this method consists of the asymmetric addition of N-methyl phenyl sulfoximine **32a** to prochiral ketones or aldehydes **61**, subsequent separation of the produced diastereomeric adducts **62**, and final desulfurization with Raney nickel to form chiral sulfur-free tertiary alcohols **63**. A wide range of diastereomeric sulfoximines **62** were prepared by this method (Scheme 4.3.2). After separation by preparative medium-pressure liquid chromatography (MPLC), the single diastereomers were converted to the enantiomerically pure tertiary alcohols listed in Table 4.3.1.

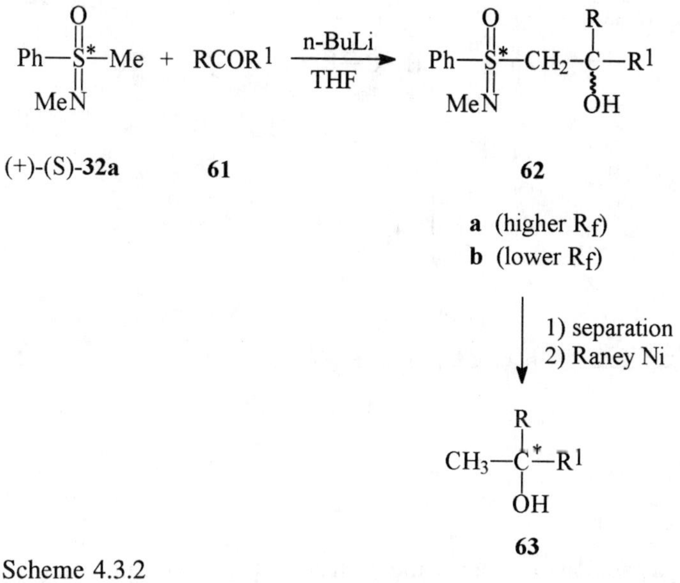

$$\mathrm{Ph{-}\overset{O}{\underset{MeN}{S^*}}{-}Me} + \mathrm{RCOR^1} \xrightarrow[\text{THF}]{\text{n-BuLi}} \mathrm{Ph{-}\overset{O}{\underset{MeN}{S^*}}{-}CH_2{-}\overset{R}{\underset{OH}{C}}{-}R^1}$$

(+)-(S)-**32a** **61** **62**

a (higher R_f)
b (lower R_f)

1) separation
2) Raney Ni

$$\mathrm{CH_3{-}\overset{R}{\underset{OH}{C^*}}{-}R^1}$$

63

Scheme 4.3.2

Table 4.3.1 Optically Active Alcohols **53** from ß-Hydroxy Sulfoximines **62** (Scheme 4.3.1)[a]

Adduct 62	R	R′	Alcohol 63 obsd $[\alpha]_D$
a	Ph	Et	$[\alpha]^{27}_D$ +16.1 (*c* 1.3, EtOH)
b			$[\alpha]^{27}_D$ −15.7 (*c* 0.8, EtOH)
a			$[\alpha]^{22}_D$ +18.0 (neat)
			$[\alpha]^{27}_D$ +15.4 (*c* 2.5, EtOH)
a	Ph	n-Pr	$[\alpha]^{25}_D$ +5.59 (*c* 0.9, acetone)
b			$[\alpha]^{25}_D$ −5.14 (*c* 1.4, acetone)
a	Ph	n-Bu	$[\alpha]^{22}_D$ +9.9 (*c* 3.2, acetone)
b			$[\alpha]^{22}_D$ −9.6 (*c* 3.1, acetone)
a	Ph	c-C_6H_{11}	$[\alpha]^{22}_D$ +20.8 (*c* 2.7, $CHCl_3$)
b			$[\alpha]^{22}_D$ −20.7 (*c* 2.1, $CHCl_3$)
a	$PhCH_2$	Et	$[\alpha]^{31}_D$ −8.61 (*c* 1.3, EtOH)
b			$[\alpha]^{31}_D$ +7.79 (*c* 1.3, EtOH)
a	*i*-Bu	Et	$[\alpha]^{21}_D$ −1.86 (neat)
			$[\alpha]^{21}_D$ −3.36 (*c* 3.3, acetone)
b			$[\alpha]^{21}_D$ +3.6 (*c* 3.3, acetone)

From Johnson, C. R., Stark, J., Jr., *J. Org. Chem.*, 47, 1193, 1982. With permission.

Pairs of diastereomeric β-hydroxy sulfoximines were also prepared from **32a** and aldehydes. Their separation by preparative-scale MPLC was, however, very poor. Therefore, they were directly converted to secondary alcohols having moderate optical purities (25–46%).[37]

An alternative approach to the preparation of chiral secondary alcohols using diastereomeric β-hydroxy sulfoximines as precursors is based on the diastereoselective reduction of optically active β-keto sulfoximines **64** prepared by butyllithium-mediated condensation of (+)-(S)-**32a** with nitriles **65**. The β-keto sulfoximines **64** thus obtained afforded on treatment with a variety of reducing agents β-hydroxy sulfoximines **66** which, after Raney nickel desulfurization, gave secondary alcohols **67** with optical purities in the range from 18 to 62% (Scheme 4.3.3).[38]

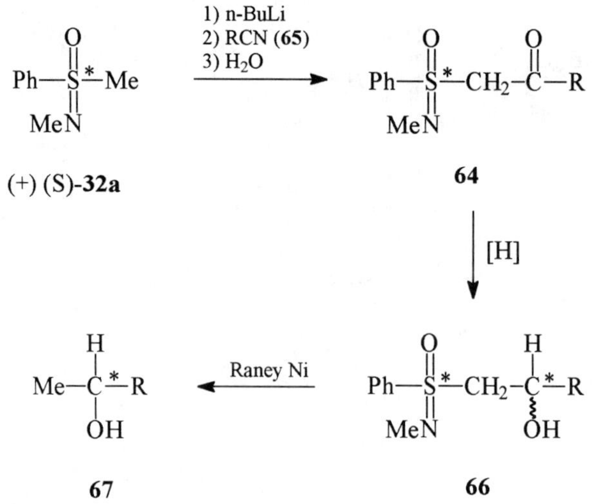

Scheme 4.3.3

Another approach to the preparation of alcohols, in which chiral β-hydroxy sulfoximines play a role of a chiral inducer, involves the asymmetric reduction of prochiral ketones with a reducing system containing chiral sulfoximine as a component. For the first time this methodology was used also by Johnson and Stark[39] as early as 1979. They found that chiral β-hydroxysulfoximines **68a,b** and **69** upon treatment in toluene solution at –78°C with 2 equivalents of gaseous diborane form complexes which were able to reduce ketones to the corresponding chiral secondary alcohols with ee values ranging from 3–82% (Scheme 4.3.4).

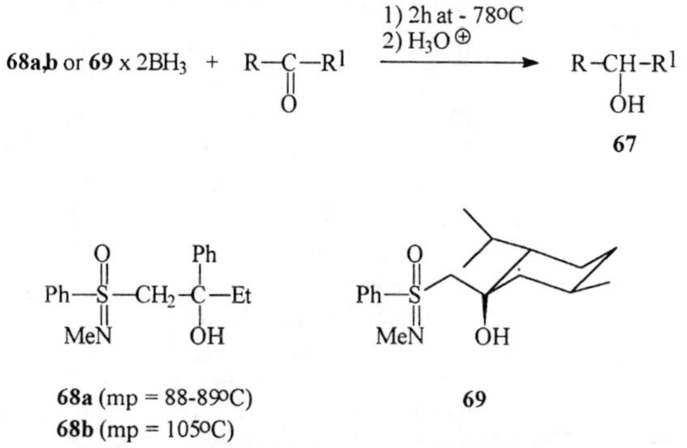

68a (mp = 88-89°C)
68b (mp = 105°C)

Sulfoximine	R	R^1	Alcohol **67** Yield [%]	Alcohol **67** ee [%]	Abs. conf.
68a	Ph	Me	80	60	S
68b	Ph	Me	92	57	R
69	Ph	Me	66	82	S
68a	Ph	Et	66	33	S
68b	Ph	Et	74	74	R
69	Ph	Et	48	29	S
68b	Ph	n-Pr	81	70	R
68b	Ph	i-Pr	53	8	R
68b	Me	n-C_6H_{13}	79	8	R

Scheme 4.3.4

Very recently, it was reported[40] that chiral β-hydroxysulfoximines **70** catalyze the borane reductions of prochiral ketones, affording chiral secondary alcohols in high yields and with good enantiomeric excesses (up to 93%) (Scheme 4.3.5).

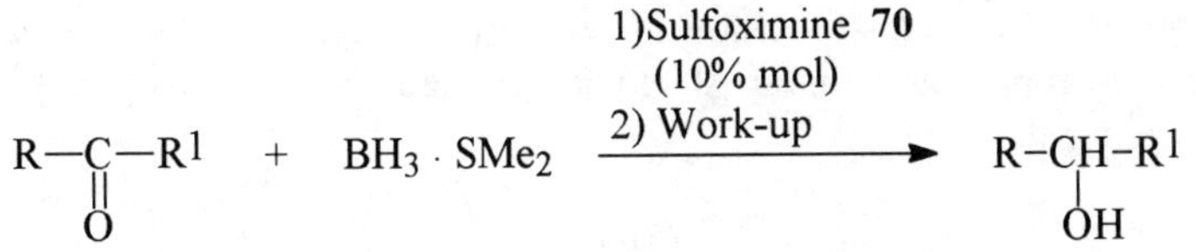

R—C(=O)—R1 + $BH_3 \cdot SMe_2$ —[1) Sulfoxime **70** (10% mol); 2) Work-up]→ R–CH(OH)–R1 **67**

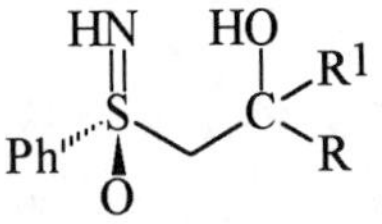

70a-e

a, R = R¹ = Ph
b, R = R¹ = Me
c, R = R¹ = i-Pr
d, R = R¹ = Adamantyl
e, R = Ph, R¹ = i-Pr

70	Alcohol 67			
	R	R¹	ee [%]	Abs. conf.
a	Ph	Me	76	R
a	Ph	Et	73	R
a	$PhCH_2$	Me	70	R
a	Ph	CH_2Cl	84	S
a	Ph	CH_2OR	93	S
b	Ph	Me	70	R
c	Ph	Me	73	R
d	Ph	Me	74	R
e	Ph	Me	61	R

Scheme 4.3.5

It was suggested that the formation of the intermediate complexes **71** and **72** is responsible for a very high enantioselectivity of the reduction under consideration. In such complexes the substituents at the chiral sulfur atom determine the conformation of the heterocycle formed in which the steric requirements of the phenyl groups and the electronic properties of the sulfoximine oxygen direct the coordination of a ketone towards the less hindered β-face of the catalyst.

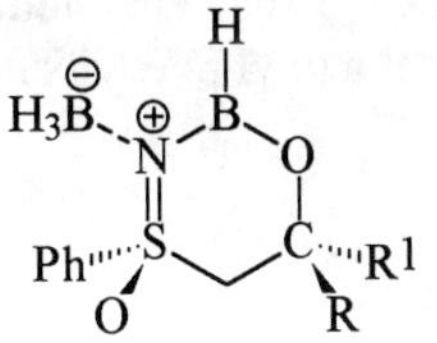

71 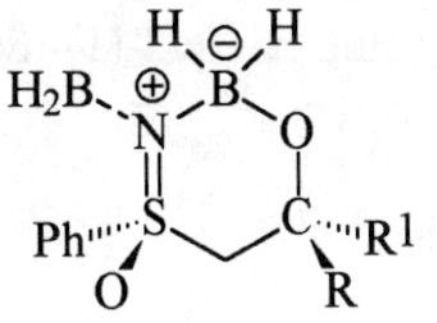 **72**

This procedure was improved by the use of sodium borohydride and trimethylsilyl chloride as a reducing mixture and chiral sulfoximine **70a** as a catalyst.[41] The results of the reduction of several ketones by this new reducing system presented in Scheme 4.3.6 show that the level of asymmetric introduction compares favorably with the optimal results collected in Scheme 4.3.5.

R—C(=O)—R1 + $NaBH_4 / Me_3SiCl$ —[1) (S)-**70a**; 2) Work - up]→ R—CH(OH)—R1

R	R¹	Yield [%]	ee [%]	Abs. conf.
Ph	Me	92	84	R
Ph	Et	86	79	R
Ph	CH_2Cl	88	88	S
Ph	CH_2Br	80	88	S
Ph	i-Pr	72	67	S
$PhCH_2$	CH_3	82	73	R

Scheme 4.3.6

Chiral β-hydroxysulfoximines **73** were also found to be effective catalysts for the enantioselective transfer of the ethyl group from diethylzinc to aldehydes giving chiral secondary alcohols with high enantioselectivity (Scheme 4.3.7).[42]

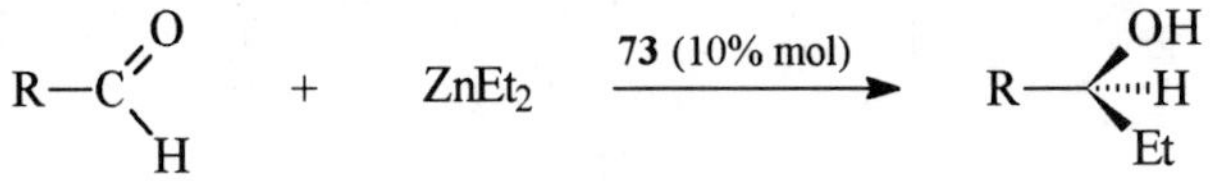

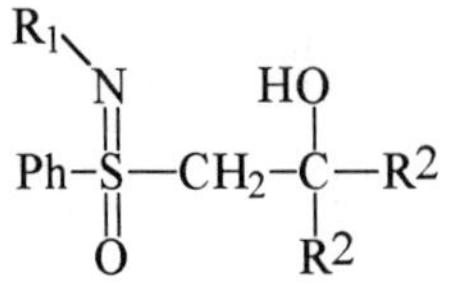

R—C(=O)H + $ZnEt_2$ —**73** (10% mol)→ R—CH(OH)Et

$Ph-S(=O)(=NR^1)-CH_2-C(OH)(R^2)-R^2$

73

a, R^1 = Me, $(R^2)_2$ = $(CH_2)_3$
b, R^1 = Me, $(R^2)_2$ = Me_2
c, R^1 = Ts, $(R^2)_2$ = Me_2
d, R^1 = t-$BuMe_2Si$, $(R^2)_2$ = Me_2

		Alcohol	
73	R	yield [%]	ee [%]
a	Ph	55	84
b	Ph	73	85
c	Ph	52	1
d	Ph	7	4
b	p-ClC_6H_4	79	79
b	p-$MeOC_6H_4$	48	75
b	$PhCH_2CH_2$	50	61

Scheme 4.3.7

Chiral β-hydroxysulfoximines combined with nickel acetylacetonate are able to catalyze the conjugate addition of diethylzinc to chalcone **74**, affording the corresponding alkylated ketone with good enantioselectivities.[43] The structure of the sulfoximine **73** was optimized and an enantiomeric excess of up to 72% was achieved using N-ethyl-(1)-hydroxycyclopentenyl methyl phenyl sulfoximine **73e** (Equation 4.3.1).

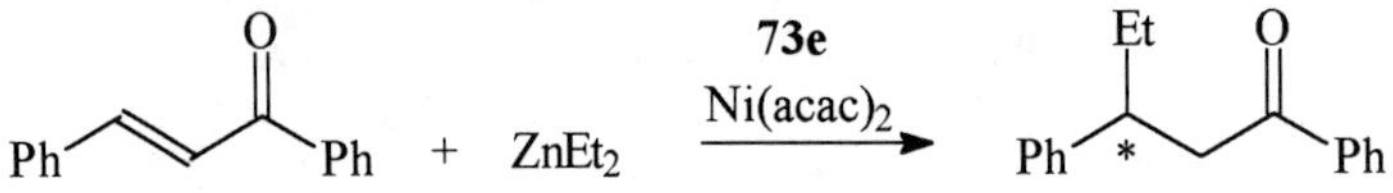

PhCH=CHC(=O)Ph (**74**) + $ZnEt_2$ —**73e**, $Ni(acac)_2$→ PhCH*(Et)CH_2C(=O)Ph (**75**)

$Ph-S(=O)(=NEt)-CH_2-C(OH)$(cyclopentyl)

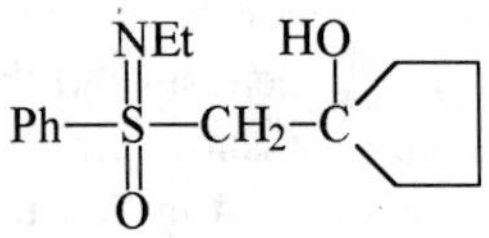

73e

(4.3.1)

A chiral titanium reagent prepared *in situ* from chiral sulfoximine **37** and titanium tetraisopropoxide was reported[24] to induce the asymmetric addition of trimethylsilyl cyanide to aldehydes giving the corresponding cyanohydrins **76** in very good yields and with high enantioselectivities (Scheme 4.3.8).

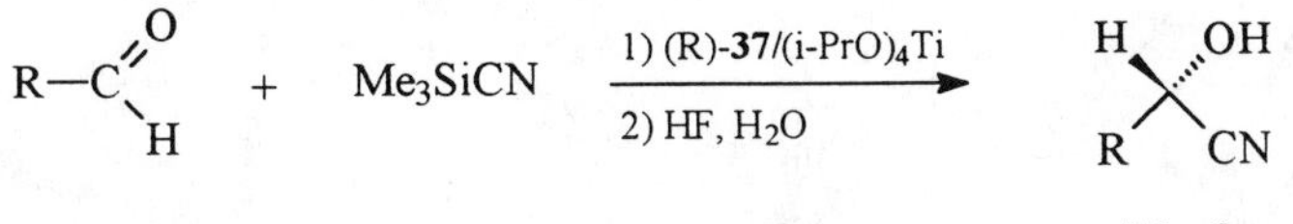

R—C(=O)H + Me_3SiCN —1) (R)-**37**/$(i\text{-}PrO)_4Ti$; 2) HF, H_2O→ H, OH / R, CN

76a-f

No	R	Yield [%]	ee [%]	Abs. conf.
a	Ph	72	91	S
b	4-$MeOC_6H_4$	60	87	S
c	2-$MeOC_6H_4$	72	74	S
d	1-Nph	92	76	S
e	t-Bu	70	81	S
f	C_5H_{11}	64	89	S

Scheme 4.3.8

It is interesting that this catalytic system is also effective in the addition of hydrogen cyanide to benzaldehyde. The cyanohydrin formation occurred at –50°C and after 16 h the chiral product **76a** was isolated in 89% yield with the ee value equal to 64%. The sequence of processes shown in Scheme 4.3.9 was proposed to explain the predominant formation of the (S)-cyanohydrins **76a–f**. According to this proposal, such a stereochemical outcome is a result of two subsequent stereo-controlled events: the first consists in selective complexation of an aldehyde to the β-face of the titanium reagent (R)-**77** to form the complex (R)-**78**, and the second is the enantioselective cyanide addition to the activated carbonyl group in the (R)-**78** to give the complex **79**.[24]

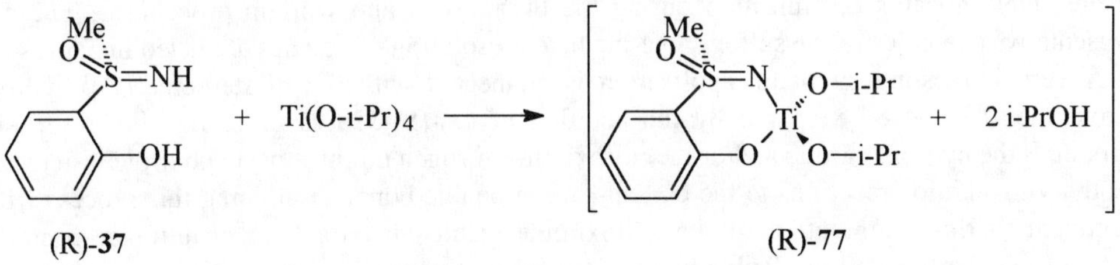

(R)-**37** + $Ti(O\text{-}i\text{-}Pr)_4$ → [(R)-**77**] + 2 i-PrOH

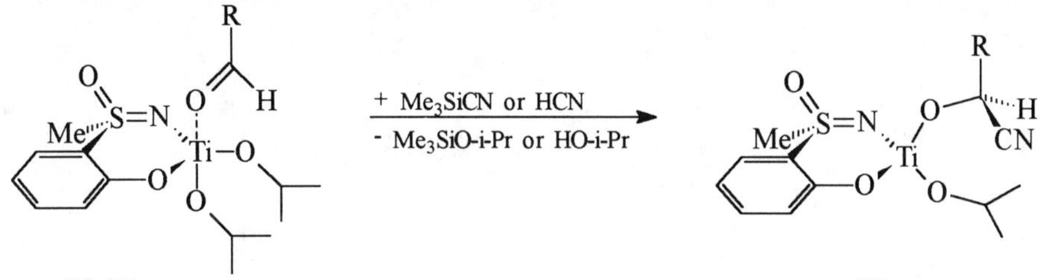

(R)-**78** —+ Me_3SiCN or HCN; - Me_3SiO-i-Pr or HO-i-Pr→ **79**

Scheme 4.3.9[24]

The diastereomeric β-hydroxysulfoximines were found to play an important role in optical resolution of ketones. This resolution procedure consists of two steps. The first is the addition of enantiomerically pure sulfoximine **32a** to selected racemic ketones **80** to form the diastereomeric β-hydroxysulfoximines **81** which, after separation into the pure diastereomers, undergo in the second step efficient thermal decomposition to regenerate the chiral ketone **80** and resolving agent **32a** (Scheme 4.3.10).[44]

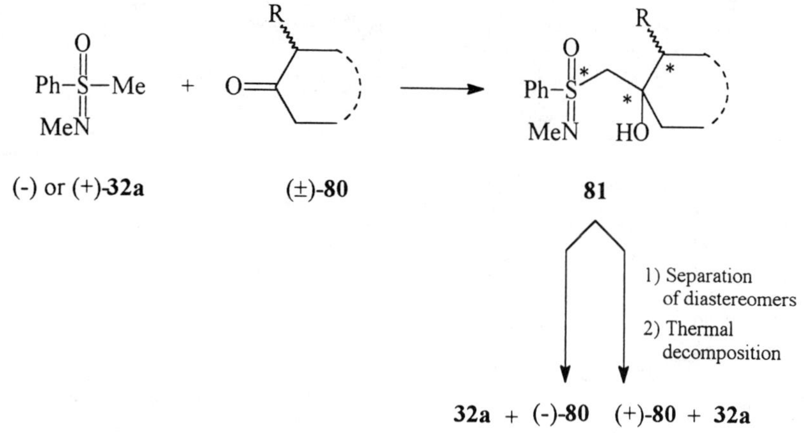

Ph–S(=O)(=NMe)–Me + (±)-80 → 81

(-) or (+)-32a (±)-80 81

1) Separation of diastereomers
2) Thermal decomposition

32a + (-)-80 (+)-80 + 32a

Scheme 4.3.10

The diastereomeric sulfoximines **81** were generally prepared by the addition of the α-lithio derivative of (+)-(S) or (–)-R-sulfoximine **32a** to selected (±)-ketones in tetrahydrofuran (THF) at 0°C followed by acid quench at that temperature. They were chromatographically separated on silica gel, and the pure diastereomers **81** were thermolyzed at ca. 130°C to produce chiral ketones and the chiral sulfoximine **32a**, which could be reused. Extensive studies of Johnson and Zeller[44a] have shown that this method can be applied to the resolution of a variety of ketones and is best suited to those ketones which exhibit high diastereoface selectivity in the addition of sulfoximine and are stable towards racemization during the thermolysis and workup procedures. The most representative examples of the sulfoximine-mediated resolution of ketones are listed in Table 4.3.2.

A very interesting synthetic application is connected with the diastereomeric β-hydroxy-sulfoximines **82** and **83** prepared by the addition of (+)-(S)-**32a** to prochiral enones. In such compounds the hydroxyl group and/or the sulfoximine nitrogen might serve to control diastereoface selectivity in addition reactions to the carbon-carbon double bond. Combining this process with a subsequent thermal elimination of the sulfoximine results, in effect, in enantioface-controlled addition to the carbon-carbon double bond of the starting enone (Scheme 4.3.11).

Table 4.3.2 Results of the Sulfoximine-Mediated Ketone Resolution (see Scheme 4.3.10)[a] (Reference 44a)

Ketone	Addition yield [%]	Diastereomers No	Diastereomers Yield [%][b]	Thermolysis yield [%]	[α] (solvent) observed	Ref.
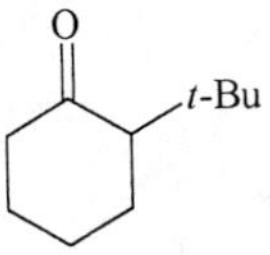 O, t-Bu	86	I	56	93	+36.0 (MeOH)	44a
		II	12	98	−35.3 (MeOH)	
		III	20	73	−35.4 (MeOH)	
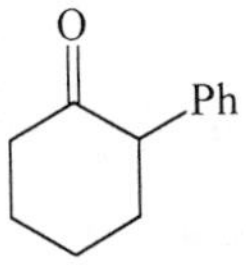 O, Ph	98	I	37	94	+112.5 (C_6H_6)	44a
		II	16	81	−59.3 (C_6H_6)	
		III	29	88	−112.0 (C_6H_6)	
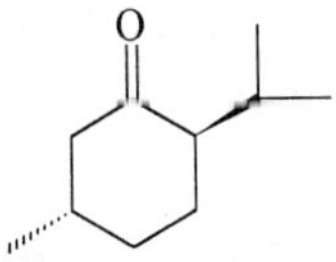 O	98	I	49	89	+26.5 ($CHCl_2$)	44a
		II	9			
		III	33	88	−27.5 ($CHCl_3$)	
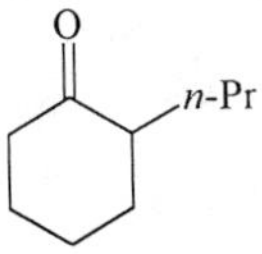 O, n-Pr	95	I	18			44a
		II	31	35	−26.9 (MeOH)	
		III	18			
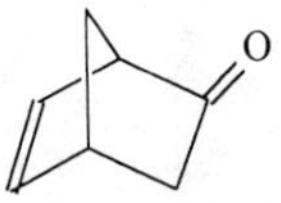 O	90	I	40			44a
		II	32	50	−1135.7 ($CHCl_3$)	
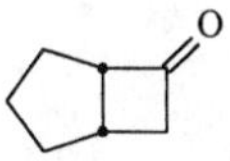 O	98	I	25	80	−64.3 ($CHCl_3$)	44a
		II	23	98	+59.6 ($CHCl_3$)	
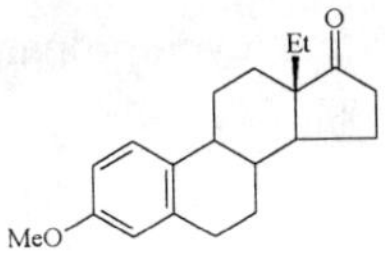 Et, O, MeO	96	I	48	48	+106.4 (1:1 MeOH/$CHCl_3$)	44a
		II	37	61	−104.4 (1:1 MeOH/$CHCl_3$)	
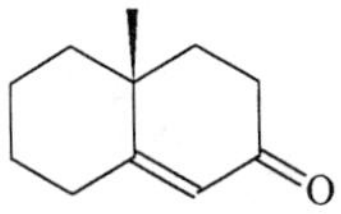 O	91	I	42	98	−207.0 (EtOH)	44a
		II	21	84	+208.3 (EtOH)	
PhC(O)CH(OMe)Ph	98	I	38	96	+52.8 (C_6H_6)	44a
		II	8			
		III	31	93	−51.5 (C_6H_6)	
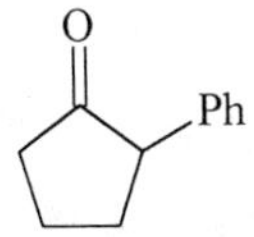 O, Ph	76	I	15	88	−23.8 (toluene)	44a
		II				
		III	21	88	+4.0 (toluene)	
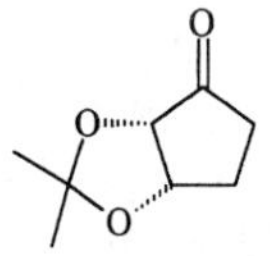 O, O, O	76	I	34	92	−70.8 ($CHCl_3$)	45
		II	42	98	+71.8 ($CHCl_3$)	
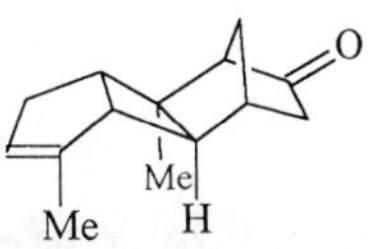 O, Me, Me, H	95	I	48	96.3	+193.2 ($CHCl_3$)	46
		II	43			

[a] Using (+)-(S)-**32a** as a chiral resolving agent.

[b] After chromatography.

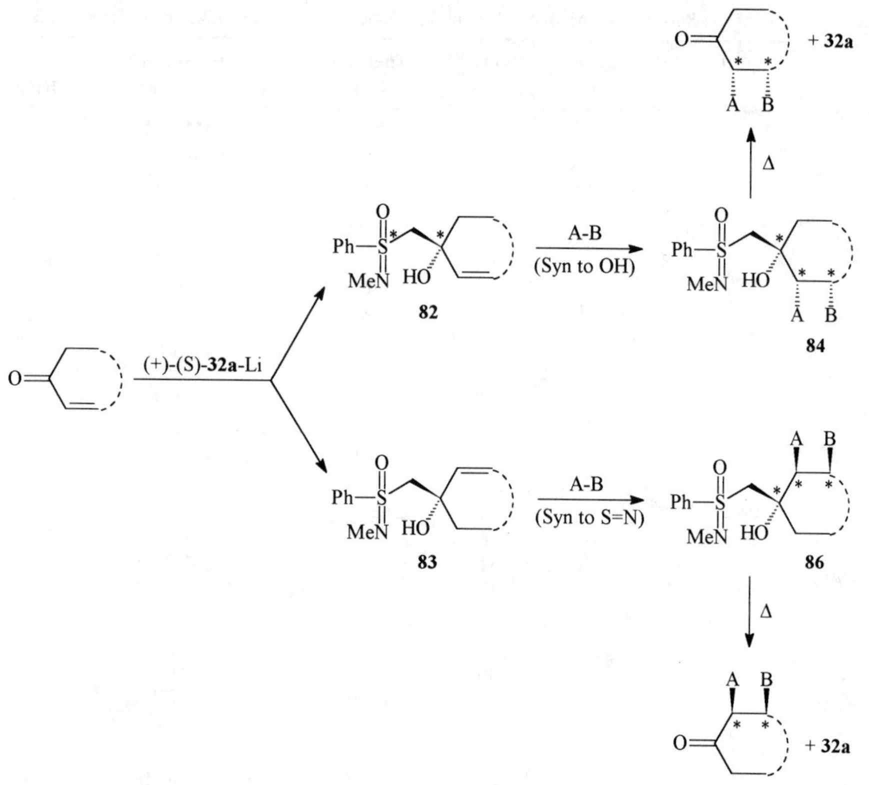

+ **32a**

Δ

Ph—S(O)(=NMe) **82** HO A-B (Syn to OH) **84** A B

O= (+)-(S)-**32a**-Li

Ph—S(O)(=NMe) **83** HO A-B (Syn to S=N) **86** A B

Δ

+ **32a**

Scheme 4.3.11

This approach was applied[47,48] in a very effective way for the Simmons-Smith cyclopropanation of a series of enones listed in Table 4.3.3 (Scheme 4.3.12).

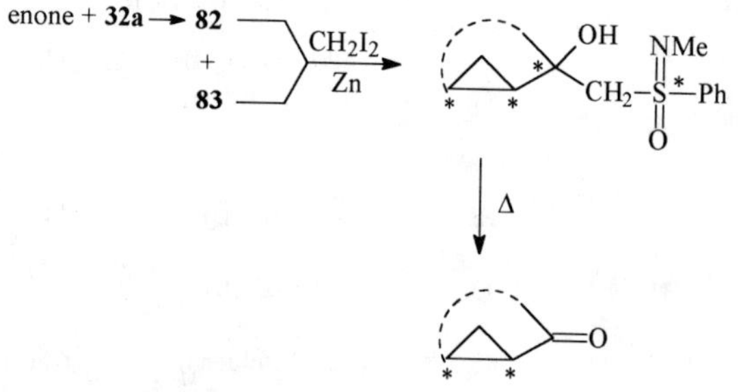

enone + **32a** → **82** + **83** CH_2I_2 / Zn → OH NMe CH_2–S*–Ph O

Δ

=O

Scheme 4.3.12

The results listed in Table 4.3.3 clearly indicate the efficiency of this methodology for the synthesis of both enantiomers of cyclopropyl ketones. In some instances (see entries 2 and 6), it was more expedient to separate the diastereomers after cyclopropanation. The usefulness of this procedure is also evident from Scheme 4.3.13 in which the sequence of reactions utilized in the synthesis of the natural and unnatural (+) and (–) enantiomers of thujopsene **87** are shown.

Table 4.3.3 Synthesis of Chiral Cyclopropyl Ketones According to Scheme 4.3.12[a]

No	Enone	Method[b]	Cyclopropyl ketone	$[\alpha]_D^{25}$ ($CHCl_3$), deg
1	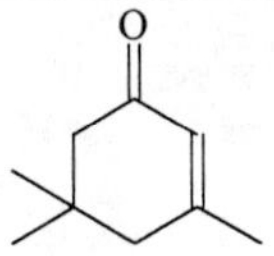	A	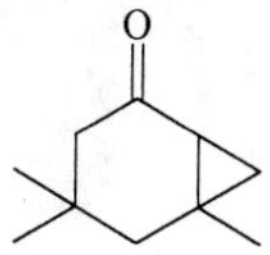	+160.6 (*c* 2.17) −164.9 (*c* 2.33)
2	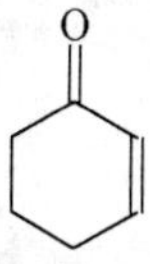	B	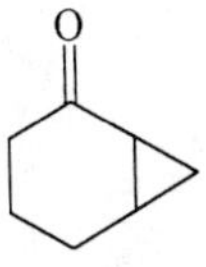	+15.3 (*c* 2.04) −15.5 (*c* 1.28)
3	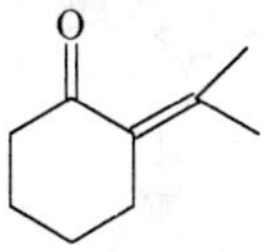	A	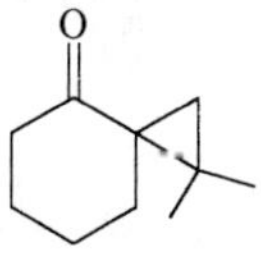	+47.3 (*c* 0.51) −50.8 (*c* 1.31)
4	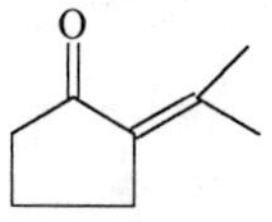	A	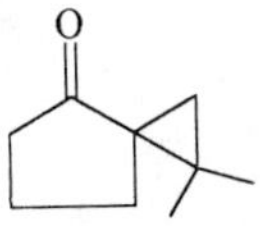	+162.3 (*c* 0.63) −171.9 (*c* 1.11)
5	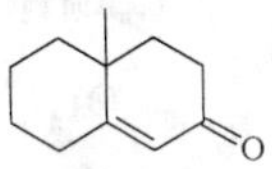	A	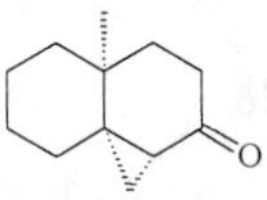	+97.7 (*c* 1.76) −95.3 (*c* 1.47)
6	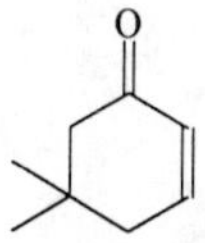	B	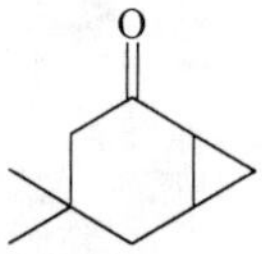	+215.3 (*c* 1.08) −214.5 (*c* 1.02)
7	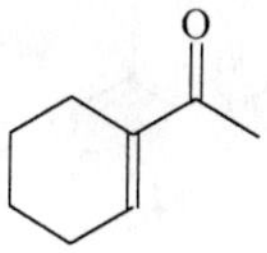	A	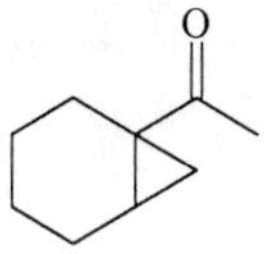	+101.4 (*c* 2.09)

[a] **Method A** sequence: (1) addition of **32a** to enone (yields ~95%); (2) separation of diastereomeric enone adducts by medium-pressure liquid chromatography on silica gel with EtOAc/hexanes (combined recovery 85-96%); (3) cyclopropanation (yields 77-96%); (4) thermal release of cyclopropyl ketone (yields 73-98%). **Method B** sequence: (1) addition of **32a** to enone (yields ~95%); (2) cyclopropanation (yields 91-98%); (3) chromatographic separation (as above) of cyclopropanated adducts (combined recovery 65-98%); (4) thermal release of cyclopropyl ketones (yields 60-75%).

From Johnson, C. R., Barbachyn, M. R., *J. Am. Chem. Soc.*, 104, 4290, 1982. With permission.

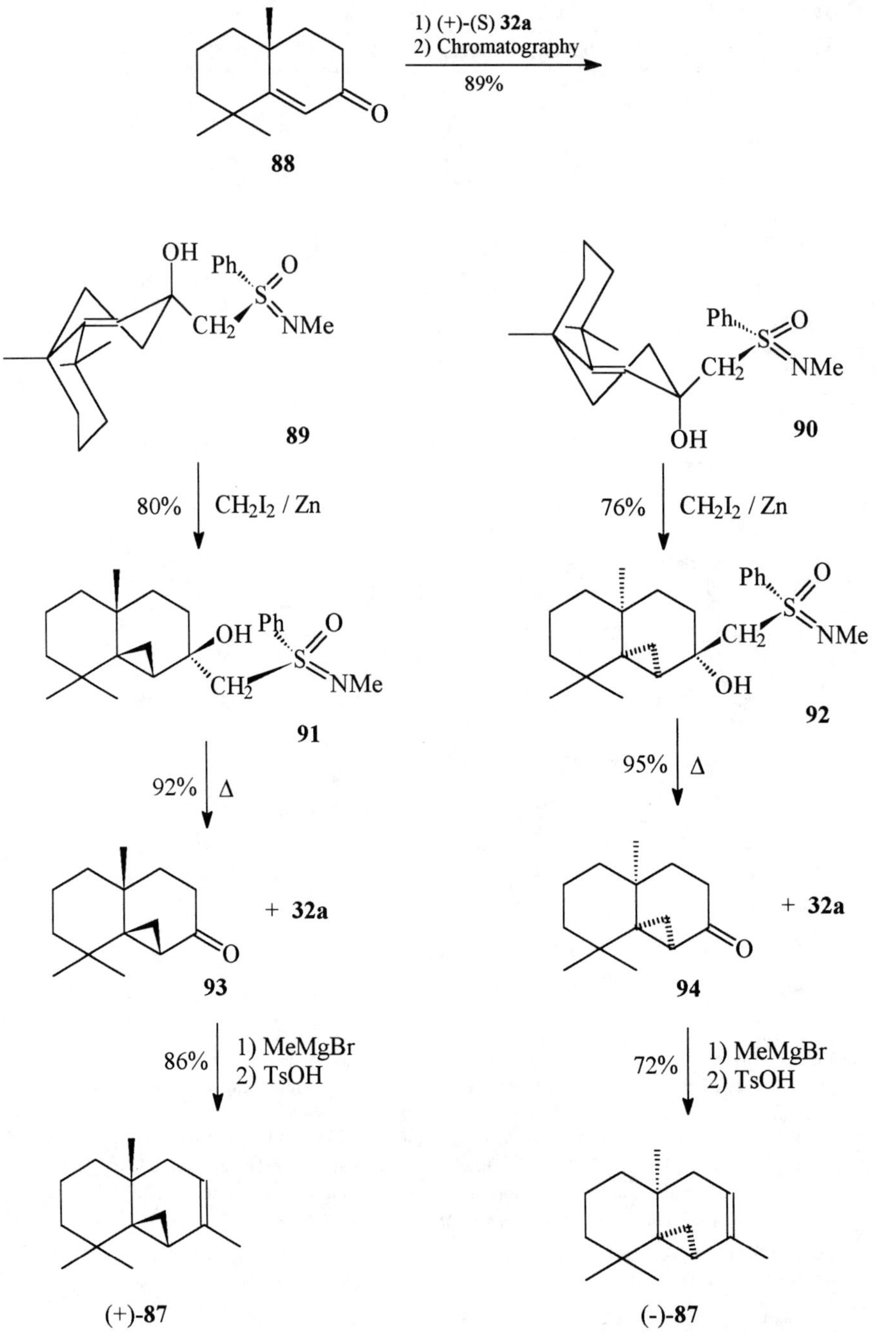

Scheme 4.3.13

A similar strategy was applied for the synthesis of enantiomerically pure dihydroxycycloalkanones (Scheme 4.3.14). It involves the direct osmylation of the diastereomerically pure sulfoximines **82** and **83** to form vicinal trihydroxysulfoximine derivatives **95** which, after thermal elimination of the sulfoximine **32a**, afford enantiomerically pure dihydroxyketones **96**.[48,49]

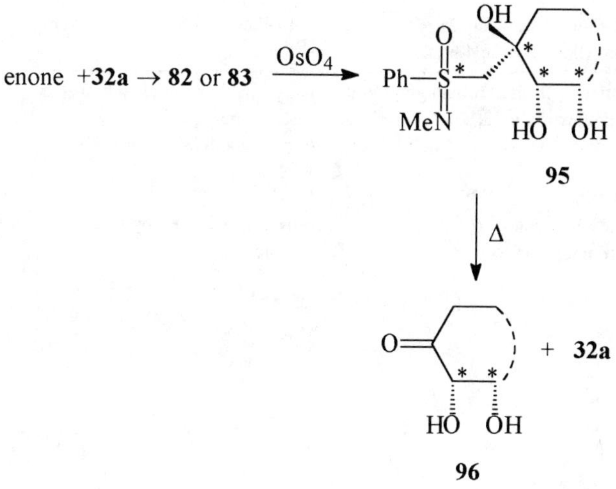

enone +**32a** → **82** or **83** $\xrightarrow{OsO_4}$ **95** $\xrightarrow{\Delta}$ **96** + **32a**

Scheme 4.3.14

The results of the most representative dihydroxylations are collected in Table 4.3.4.

Table 4.3.4 Preparation of Enantiomerically Pure Dihydroxycycloalkanones from Cycloalkenones via Sulfoximine Directed Osmylation Procedure (See Scheme 4.3.14)

Ketone	Adduct 82 and 83[a] Yield [%][d]	No[b]	Osmylation yield [%][c]	Thermolysis yield [%]	Product 96	$[\alpha]^{25}_D$ (conc) $CHCl_3$
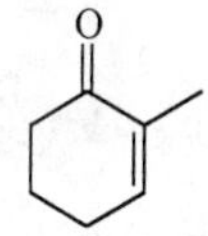	84	(+)-I[d] (−)-I[e]	(74) (73)	81 70	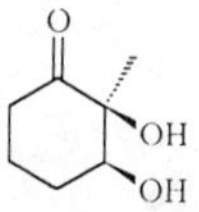	+0.7° (1.01)[h] -0.8° (1.06)
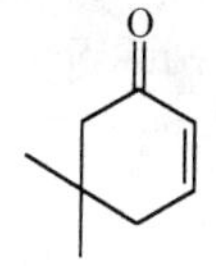	94	ef	87[f]	85 76	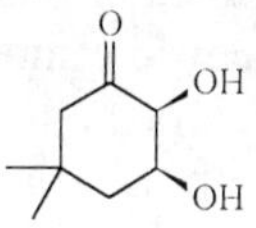	−21.2° (1.08)[i] +21.6° (0.36)[j]
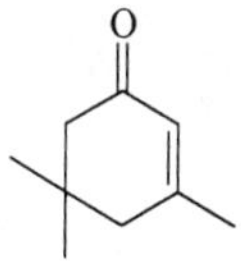	96	(+)-I[e] (−)-II[e]	87 90	83 97	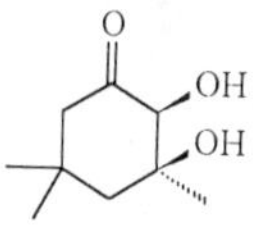	−29.3° (1.09)[k] +28.6° (1.05)[l]
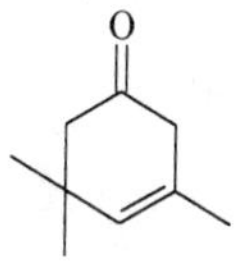	47	(+)-I[e] (+)-II[e]	94 98	82 89	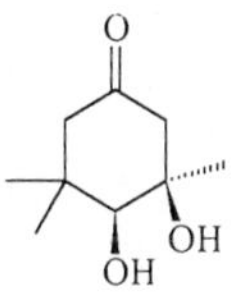	−24.2° (0.39)[m] +24.0° (0.92)[n]
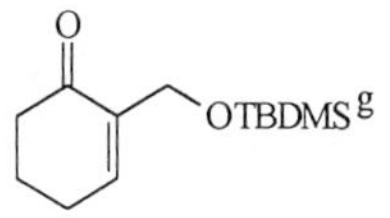 OTBDMS[g]	92	(+)-I[d]	87 (78) 89 (78)	96 65	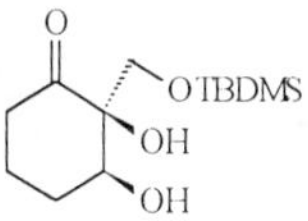 OTBDMS	−52.3° (1.34) +51.2° (1.01)
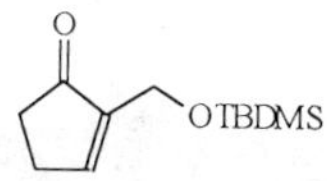 OTBDMS	83	(−)-I[e] (+)-II[e]	97 84	96 95	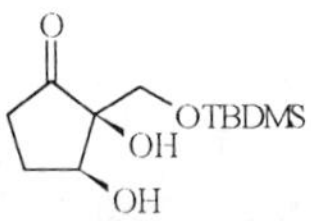 OTBDMS	+35.2° (1.04) -34.7° (1.00)

Table 4.3.4 (continued) Preparation of Enantiomerically Pure Dihydroxycycloalkanones from Cycloalkenones via Sulfoximine Directed Osmylation Procedure (See Scheme 4.3.14)

[a] Combined yield of adducts after purification by flash chromatography on silica gel with hexane/ethyl acetate.
[b] Entry designates sign of rotation of diastereomer; I indicates faster eluting diastereomer.
[c] Yields in parentheses were obtained in catalytic procedures using trimethylamine *N*-oxide.
[d] Adducts from (–)-(R)-**32a**.
[e] Adducts from (+)-(S)-**32a**.
[f] A mixture of diastereomers was oxidized. Separation of the diastereomeric triols was achieved by silica gel chromatography; a small amount of a third diastereomer was formed in the osmylation reaction.
[g] TBDMS = *tert*-butyldimethylsilyl.
[h] Colorless oils unless otherwise stated.
[i] mp 47–48°C.
[j] mp 48–49°C.
[k] mp 57–58°C.
[l] mp 65–67°C.
[m] mp 129–130°C.
[n] mp 132–133°C.

From Johnson, C. R., Barbachyn, M. R., *J. Am. Chem. Soc.*, 106, 2459, 1984. With permission.

β-Hydroxysulfoximines obtained by the addition of **32a** to aldehydes and ketones, having the general structure **81**, undergo reductive elimination to give alkene upon treatment with aluminum/mercury amalgam in a mixture of THF, water, and acetic acid (Equation 4.3.2).[47,48]

Ph–S(O)(=NMe)–CH$_2$–C(OH)< $\xrightarrow[\text{THF, AcOH}]{\text{Al(Hg) / H}_2\text{O}}$ CH$_2$=C<

81 (4.3.2)

The combination of this olefination procedure with the chromatographic separation of sulfoximine diastereomers constitutes the methodology which allows ketone methylenation with optical resolution to be carried out. The application of this technique for the total synthesis of the ginseng sesquiterpene (–)-β-panasinsene **97** and its enantiomer is pictured in Scheme 4.3.15.[50]

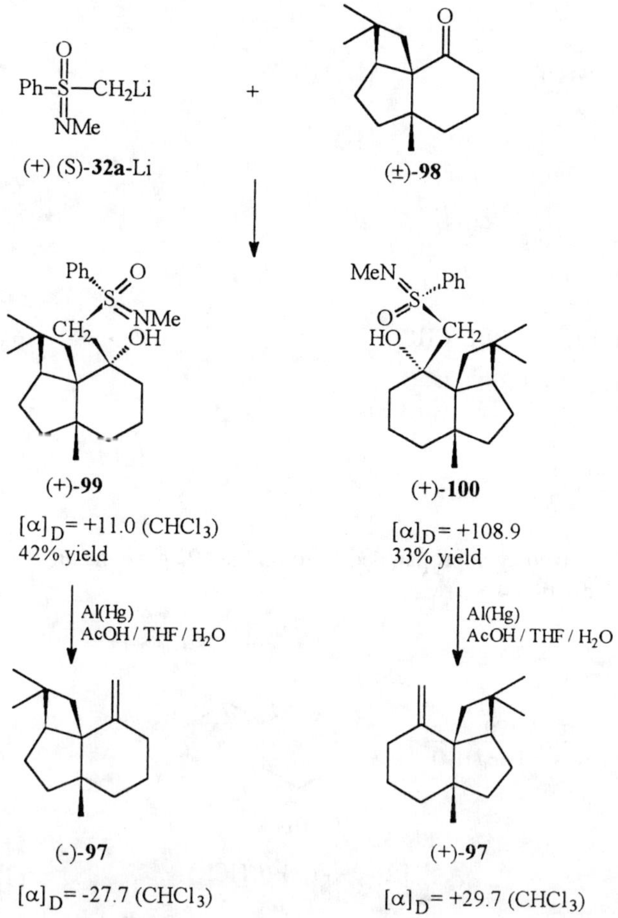

O
Ph–S–CH_2Li
NMe
(+) (S)-**32a**-Li

+

O

(±)-**98**

Ph O S NMe CH_2 OH

MeN Ph S O HO CH_2

(+)-**99**
$[\alpha]_D$= +11.0 ($CHCl_3$)
42% yield

(+)-**100**
$[\alpha]_D$= +108.9
33% yield

Al(Hg)
AcOH / THF / H_2O

Al(Hg)
AcOH / THF / H_2O

(-)-**97**
$[\alpha]_D$= -27.7 ($CHCl_3$)

(+)-**97**
$[\alpha]_D$= +29.7 ($CHCl_3$)

Scheme 4.3.15

The addition of the lithium salts of (+)-(S)-**32** to (±) **98** gave two diastereomeric sulfoximines (+)-**99** and (+)-**100** which were readily separated by flash chromatography. Treatment of the diastereomer **99** with aluminum amalgam gave (–)-β-panasinsene **97** in 96% yield. Under the same conditions from the diastereomer (+)-**100**, the unnatural enantiomeric (+)-β-panasinsene **97** was obtained in 92% yield.[50]

It is interesting to note that the Raney nickel desulfurization of both diastereomeric sulfoximines **99** and **100** gave the tertiary alcohols (–) and (+)-**101** (Scheme 4.3.16). The relative configuration at the chiral carbinol center in enantiomers of **101** is, however, not known.

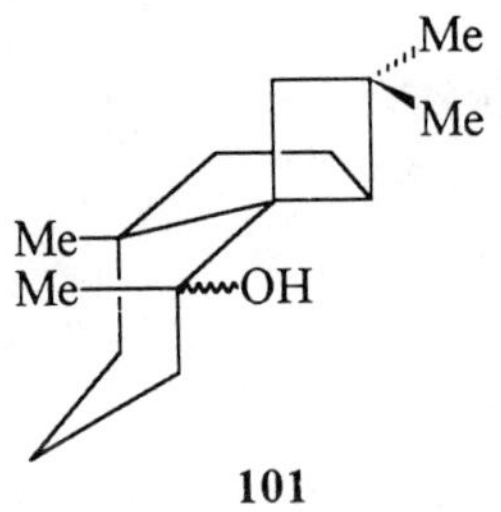

Me
Me
Me
Me
OH

101

(+) **99** $\xrightarrow[78\%]{\text{Raney nickel}}$ (-) **101**, $[\alpha]_D$ = -52.1 ($CHCl_3$)

(+) **100** $\xrightarrow[54\%]{\text{Raney nickel}}$ (+) **101**, $[\alpha]_D$ = +53.0 ($CHCl_3$)

Scheme 4.3.16

A similar desulfurization of the resolved cyclopropanated sulfoximine **102** gave the unsaturated spiro-compound **103** in 75% yield (Equation 4.3.3).[47]

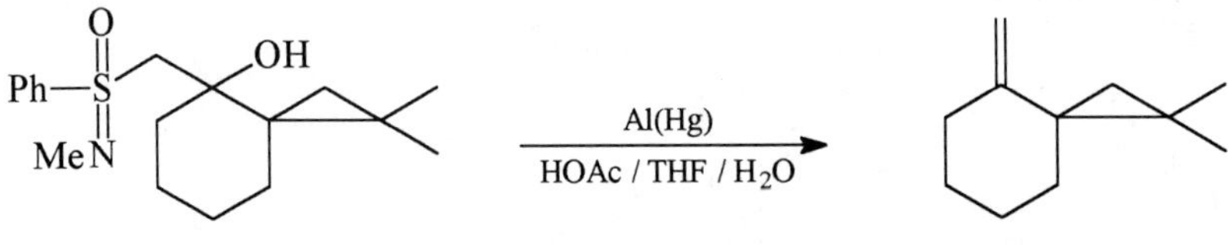

102 $\xrightarrow[\text{HOAc / THF / } H_2O]{\text{Al(Hg)}}$ **103**

102 $[\alpha]_D$ = +49.1 ($CHCl_3$)

103 $[\alpha]_D$ = -12.1 ($CHCl_3$) (4.3.3)

The multifaceted role of diastereomeric β-hydroxysulfoximines is evidently seen in Scheme 4.3.17, which outlines the total synthesis of the nonhead-to-tail monoterpene, (–)-rothrockene **104**.[51]

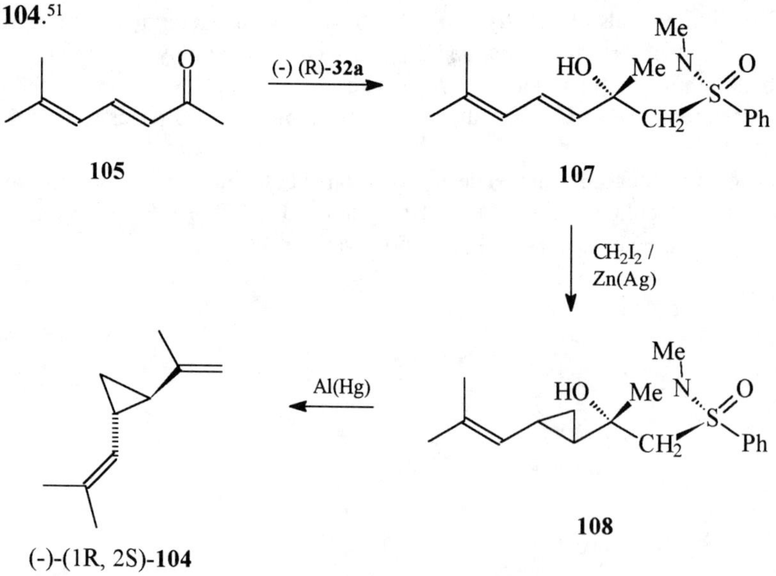

105 $\xrightarrow{\text{(-) (R)-}\mathbf{32a}}$ **107**

107 $\xrightarrow{CH_2I_2 / Zn(Ag)}$ **108**

108 $\xrightarrow{\text{Al(Hg)}}$ (-)-(1R, 2S)-**104**

(-)-(1R, 2S)-**104**
$[\alpha]_D$-65.5

Scheme 4.3.17

First of all, the presence of the β-hydroxysulfoximine moiety allowed the resolution of the precursor **107**. Moreover, due to the presence of this group ensuring a regio- and diastereoface selectivity, the cyclopropanation of **107** afforded **108**, the absolute configuration of which was easily determined by X-ray analysis by relating the carbon stereochemistry to the known absolute stereochemistry at sulfur. Finally, the β-hydroxysulfoximine moiety allowed introduction of the second carbon-carbon double bond in **104**.[51]

Asymmetric induction as high as 90% ee was observed during the addition of enolate nucleophiles, containing chiral sulfoximine auxiliaries **109** to diene-molybdenum and dienyliron complexes **110–113** (Equations 4.3.4–4.3.7).[29]

MeO_2CCH_2–S(=O)(=NR)–Ph

(R)-**109a-d**

a, R = Ts
b, R = Me
c, R = TBDMS
d, R = DMTS

110 $\oplus$Mo(CO)$_2$Cp $PF_6^{\ominus}$ —109/base→ **114** Me(CO)$_2$Cp, CO_2Me, O=S=NR, Ph → **118** Me(CO)$_2$Cp, CO_2Me (4.3.4)

111 $\oplus$Mo(CO)$_2$Cp $PF_6^{\ominus}$ —109/base→ **115** Me(CO)$_2$Cp, CO_2Me, O=S=NR, Ph → **119** Me(CO)$_2$Cp, CO_2Me (4.3.5)

112 $\oplus$Fe(CO)$_2$P(OPh)$_3$ $PF_6^{\ominus}$ —109/base→ **116** Fe(CO)$_2$P(OPh)$_3$, CO_2Me, O=S=NR, Ph → **120** Fe(CO)$_2$P(OPh)$_3$, CO_2Me (4.3.6)

113 Fe(CO)$_3$ $\oplus$ $PF_6^{\ominus}$ —109/base→ **117** Fe(CO)$_3$, CO_2Me, O=S=NR, Ph → **121** Fe(CO)$_3$, CO_2Me (4.3.7)

The enolate anions of **109a–d** were generated *in situ* by the addition of a base and were found to react satisfactorily with diene-$Mo(CO)_2Cp$ **110** and **111**, and dienyl-$Fe(CO)_2L$ complexes **112** and **113**. Examinations of the ^1H NMR spectra of the crude products **114–117** showed that they constitute a mixture of diastereomers, usually with one of them predominating. Since no information regarding asymmetric induction could be deduced from these data, each adduct was desulfoximinated using sodium or aluminum amalgam to give monoester derivatives **118–121** which showed optical activity. Their enantiomeric purities were determined by the measurement of the ^1H NMR spectra in the presence of the chiral lanthanide shift reagent $[Eu(hfbc)_3]$ (see Table 4.3.5).

Table 4.3.5 Asymmetric Induction Observed during the Addition of Sulfoximinyl Ester Enolates Derived from **109a–d** to Complexes **110–113**

Starting complex	Substituent (R)[a] on 109	Enolate countercation	Monoester product (yield, %)[b]	$[\alpha]_D$	ee [%][c]
110	Me (+)	Na	**118** (45)	+15.4	35
110	TBDMS (+)	Li	**118** (75)	+16	49
110	TBDMS (+)	Na	**118** (75)	+39	75
110	TBDMS (+)	K	**118** (80)	+38	78
110	DMTS (–)	Li	**118** (77)	–29	55
110	DMTS (–)	Na	**118** (83)	–38	75
110	DMTS (–)	K	**118** (80)	–43	80
111	Ts (+)	K	**119** (70)	+28	49
111	TBDMS (+)	Li	**119** (75)	+30	50
111	TBDMS (+)	Na	**119** (77)	+51	86
111	TBDMS (+)	K	**119** (80)	+49	84
111	DMTS (–)	Li	**119** (78)	–42	70
111	DMTS (–)	Na	**119** (83)	–56	89
111	DMTS (–)	K	**119** (80)	–53	85
113	Ts (+)	Li	**121** (72)	+3.2	20
113	Ts (+)	K	**121** (81)	+3.9	25
113	Ts (+)	K (DME)[d]	**121** (79)	+5.5	30

[a] Ts = 4-toluenesulfonyl-, TBDMS = *tert*-butyldimethylsilyl-, DMTS = dimethylhexylsilyl.
[b] Overall yield after desulfonylation.
[c] Determined by 200-MHz ^1H NMR as outlined in text. Precision of measurement of the order ±5% of value quoted.
[d] Run in presence of 18-crown-6-ether.

From Pearson, A. J., Blustone, S. L., Nar, H., Pinkerton, A. A., Roden, B. A., Yoon, J., *J. Am. Chem. Soc.*, 111, 134, 1989. With permission.

Earlier, it was reported[32] that the chiral vinyl sulfoximines **46** and **47a–d** undergo highly asymmetric conjugate addition reactions with organometallic reagents (Scheme 4.3.18 and Table 4.3.6).

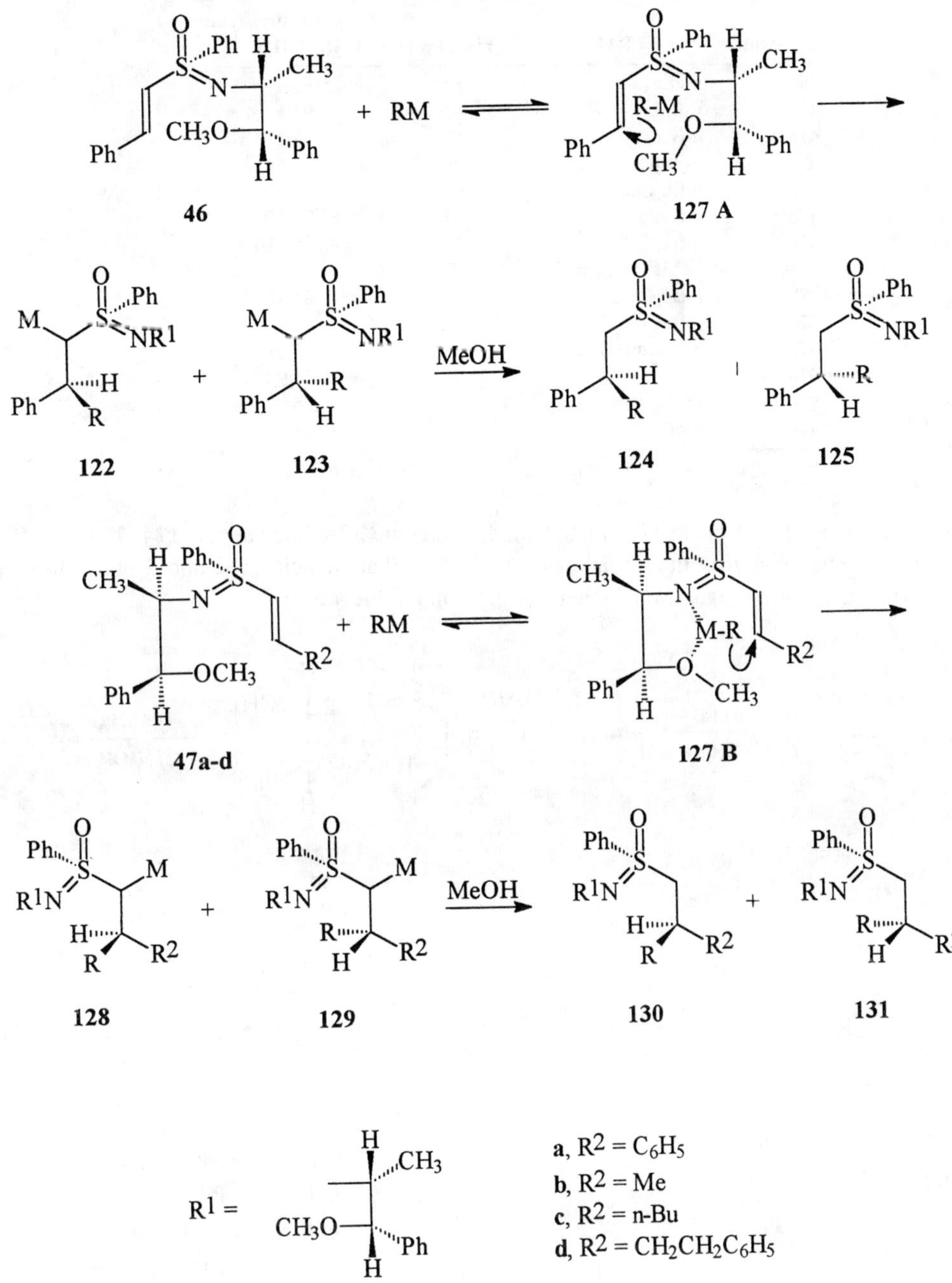

+ RM ⇌ → MeOH →

46 127 A

122 123 124 125

47a-d 127 B

128 129 130 131

R^1 =

a, $R^2 = C_6H_5$
b, $R^2 = Me$
c, $R^2 = n\text{-}Bu$
d, $R^2 = CH_2CH_2C_6H_5$

Scheme 4.3.18

Table 4.3.6 Conjugate Addition Reactions of Vinyl Sulfoximines **46** and **47a–d** with Organometallic Reagents RM

Substrate	RM	Yield [%]	Diastereomeric ratio 124:125 (130:131)
46	MeLi	85	96:4
46	n-BuLi	82	95:5
47a	n-BuLi	69	73:27
47a	n-Bu_2CuLi	76	86:14
47a	n-$Bu_2CuCNLi_2$	69	82:18
47a	n-BuCu	71	4.5:95.5
47a	n-Bu_2CuLi +LiI (5 equiv)	65	85:15
47b	n-Bu_2CuLi	77	81:19
47b	n-BuCu	81	5:95
47b	n-BuCu (LiI "Free")	68	33:67
47a	MeCu	78	4:96
47c	MeCu	72	3.5:96.5
47d	MeCu	75	4.5:95.5

From Pyne, S. G., *J. Org. Chem.*, 51, 81, 1986. With permission.

The stereochemistry of the newly created chiral center at C-2 in the adducts **124**, **130**, and **131** was determined chemically by their conversion to 3-alkylalkanoic acids **132** and 2-phenylhexane **133** of known absolute configuration (Scheme 4.3.19 and Table 4.3.7).[32]

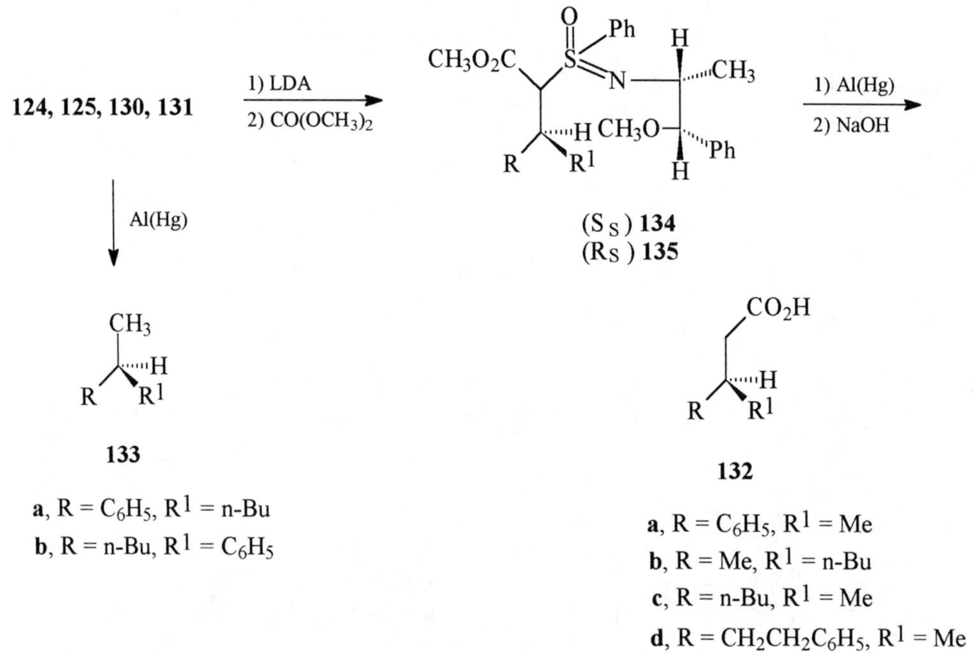

124, 125, 130, 131 — 1) LDA, 2) $CO(OCH_3)_2$ → (S_S) **134**, (R_S) **135** — 1) Al(Hg), 2) NaOH → **132**

124, 125, 130, 131 — Al(Hg) → **133**

133

a, R = C_6H_5, R^1 = n-Bu
b, R = n-Bu, R^1 = C_6H_5

132

a, R = C_6H_5, R^1 = Me
b, R = Me, R^1 = n-Bu
c, R = n-Bu, R^1 = Me
d, R = $CH_2CH_2C_6H_5$, R^1 = Me

Scheme 4.3.19

For example, the adduct **124**(R=Me) was converted to (–)-(R)-3-phenylbutanoic acid **132a** (90% ee) by the following sequence of reactions: (a) methoxycarbonylation with lithium diisopropylamide (LDA) in tetrahydrofuran (THF) and dimethyl carbonate, (b) reduction with aluminum amalgam in aqueous HOAc and THF, and (c) basic hydrolysis.

Table 4.3.7 Optically Active Acids **132a–d** and Hydrocarbons **133a,b** Prepared via the Conjugate Addition of **46** and **47a–d** with Organometallic Reagents RM

		Chiral	$[\alpha]_D$, (solvent, *c*)		
Substrate	**RM**	**deriv.**	**obsd**	**ee [%]**	**Abs. config.**
46	MeLi/n-BuLi	**132**	–51 (PhH, 0.02)	90	R
46	n-BuLi	**133**	+20.6 (hexane, 0.01)	89	S
47a	n-BuCu	**133**	–10.4 (hexane, 0.01)	45	R
47a	n-Bu_2CuLi	**133**	+20.6 (hexane, 0.015)	89	S
47b	n-BuCu	**132**	–2.5 (acetone, 0.01)	59	S
47b	MeCu	**132**	+4.6 (acetone, 0.01)	≥90	R
47b	MeCu	**132**	–53.4 (PhH, 0.02)	93	R
47c	MeCu	**132**	–4.7 (acetone, 0.03)	≥92	S
47d		**132**	–66.1 (PhH, 0.01)	≥91	S

From Pyne, S. G., *J. Org. Chem.*, 51, 81, 1986. With permission.

The observed stereoselectivity in the reaction of the sulfoximine **46** with alkyllithium reagents was rationalized in terms of the intermediate **136** assuming prior complexation of RM to the sulfoximine nitrogen of **46** (Scheme 4.3.20).[32]

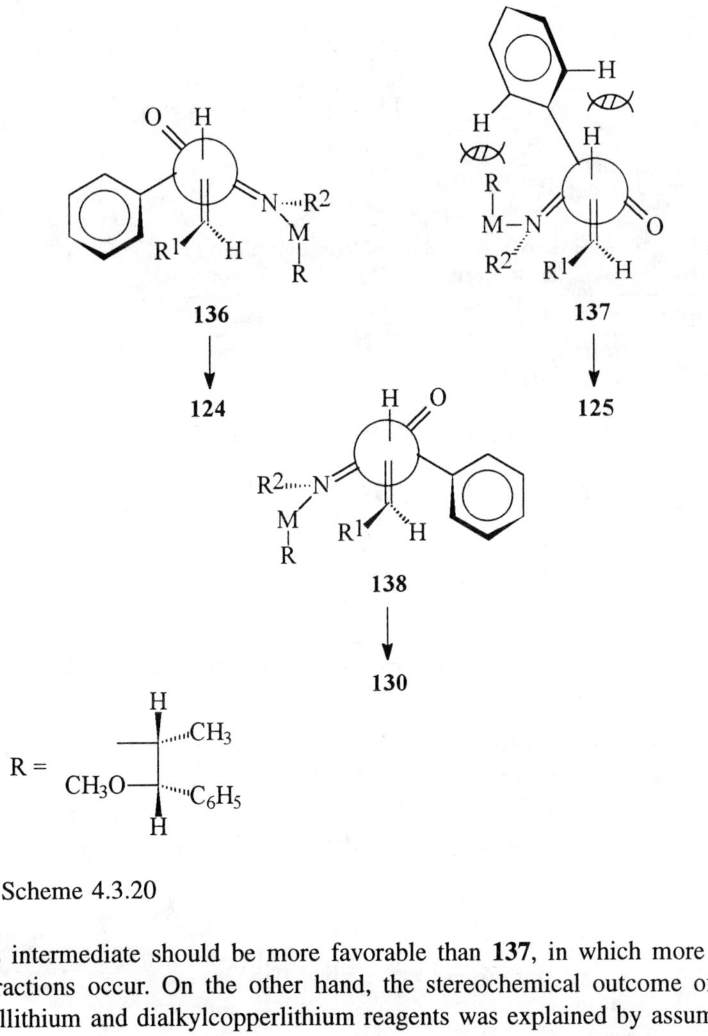

Scheme 4.3.20

This intermediate should be more favorable than **137**, in which more severe nonbonding steric interactions occur. On the other hand, the stereochemical outcome of the reaction of **47** with alkyllithium and dialkylcopperlithium reagents was explained by assuming that the reaction proceeds via the conformation **138**.

The results of the reaction of **47b** with LiI "free" n-BuCu, in which **131b** (R = n-Bu) was formed as a major diastereomer, suggested the involvement of the LiI-chelated species **139**.[32]

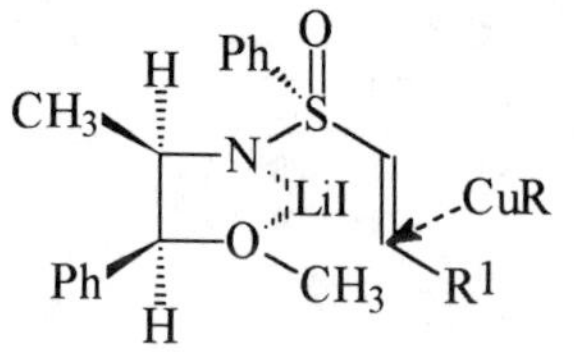

139 → **131b**

Similar asymmetric conjugate addition of organocopper reagent to sulfoximines **50a,b** and **51a,b** proceeded with modest selectivity (Scheme 4.3.21 and Table 4.3.8) and afforded a mixture of the diastereomeric sulfoximines **140** and **141**.[33]

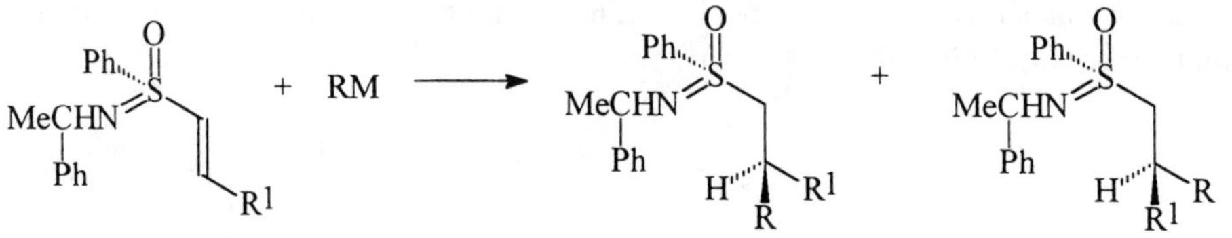

(R_S)-**50a,b** + RM → **140** + **141**

Scheme 4.3.21

Table 4.3.8 Conjugate Addition Reactions of Vinyl Sulfoximines **50a,b** with Organometallic Reagents

Sulfoximine	RM	Yield [%]	Ratio 140:141
50a	$(CH_3)_2CuLi$	60	23:77
50a	$(CH_3)_2CuLi$ LiI "free"	69	90:10
50a	$(CH_3)_2CuLi + ZnBr_2$ (1.1 equiv.)	65	12:88
50a	CH_3Cu	79	21:79
50a	CH_3Cu LiI "free"	80	13:87
50b	n-Bu_2CuLi	59	12:88
50b	n-$Bu_2CuLi + ZnBr_2$ (1.1 equiv.)	62	56:44
50b	n-BuCu	69	36:64

From Pyne, S. G., *Tetrahedron Lett.*, 27, 1691, 1986. With permission.

Another type of reactivity exhibited by vinyl sulfoximines is the nickel-catalyzed cross coupling with organometallics. This conversion schematically described by Equation 4.3.8 affords the products of α-substitution in a vinyl moiety **143** and the sulfinamide derivatives **144**. Therefore, it can be considered as a very useful way for the stereoselective synthesis of chiral unsaturated compounds.

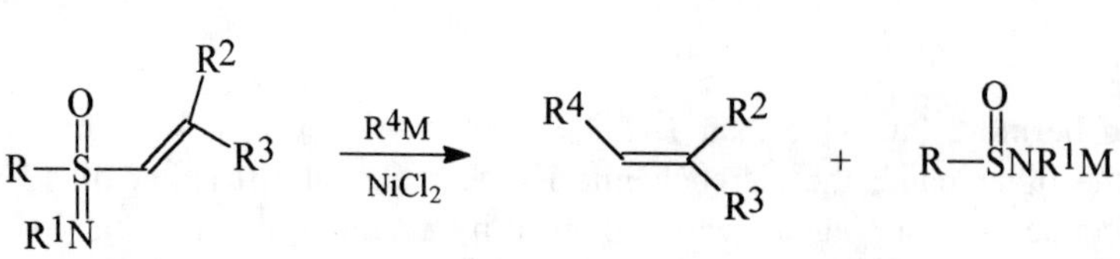

142 $\xrightarrow[NiCl_2]{R^4M}$ **143** + R–S(O)NR^{1}M **144** (4.3.8)

Recently, Gais and co-workers[56] have carried out extensive studies on this reaction and applied it for the synthesis of a few complex unsaturated systems in enantiomerically pure forms.

In the first report[52] from this group the cross coupling of the organozinc reagents **145** with the sulfoximine **146** in the presence of $MgBr_2$ (1 equiv.), LiBr (1 equiv.), $ZnCl_2$ (2 equiv.), or $NiCl_2$ (dppp) [dppp = $Ph_2P(CH_2)_3PPh_2$] as catalysts was described to give the corresponding aryl alkenes **147** in high yields and with 99:1 dr (Scheme 4.3.22). It was observed that without inorganic salts added as co-catalysts, practically no coupling occurs.

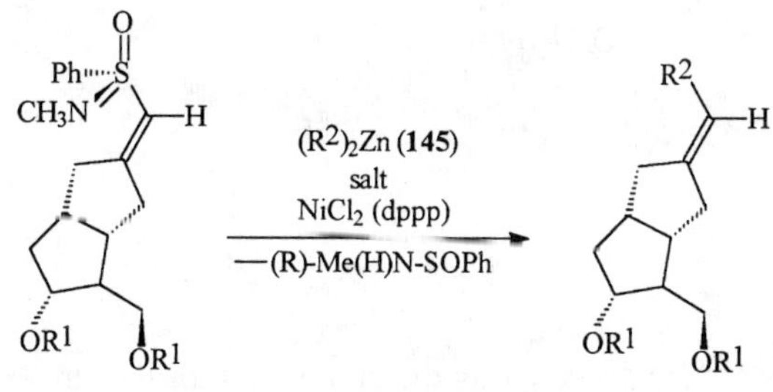

$(R^2)_2Zn$ (**145**)
salt
$NiCl_2$ (dppp)
—(R)-Me(H)N-SOPh

146 **147**

R^1 = t-BuMe$_2$Si
R^2 = t-BuMe$_2$Si

No	R^2	Yield [%]	dr
a	Ph	83	99:1
b	m-(R^1OCH_2)C_6H_4	89	99:1
c	$(CH_2)_4OR^3$	70	99:1

Scheme 4.3.22

This cross coupling was also successfully applied for the synthesis of the enantiomerically pure alkene **148** (Equation 4.3.6).

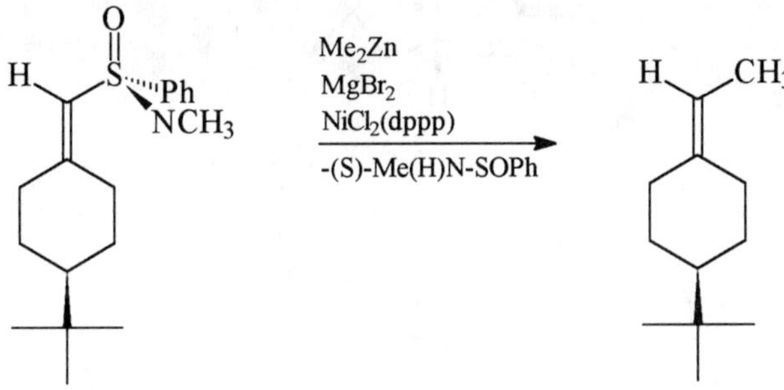

Me_2Zn
$MgBr_2$
$NiCl_2$(dppp)
-(S)-Me(H)N-SOPh

148 (4.3.9)

Of interest is that the coupling of **146** with more basic organomagnesium or lithium reagents gave unexpectedly the cross-coupling products **147** with complete loss of olefinic stereochemistry. It was proved that the formation of the alkenyl metal derivatives **149** (Equation 4.3.10) is responsible for this loss of stereochemical integrity in the produced **147**.[53]

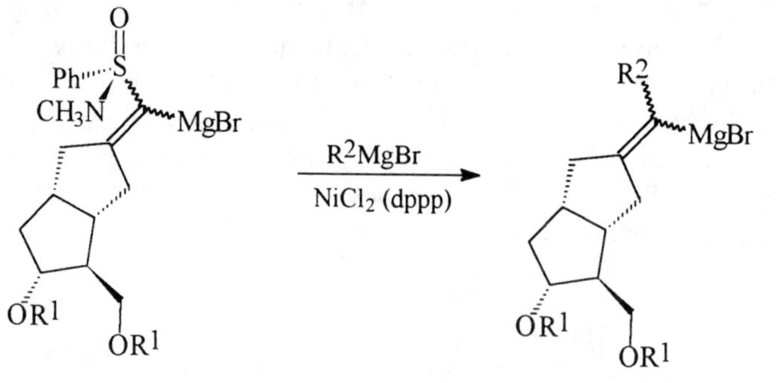

R²MgBr
NiCl₂ (dppp)

149

(4.3.10)

a, $R^2 = Ph$
b, $R^2 = m\ C_6H_4CH_2OR^1$
c, $R^2 = (CH_2)_4OR^3$

All vinyl sulfoximines discussed above were prepared by the elimination of the suitable functionalized α-hydroxysulfoximines obtained in the addition reaction of the lithiated chiral sulfoximine **32a** with the appropriate carbonyl components. In this context it should be noted that this type of conversion was also used[52] for a highly stereoselective asymmetric synthesis of 1-alkenylsulfoximines with axial and central chirality. For example, selective *exo*-addition of the lithioderivative of **32a** to the ketone **150** gave the β-hydroxysulfoximine **151a** in 96% yield and with dr ≥ 98:2; smooth silylation of the lithium alkoxide **151b** with Me_3SiCl afforded the sulfoximine **151c** in 95% yield. The latter, upon treatment with n-BuLi, gave the (S, α-R) alkenylsulfoximine **152a**, which was isolated in 91% yield and with dr ≥ 98 (Scheme 4.3.23).[52]

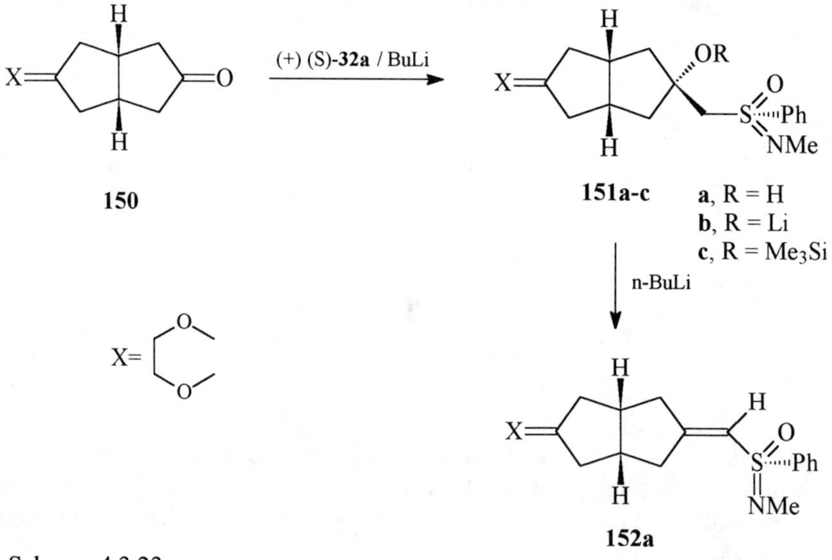

(+) (S)-**32a** / BuLi

150

151a-c **a**, R = H
b, R = Li
c, R = Me₃Si

n-BuLi

152a

Scheme 4.3.23

A similar sequence of reactions was used for the selective synthesis of (Z)- and (E)-alkenylsulfoximines **153** and **154** starting from the ketone **155** (Scheme 4.3.24).[52]

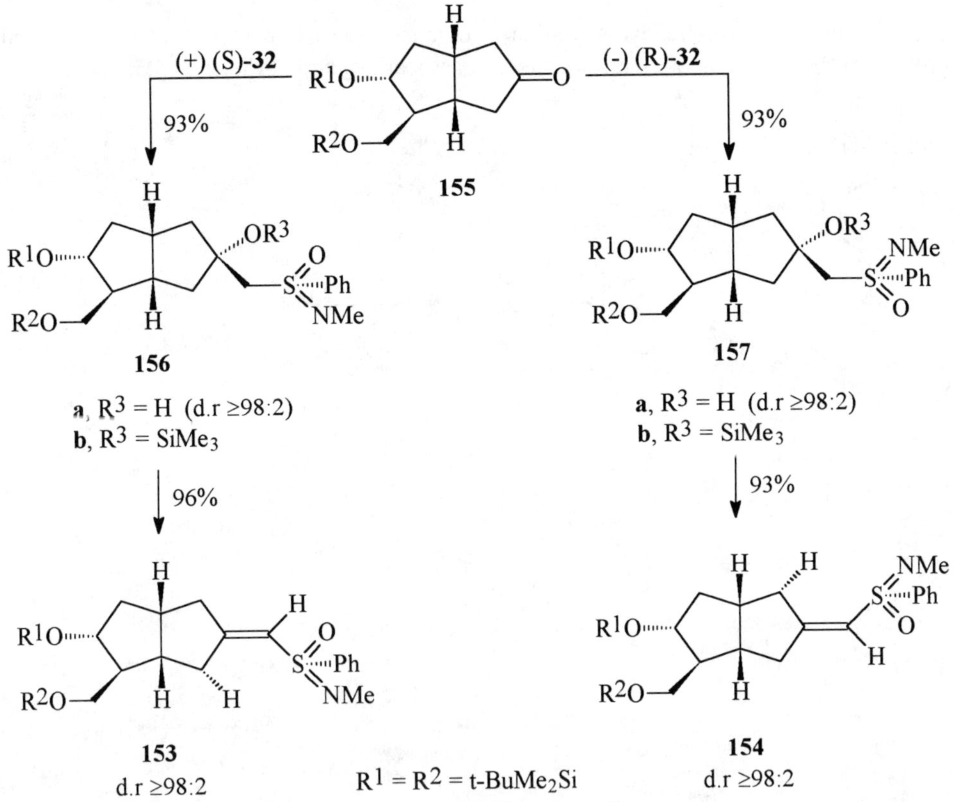

156 a, $R^3 = H$ (d.r ≥98:2); b, $R^3 = SiMe_3$

157 a, $R^3 = H$ (d.r ≥98:2); b, $R^3 = SiMe_3$

153 d.r ≥98:2

$R^1 = R^2 = t\text{-}BuMe_2Si$

154 d.r ≥98:2

Scheme 4.3.24

In the case of allyl sulfoximines **158** the transition metal-mediated substitution with organometallic or carbanion salts[54,55] can occur either at the α- and γ-position of the allyl moiety to afford the nonrearranged **159** and rearranged **160** unsaturated hydrocarbons and the appropriate sulfinamide derivatives **161** (Equation 4.3.11).

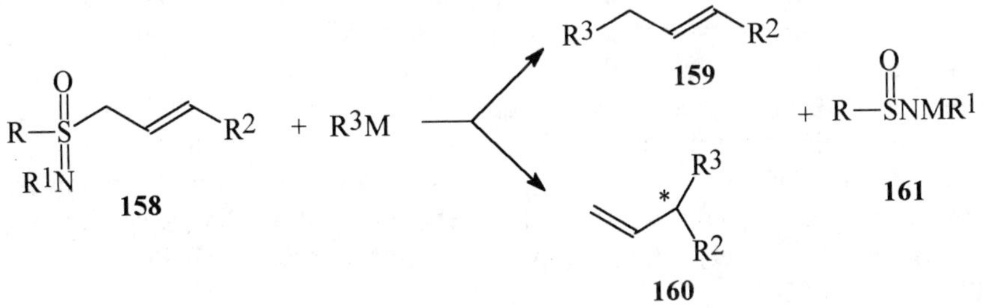

(4.3.11)

It is evident that the substitution reaction at the γ-position is accompanied by transfer of chirality from the sulfoximidoyl sulfur atom to the newly created chiral carbon atom of the allyl moiety. On the other hand, substitution at the α-position affords products which should preserve stereochemical integrity. Therefore, this type of chiral sulfoximines can be used either as substrates for the asymmetric synthesis of the chiral rearranged allylic hydrocarbons or precursors of the chiral alkenes.

The regio- and enantioselective allylation of organocopper reagents with chiral sulfoximines was utilized in the synthesis of chiral precursors of isocarbacyclins[53] (Scheme 4.3.25).

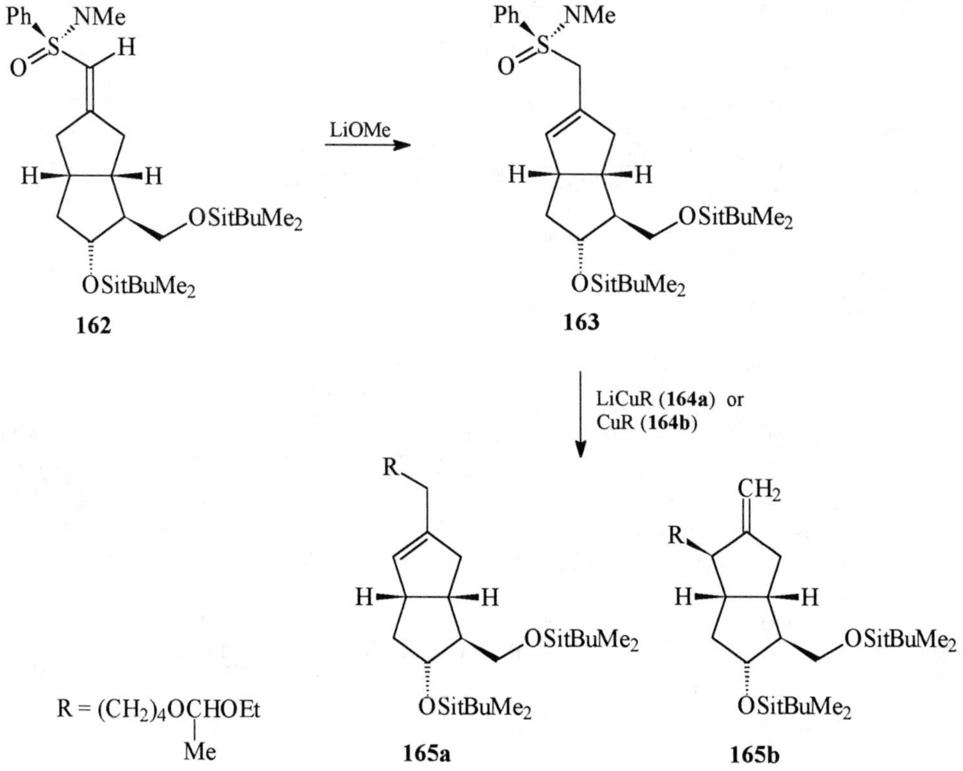

162 —LiOMe→ **163**

163 —LiCuR (**164a**) or CuR (**164b**)→ **165a** + **165b**

R = $(CH_2)_4OCH(Me)OEt$

Scheme 4.3.25

Thus, regioselective isomerization of the vinyl sulfoximine **162** upon treatment with lithium methoxide and subsequent aqueous workup gave the allylic analog **163** in 97% yield. After reaction with 3 equivalents of the homocuprate **164a**, it delivered a 96% yield of the endocyclic alkene **165a** with less than 0.5% of the γ-substitution product **165b**. Regioselective substitution in the γ-position, on the other hand, was achieved by treatment of the allylic sulfoximine **162** with the organocopper reagent **164b** in the presence of an equimolar amount of $BF_3 \cdot OEt_2$. The exocyclic alkene **165b** was isolated in 86% yield as a single diastereomer together with 2% of **165a**.

Asymmetric induction, which occurs at the γ-carbon atom of the allylic sulfoximines upon treatment with organocopper reagents, was recently nicely exploited for the preparation of chiral exocyclic alkenes using a variety of optically active sulfoximines **167–174** as substrates (Scheme 4.3.26 and Table 4.3.9).[53,56]

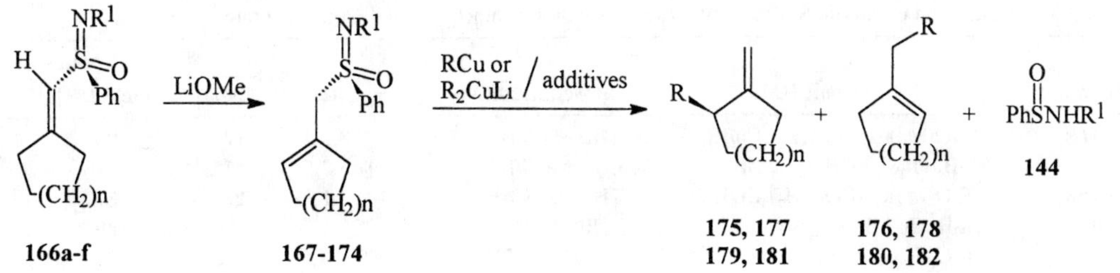

NR^1 — H, S=O, Ph — $(CH_2)n$ **166a-f** → LiOMe → NR^1, S=O, Ph — $(CH_2)n$ **167-174** → RCu or R_2CuLi / additives → R — $(CH_2)n$ **175, 177 179, 181** + R — $(CH_2)n$ **176, 178 180, 182** + $PhS(O)NHR^1$ **144**

No	R^1	n
167	Me	0
168	Me	1
169	Me	2
170	Me	3
171	t-BuPh$_2$Si	1
172	H	1
173	SO$_2$Tol-o	1
174	SO$_2$CF$_3$	1

n	No
0	**179, 180**
1	**175, 176**
2	**177, 178**
3	**181, 182**

Scheme 4.3.26

The most representative examples are presented in detail in Table 4.3.9, and the most characteristic features can be summarized as follows:

a) Reaction of the allylic sulfoximines **168**, **173**, and **174** with the organocuprate reagents RCu/LiI led with α-selectivities of 92:8 to 99:1 to the endocyclic alkanes **176** in good to high yields;
b) Reaction of the sulfoximines **167–170**, **173**, and **174** with the organocopper reagents RCuLi in the presence of $BF_3 \cdot OEt_2$ resulted in the formation of the exocyclic alkenes **175**, **177**, **179**, and **181**, respectively, in good to high yields with γ-selectivities of 80:20 to 99:1;
c) Enantiomeric excess of the chiral exocyclic alkenes is in the range of 27–90% and the (S) sulfonimidoyl group leads to a preferential bond formation from the *si* side of the double bond;
d) The N-methylsulfoximine **168** and the N-methylsulfoximine **167** gave the highest and the lowest enantioselectivity with the butylcopper reagents n-BuCu/LiI;
e) The enantioselectivity of the γ-substitution was considerably lower for the N-sulfonylsulfoximines **173** and **174**.

Vinyl sulfoximines function also as chiral dienophiles in the Diels-Alder reaction. For example, p-tolyl vinyl N-phthalimidosulfoximine **33e** with cyclopentadiene was found to produce a mixture of four cycloadducts **183a–d** in 95% yield (Scheme 4.3.27). The ratio of the products **183a**:(**183b** + **183c**):**183d** was approximately 1:4:4. The major product was found to be **183d** by a single-crystal X-ray analysis.

Table 4.3.9 Substitution of Allylic Sulfoximines **167–174** with Organocopper and Organocuprate Reagents

Sulfoximine	Reagent RM	Solvent/additives	γ:α	Product No	Product Yield [%]	Product ee [%]
168	$EtO(Me)C(H)O(CH_2)_4Cu/LiI$	$THF/Me_2S/BF_3$	98:2	**175**	90	71
168	$t\text{-}BuPh_2SiO(CH_2)_4Cu/LiI$	$Et_2O/Me_2S/BF_3$	97.5:2.5	**175**	88	90
168	$EtO(Me)C(H)O(CH_2)_3Cu/LiI$	$THF/Me_2S/BF_3$	87:13	**175**	87	63
168	n-BuCu/LiI	$THF/Me_2S/BF_3$	99:1	**175**	89	72
168	$n\text{-}C_{11}H_{23}Cu/LiI$	$THF/Me_2S/BF_3$	99.5:0.5	**175**	95	67
169	$EtO(Me)C(H)O(CH_2)_4Cu/LiI$	$THF/Me_2S/BF_3$	91:9	**177**	91	60
169	$t\text{-}BuPh_2SiO(CH_2)_4Cu/LiI$	$THF/Me_2S/BF_3$	96:4	**177**	81	63
169	n-PrCu/LiI	$THF/Me_2S/BF_3$	99:1	**177**	68	72
169	n-PrCu/LiI	$THF/Me_2S/BF_3$	99:1	**177**	83	60
169	$Ph(CH_2)_2Cu/LiI$	$THF/Me_2S/BF_3$	97:3	**177**	91	61
173	n-BuCu/LiI	THF/BF_3	99:1	**175**	70	27
174	n-BuCu/LiI	THF/BF_3	98:2	**175**	84	33
168	n-BuCu/LiI	THF/BF_3	99:1	**175**	86	65

From Gais, H. J., Müller, H., Bund, J., Scommoda, M., Brandt, J., Raabe, G., *J. Am. Chem. Soc.*, 117, 2453, 1995. With permission.

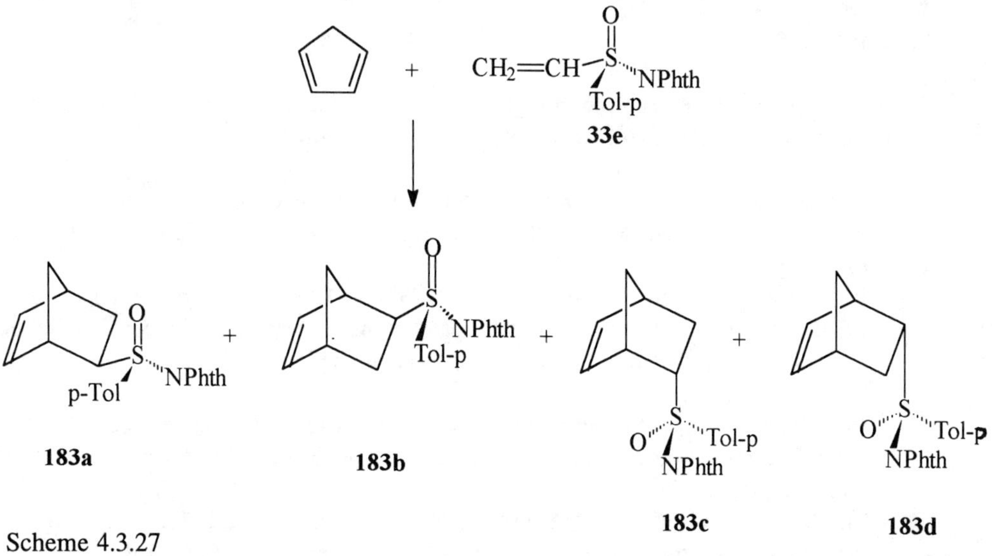

Scheme 4.3.27

Very high asymmetric induction was observed in the alkylation reaction of diastereomerically pure α-sulfinylsulfoximines **184a** and **184b** with alkyl halides under phase transfer catalytic conditions (Equation 4.3.12 and Table 4.3.10).[58]

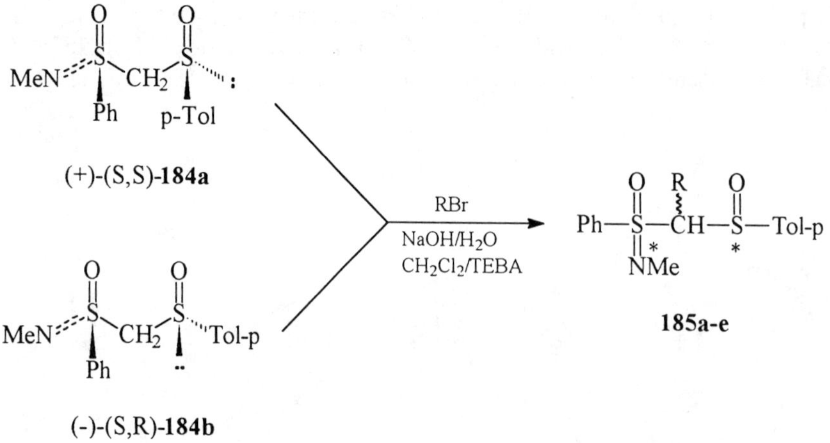

(4.3.12)

Table 4.3.10 Alkylation of (+)-(S,S)-184a and (S,R)-184b with Alkyl Halides under PTC Conditions

		Product			
Sulfoximine	**Alkyl halide**	**No**	**Yield [%]**	$[\alpha]_D^{25}$	**Diastereomeric ratio**
(+)-(S,S)-**184a**	CH_2=$CHCH_2Br$	**a**	93	+97.7°	100:0
(–)-(S,R)-**184b**	CH_2=$CHCH_2Br$	**a**	87	–49.0°	83:17
(+)-(S,S)-**184a**	$PhCH_2Br$	**b**	77	+86.3°	100:0
(–)-(S,R)-**184b**	$PhCH_2Br$	**b**	79	–34.4°	80:20
(+)-(S,S)-**184a**	HC≡C–CH_2Br	**c**	91	+80.0°	100:0
(–)-(S,R)-**184b**	HC≡C–CH_2Br	**c**	73	–28.5°	100:0
(+)-(S,S)-**184a**	EtBr	**d**	80	+103.0°	100:0
(–)-(S,R)-**184b**	EtBr	**d**	82	–18.2°	80:20
(+)-(S,S)-**184a**	EtI	**d**	70	+103.0°	100:0
(+)-(S,S)-**184a**	n-BuBr	**e**	30	+84.6°	100:0

From Annunziata, R., Cinquini, M., Cozzi, F., *Synthesis*, 767, 1982. With permission.

The results collected in Table 4.3.10 show that the stereoselectivity of the alkylation reaction leading to the compounds **185a–e** was higher when the (+)-(S,S)-**184a** was used as a substrate, since only one stereoisomer was formed.

4.4 SELECTED EXPERIMENTAL PROCEDURES

4.4.1 (–) o-Substituted Diphenyl N-(Substituted) Sulfilimines 6a–g from Sulfides and Amide Anions[6]

(A) An equimolar mixture (20 mmol) of a sulfide, (–)-menthol, and pyridine was dissolved in 70 mL of dry acetonitrile and the solution was cooled to –25°C. To the solution was added over 20 min stirring an equimolar amount of t-butyl hypochlorite dissolved in 20 mL of the same solvent. To the above solution, amide anion (22 mmol) (i.e., crystals of sodium arenesulfonamide prepared from the amide and NaOMe in MeOH, or a suspension of sodium benzamide in t-BuONa-t-BuOH) was added over a 30-min period under vigorous stirring at –25°C. The cooling bath was removed, while stirring was continued for an additional 10 min. The mixture was poured into dilute aqueous $Na_2S_2O_3$, then was extracted with chloroform. The chloroform layer was shaken with 10% aqueous NaOH and water, dried over $MgSO_4$, and the solvent was distilled off below 50°C. The residue was recrystallized from acetone-hexane (N-arylsulfonylsulfilimines) or benzene-hexane (N-benzoylsulfilimine).

(B) The recrystallization was repeated two or three times until the specific rotation became constant.

(C) The residue (2 mmol) from the chloroform solution prepared by method (A) was dissolved in 5 mL of acetone, and finely powdered $KMnO_4$ (2 mmol) was added and the whole mixture was stirred for 1 h at room temperature. To the mixture was added 1 mL of MeOH, and stirring was continued for 1 h. The insoluble product of the mixture, after being added into 30 mL of chloroform, was filtered off and the filtrate was washed with water. The solution was dried over $MgSO_4$ and the solvent was distilled off below 50°C. Chromatography on silica gel with chloroform gave the pure sulfilimines, which gave identical IR and NMR spectra to those of the authentic samples.

(D) The sulfilimines obtained above were recrystallized from acetone-hexane (N-arylsulfonyl-sulfilimines) or benzene-hexane (N-benzoylsulfilimine).

(E) The solvent was evaporated from the filtrate after the workup of method (D), and the residue was again recrystallized from the same solvent.

The yields and physical properties of the sulfilimines obtained thus are listed in Table 4.1.1.

4.4.2 (–)-o-Substituted Aryl Phenyl N-(Unsubstituted) Sulfilimines 10a,b[6]

Compound **6a** (4.2 g, 11 mmol) was added step by step over 10 min into 15 mL conc. H_2SO_4 with vigorous stirring at 5°C, and stirring was continued for additional 5 min after removing the cooling bath. The resulting light green solution was poured into 300 mL ice water, and the aqueous solution was filtered after the addition of active charcoal. The acidic solution was made alkaline (pH 10) by addition of 50% aqueous NaOH, together with crushed ice, below 5°C. The mixture was extracted with three 150-mL portions of benzene. The combined benzene extract was washed with water, dried over $MgSO_4$, and the solvent was evaporated under reduced pressure below 50°C. The light yellow oil thus obtained crystallized upon cooling. The product was then recrystallized from benzene-hexane, IR ν_{max} (KBr) 3170 and 930 (907 in CCl_4) cm^{-1}, NMR δ ($CDCl_3$) 1,45 (1H, s, NH), 3.74 (3H, s, *o*-CH_3O) and 6.86–7.91 (9H, m, aromatic). Derivative **10b**, prepared by a similar treatment, gave IR and NMR spectra identical to those of the authentic sample. These free sulfilimines were stored in a refrigerator, avoiding moisture, since they are somewhat unstable at room temperature.

4.4.3 Preparation of (+)-(R)-N-p-Toluenesulfonyl-S-Methyl-S-p-Tolyl-Sulfilimine, (+)-13a in Benzene at 27°C[7c]

Sulfoxide (+)-**12a** (0.2095 g, 1.36 mmol, 99% optically pure) and N,N′-bis-(p-toluenesulfonyl)sulfur diimide, DIM (0.8006 g, 2.16 mmol) were dissolved in 15 mL of benzene and allowed to stir at 27°C in a drybox for 2.75 h, after which 10 mL of water and 20 mL of chloroform were added. The product was isolated as in other runs to yield 0.3528 g of a yellow oil that crystallized on standing. The mixture was chromatographed on 30 g of silica gel (ethyl acetate, 30-ml fractions). Pure (+)-**13a** was recovered from fractions 3–7, 229.1 mg (54.8%, mp. 114–116°C, $[\alpha]^{25}_{546}$ +302.4 (c 0.93, acetone). The product was therefore 93% optically pure and the reaction proceeded with 94% stereospecificity.

4.4.4 (+)-(S)-S-Methyl-S-Phenylsulfoximine 24[19]

A hot solution of (–)-d-camphorosulfonic acid (15.7 g, 68 mmol, recrystallized from ethyl acetate) in 100 mL of acetone (HPCL or spectrophotomeric grade, 1.5 mL/mmol) was carefully added with swirling to a hot solution of (±)-**24** (10.0 g, 64.5 mmol) in 195 mL of acetone (3.0 mL/mmol) in a 500-mL Erlenmeyer flask. Because a vigorous reaction occurs on mixing, the clear boiling solutions were removed from the hot plate immediately prior to the addition. Solvent was evaporated on a hot plate under a stream of argon until crystals formed, usually after 1–2 min, and the mixture was allowed to cool. After 2 h, the crystals were filtered, washed with acetone, and suction filtered to dryness, furnishing 10.4 g of the salt as white needles (26.9 mmol, 83% yield), mp 182–183°C. Derivatization and GC analysis indicated that the isomeric purity of this salt was 98.8% de. For recrystallization, 10.4 g (26.9 mmol) of the salt was dissolved in 250 mL of hot, dry acetonitrile (24 mL/g). After hot filtration, the solution was slowly cooled to room temperature under argon. Further cooling at –20°C for 12 h enhanced the yield without compromising diastereomeric purity. After filtration, the crystals were washed with acetonitrile and suction filtered to dryness, furnishing 9.2 g of salt (23.7 mmol, 73% yield overall), mp 182–183°C, 99.7% de by derivatization and GC analysis. This material was recrystallized as above from 22 mL of acetonitrile to afford 8.3 g of the salt (21.4 mmol, 66% yield overall), mp 182–183°C, $[\alpha]^{25}_D$ –43.2 (c 3.60, MeOH), –28.8°C (c 7.57, H_2O). After derivatization, GC analysis indicated that the diastereomer ratio was 3550:1, greater than 99.9% de. A third recyclization gave salt of higher de, but analysis still revealed traces of material with the retention time of minor N-acyl diastereomer.

4.4.5 (–)-(R)-N-(p-Toluenesulfonyl)-S-Methyl-S-(p-Tolyl)Sulfoximine-27[17a]

To a solution of 10 g (0.065 mol) of (+)-(R)-methyl-p-tolyl sulfoxide **12a**, $[\alpha]_D$ +149.9 (c 1.2 acetone), and 25.6 g (0.130 mol) of p-toluenesulfonyl azide in 50 mL of absolute methanol was added to Raney copper (ca. 3 g) wet from a methanol suspension. The mixture was gently refluxed and the decomposition of the azide was monitored by observing the 2130-cm^{-1} band in the infrared spectrum. After 24 h, more copper (ca. 1 g) was added, and the mixture was refluxed for an additional 24 h. The mixture was then transferred to a 1 liter Erlenmeyer flask, using 300 mL of methylene chloride and 200 mL of a saturated Na_2EDTA solution. After 30 min of vigorous stirring, the mixture was filtered through Celite; the residue was washed with several portions of water and methylene chloride. The organic layer was separated and the aqueous layer was further extracted with methylene chloride. The combined organic extracts were washed with 5% sodium hydroxide solution in order to remove any p-toluenesulfonamide. The solution was then dried over $MgSO_4$ and evaporated. The residue was recrystallized from the methylene chloride/ether, yielding 18.4 g (90.3%) of a white solid: mp 158–161°C; $[\alpha]_D$ –144 (c 1.15, acetone).

4.4.6 (–)-(R)-S-Methyl-S-(p-Tolyl)Sulfoximine 28[17a]

A solution of 17.7 g (0.0568 mol) of (–)-(R)-N-(p-toluenesulfonyl)-S-methyl-S-(p-tolyl)sulfoximine **27**, $[\alpha]_D$ –144 (c 1.15, acetone), in 27 mL of concentrated sulfuric acid was heated on a steam bath for 15 min. The mixture was then carefully poured into 100 mL of water. Upon cooling, crystals began to form, but they were not filtered. The mixture was made slightly alkaline with 20% sodium hydroxide solution and extracted with methylene chloride. The solution was dried over magnesium sulfate and evaporated, giving 8.2 g (85%) of a light brown oil. The material crystallized upon standing, mp 58–62°C; $[\alpha]_D$ –33.8 (c 1.13, acetone).

4.4.7 (+)-(S)-N,S-Dimethyl-S-Phenylsulfoximine 32a[17a]

A 0.775 g (5.00-mmol) sample of (+)-(S)-S-methyl-S-phenylsulfoximine **24a**, $[\alpha]_D$ +35.6 (c. 0.97, acetone) (97.5% optically pure), was heated with 8 mL of 37% aqueous formaldehyde and 40 mL of 98% formic acid on a steam bath in an open flask. After 48 h, the mixture was poured into an evaporating dish and heated on a steam bath to an oil. The almost colorless material was dissolved in 30 mL of 2 M sulfuric acid and extracted with methylene chloride. The aqueous layer was neutralized with solid sodium carbonate. Methylene chloride extraction, drying ($MgSO_4$), and removal of the solvent *in vacuo* gave a nearly colorless oil (0.780 g, 92%), $[\alpha]_D$ +177.4 (c 1.16, acetone).

4.4.8 (–)-(R)-Dimethylamino-Methyl-p-Tolyloxosulfonium Fluoroborate 35[17a]

To a solution of 7.5 g (0.044 mmol) of (–)-(R)-S-methyl-S-(p-tolyl)sulfoximine, $[\alpha]_D$ –33.8 (c 1.13, acetone), in 80 mL of methylene chloride was added 7.6 g (0.050 mmol) of trimethyloxonium fluoroborate all at once. This mildly exothermic reaction was cooled initially in a water bath (15–20°C). After 1 h, 24 g (0.220 mol) of anhydrous carbonate was added, and the mixture was allowed to stir for 3 h. Then another 7.6 g (0.050 mol) of trimethyloxonium fluoroborate was added and the stirring continued for an additional 3 hr. Finally, another 0.5 equiv. (3.8 g) of the fluoroborate was added and the mixture stirred for one more hour. The inorganic salts were then filtered and were washed with methylene chloride and hot ethanol; 9.1 g (73%) of a white solid was obtained, mp 65–67°C; $[\alpha]_D$ –4.6 (c 2.0, acetone).

4.4.9 Synthesis of Vinyl Sulfoximines 46, 47 (R′ = C_6H_5, CH_3, n-Bu, $C_6H_5CH_2CH_2$)

***(S_S,1S,2R)-N-(1-Methoxy-1-phenyl-2-propyl)-S-(2-phenylethenyl)-S-phenylsulfoximine 46. A Typical Procedure*[32]**

A solution of **44** (400 mg, 1.32 mmol) in THF (2.5 mL) was added dropwise to a solution of LDA (1.58 mmol) in THF (5 mL) at 0°C. After 0.5 h, the solution was cooled to –78°C and benzaldehyde (1.58 mmol, 0.161 mL) was added. After 10 min the reaction mixture was warmed slowly (45 min) to 0°C and then quenched with aqueous saturated NH_4Cl (10 mL). The product was extracted with ether (2 × 15 mL) and the combined extracts were dried and concentrated. Purification of the crude product on silica gel (20% EtOAc/hexane) (R_f 0.55, EtOAc/hexane 1:1) gave **48** (380 mg, 70%) as a colorless oil: ^1H NMR showed ca. 1:1 mixture of diastereomers.

A solution of **48** (380 mg, 0.929 mmol) and triethylamine (4.65 mmol, 0.648 mL) in CH_2Cl_2 (5 mL) at 0°C was treated dropwise with methanesulfonyl chloride (2.79 mmol, 0.216 mL) over 10 min. After 0.5 h the mixture was treated with 1,8-diazabicyclo-[5.4.0]undec-7-ene (5.57 mmol, 0.833 mL) and stirring was continued at 23°C for 12 h. The reaction mixture was diluted with ether (25 mL), and the solution was washed with water, aqueous saturated NH_4Cl, and aqueous 10% Na_2CO_3, dried, and concentrated. Purification of the crude product by chromatography on silica gel (25% EtOAc/hexane) gave **46** (210 mg, 58%) as a colorless oil.

4.4.10 Sulfoximine-Mediated Synthesis of Chiral Alcohols (Scheme 4.3.2 and Table 4.3.1)[37]

a) β-Hydroxy Sulfoximines from the Reaction of N,S-Dimethyl-S-Phenylsulfoximine with Ketones

A known quantity of N,S-dimethyl-S-phenylsulfoximines **32a** was added to a flame-dried flask fitted with a stirring bar, an inlet, and an outlet for nitrogen. THF was added to dissolve the sulfoximine and the solution was cooled to 0°C. An equivalent of butyllithium was added slowly while the solution was stirred. The cooling bath was removed and the mixture allowed to stir for 15 min; a yellow suspension formed. A THF solution of an equivalent amount of the ketone was added dropwise at room temperature. The mixture was allowed to stir under a positive nitrogen pressure overnight. An equal volume of aqueous 6 N H_2SO_4 was then added. The mixture was stirred for 15 min and then transferred to a separatory funnel. The aqueous layer was removed and the THF layer was extracted once with a small portion of aqueous 6 N H_2SO_4. (If a third oily layer forms, it should be combined with an aqueous layer.) The aqueous layers were combined and shaken twice with hexane; the hexane and THF layers were discarded. Aqueous 10% NaOH was added to the combined aqueous layers until a permanent turbidity was produced. The turbid solution was then extracted several times with dichloromethane. The organic extracts were washed once with saturated aqueous $NaHCO_3$, dried with $MgSO_4$, and concentrated on a rotary evaporator to yield the pure β-hydroxy sulfoximines as mixtures of diastereomers.

b) Synthesis of β-Hydroxy Sulfoximines from the Reaction of N,S-Dimethyl-S-Phenylsulfoximine with Aldehydes

A known quantity of N,S-dimethyl-S-phenylsulfoximine **32a** was added to a flame-dried flask fitted with a stirring bar, an inlet, and an outlet for nitrogen. THF was added to dissolve the sulfoximine and the solution was cooled to 0°C. An equivalent of butyllithium was added slowly while the solution was stirred. The cooling bath was removed and the mixture allowed to stir for 15 min; during this time a yellow suspension formed. This mixture was then cooled to –78°C and an equimolar amount of an aldehyde, dissolved in THF, was added dropwise. The mixture was allowed to stir for 1 h at –78°C and then 1 h at ambient temperature. After this time, the reaction mixture was added to an equal volume of 6 N H_2SO_4. Isolation of the pure β-hydroxy sulfoximine (obtained as a mixture of stereoisomers) proceeded in a fashion analogous to that described for the isolation of the hydroxy sulfoximines obtained from ketones.

c) *Hydrogenolysis of β-Hydroxy Sulfoximines with Raney Nickel W-2*

The β-hydroxy sulfoximine was added to a two-neck flask fitted with a mechanical stirrer and dissolved in a minimal amount of water-saturated diethyl ether. (Caution: Failure to use water-saturated ether can result in fires.) To the sulfoximine solution were transferred 20 equivalents of Raney nickel W-2 (measured as per directions in "Organic Syntheses"). The reaction mixture was vigorously stirred at room temperature and the progress of the reaction was monitored by TLC. When no starting material remained, stirring was stopped and the solvent was carefully decanted. The residue was washed twice by suspension and the organic layers were combined. The combined organic layers were dried ($MgSO_4$), filtered through Celite, and concentrated by distillation. The product alcohol was obtained from the residue in the high purity by short-path distillation or chromatography.

4.4.11 Resolution of Ketones via Diastereoisomeric Sulfoximines Derived from Optically Pure N,N-Dimethyl-S-Phenylsulfoximine 32a[44]

a) *General Procedure for the Preparation of β-Hydroxy Sulfoximines*

A solution of n-BuLi in n-hexane (10 mmol, 1.6 M) was added to a solution of N,S-dimethyl-S-phenylsulfoximine (1.7 g, 10 mmol) in dry THF (35 mL) maintained at 0°C. After stirring at room temperature for 15 min, the yellow solution was cooled to –78°C and the ketone (10 mmol) in dry THF (10 mL) was added over 5 min. The mixture was allowed to stir for 45 min at –78°C. The cold mixture was poured into a mixture of diethyl ether (50 mL) and saturated NH_4Cl aqueous (25 mL). The mixture was shaken vigorously and the layers were separated. The aqueous layer was extracted twice with 10-mL portions of diethyl ether and the combined organic layers were dried over $MgSO_4$, filtered, and concentrated on a rotary evaporator.

The diastereomeric hydroxysulfoximines resulting from addition of sulfoximine to chiral ketones will be referred to throughout the experimental section as follows: diastereomer I — the fastest-moving diastereomer on silica gel; diastereomer II — the second fastest-moving diastereomer, etc.

b) *Chromatography of β-Hydroxysulfoximines*

Analytical TLC was performed on silica gel plate (0.25 mm, EM Reagents). Flash chromatography was effected using Silica Gel 60 (230–400 mesh, EM Reagents). MPLC was performed with the above silica gel using various sizes of Michel-Miller columns (Ace Glass) and Milton Roy pump with flow rates up to 50 mL/min. Analytical HPLC was accomplished using a Varian Associates Model 5000 HPL with a duPont Zobax Sil 5- to 6-μm steel column (4.6 mm × 26 cm).

c) *General Procedures for the Thermolysis of β-Hydroxysulfoximines*

A. Volatile ketones — The hydroxysulfoximine was added to a flask fitted with a Kuglerohr collection tube and attached to a vacuum line. The collection tube was cooled to –78°C and the apparatus was placed in a Kuglerohr oven preheated to the desired temperature (usually 130°C). The products distilled as they were formed and were separated by extraction of the N,S-dimethyl-S-phenyl sulfoximine into aqueous 6 N H_2SO_4, aqueous copper (II) nitrate, or by percolation through silica gel.

B. Nonvolatile ketones — The hydroxysulfoximine was refluxed in 2-BuOH until TLC analysis indicated complete thermolysis. The solvent was removed on the rotary evaporator and the sulfoximine and ketone were separated as described under Method A.
C. Nonvolatile ketones — The hydroxysulfoximine, in a Kuglerohr tube under an argon atmosphere, was placed in an oven preheated to 130°C until thermolysis was complete (5–10 min). Workup was completed as described under Method A.

Resolution of 2-t-Butylcyclohexanone. The addition of (+)-(S)-N,S-dimethyl-S-phenylsulfoximine (1.0 g, 5.85 mmol, 99% ee) to 2-t-butylcyclohexanone (0.9 g, 5.85 mmol) yielded a mixture of diastereomers (1.63 g, 86%). Analytical HPLC (10% EtOAc in n-hexane, 2 mL/min) revealed three diastereomers: diastereomer I, 5.1 min, 61%; diastereomer II, 5.8 min, 22%; diastereomer III, 13.5 min, 17%. MPLC with elution with 8% EtOAc in hexanes yielded 1.3 g of a mixture of diastereomer I and II and 0.33 g of pure diastereomer III. Rechromatography of the mixture using 49% dichloromethane/49% n-hexane/2% EtOAc gave 0.92 g of diastereomer I and 0.21 g of diastereomer II.

Diastereomer I: mp 123–125°C; ^{13}C NMR ($CDCl_3$); δ 66.09 (S-CH_2); $[\alpha]_D^{25}$ = +57.8 (c 1.0, EtOH).
Diastereomer II: mp 121–122°C; δ 57.89 (S-CH_2); $[\alpha]_D^{24.5}$ = +45.9 (c 0.95, EtOH).
Diastereomer III: mp 87–88°C; $[\alpha]_D^{24.5}$ = +12.6 (c 1.56, EtOH).

Thermolysis of the above diastereomers using Method A and separation of the products by flash chromatography gave optically active 2-t-butylcyclohexanone. From diastereomer I: 93%; $[\alpha]_D^{24.5}$ = +36.0 (c 1.0, MeOH). From diastereomer II: 98%; $[\alpha]_D^{24.5}$ = –35.3 (c 1.0, MeOH). From diastereomer III: 73%; $[\alpha]_D^{24.5}$ = –35.4 (c 1.0, MeOH). Spectroscopic data and chromatographic behavior of the optically active samples were identical to those of the racemic ketone.

4.4.12 (+)-(S)-S-(1-Cyclopenten-1-ylmethyl)-N-Methyl-S-Phenylsulfoximine 168[56]

To a solution of 1,10-phenanthroline (1 mg) in THF (10 mL) was added n-BuLi (77 mmol, 50 mL of 1.55 M in n-hexane). The resulting dark red solution was cooled to –50°C, and MeOH was added dropwise until the suspension turned yellow. The thus-formed suspension of LiOMe was treated with **166b** (9.0 g, 38 mmol) in THF (10 mL) and toluene (20 mL) at –50°C. After the suspension was stirred for 1 h, the cooling bath was removed, and stirring was continued for 72 h at room temperature. After the addition of saturated aqueous NH_4Cl (100 mL), the mixture was extracted with EtOAc (4 × 100 mL). The combined organic extracts were dried ($MgSO_4$) and concentrated *in vacuo*. Purification of the residue by chromatography (20% n-hexane-EtOAc) gave **168** (8.5 g, 95%) as a colorless oil.

4.4.13 (+)-(R)-1-[4-(1-Ethoxyethoxy)butyl]-2-Methylenecyclopentane 175a[56]

To a solution of CuI (381 mg, 2.0 mmol) in THF (15 mL) and Me_2S (3 mL) was added EtO(Me)C(H)O$(CH_2)_4$Li (2.0 mmol, 1.5 mL of 1.33 M in Et_2O) at –50°C. The solution was stirred for 45 min and then cooled to –78°C. The resulting brown, homogeneous solution was treated with BF_3OEt_2 (246 μL, 2.0 mmol) and stirred for 10 min. Then the solution was cooled to –100°C, and **168** (117 mg, 0.5 mmol) in THF (15 mL) was added. The reaction mixture was allowed to warm to –78°C during 4 h. Aqueous workup and purification of the crude product by chromatography (9% EtOAc-n-hexane) gave a mixture of **175a** and **176a** (103 mg, 91%) in a ratio of 98:2 as a colorless oil with an ee value of 71% for **175a**: $[\alpha]_D$ +38.0 (c 0.88, THF); ^1H NMR (400 MHz, $CDCl_3$).

REFERENCES

1. Clarke, S. G., Kenyon, J., Phillips, H., *J. Chem. Soc.*, 188, 1927.
2. Holloway, J., Kenyon, J., Phillips, H., *J. Chem. Soc.*, 3000, 1928.
3. Kresze, G., Wustrow, B., *Chem. Ber.*, **95**, 2652, 1962.
4. Bohman, O., Allenmark, S., *Chem. Scripta.*, **4**, 202, 1973.
5. Christensen, B.W., Kjaer, A., *J. Chem. Soc., Chem.Commun.*, 784, 1975.
6. a) Moriyama, M., Yoshimura, T., Furukawa, N., Numata, T., Oae, S., *Tetrahedron*, **32**, 3003, 1976;
   b) Moriyama, M., Oae, S., Numata, T., Furukawa, N., *Chem. & Ind.*, 163, 1976;
   c) Moriyama, M., Kurijama, K., Iwata, T., Furukawa, N., Numata, T., Oae, S., *Chem. Lett.*, 363, 1976.
7. a) Day, J., Cram, D.J., *Am. Chem. Soc.*, **87**, 4398, 1965;
   b) Rayner, D.R., von Schriltz, D.M., Day, J., Cram, D.J., *J. Am. Chem. Soc.*, **90**, 2721, 1968;
   c) Garwood, D.C., Cram, D.J., *J. Am. Chem. Soc.*, **92**, 4575, 1970;
   d) Yamagishi, F.G., Rayner, D.R., Zwicker, E.T., Cram, D.J., *J. Am. Chem. Soc.*, **95**, 1916, 1973.
8. a) Christensen, B.W., Kjaer, A., *J. Chem. Soc., Chem. Commun.*, 934, 1969;
   b) Christensen, B.J., *J. Chem. Soc., Chem. Commun.*, 597, 1971.
9. a) Maricich, T.J., Hoffman, V.L., *J. Am. Chem. Soc.*, **96**, 7770, 1974;
   b) Maricich, T.J., Hoffman, V.L., *Tetrahedron Lett.*, 5309, 1972.
10. Sabol, M.A., Davenport, R.W., Andersen, K.K., *Tetrahedron Lett.*, 2159, 1968.
11. a) Williams, T.R., Booms, R.E., Cram, D.J., *J. Am. Chem. Soc.*, **93** 7338, 1971;
    b) Williams, T.R., Nudelman, A., Booms, R.E., Cram, D.J., *J. Am. Chem. Soc.*, **94**, 4684, 1972;
    c) Nudelman, A., *Diss.Abstr.*, **29**, 1049B, 1969; *Chem.Abstr.*, **72**, 131905h, 1970.
12. a) Bentley, H.R., McDermott, E.E., Moran, T., Pace, J., Whitehead, J.K., *Proc. R. Soc.*, London, Ser. B, **137**, 402, 1950;
    b) Kennewell, P.D., Taylor, J.B., *Chem. Soc. Rev.*, **9**, 477, 1980.
13. a) Meister, A., in *Enzyme-Activated Irreversible Inhibitors*, Geiler, N., Jung, M. and Koch-Wesser, J., eds., Elsevier, Amsterdam, 1978, p. 187.
    b) Hwang, K.J., Logusch, W., Brannigan, L.M., Thompson, M.R., *J. Org. Chem.*, **52**, 3435, 1987.
14. a) Preuss, D.L., Scanell, J.P., Ax, H.A., Kellet, M., Weiss, F., Demny, T.C., Stempel, A., *J. Antibiot.*, **26**, 261, 1973;
    b) Scannel, J.P., Preuss, D.L., Demny, T.C., Ax, H.A., Weiss, F., Williams, T., Stempel, A., *The Chemistry and Biology of Peptides*, Meienhofer, J., ed., Ann Arbor Science, Ann Arbor, MI, 1972, 415.
15. Barash, M., *Nature*, **187**, 591, 1960.
16. Fusco, R., Tenconi, F., *Chem.Ind. (Milan)*, **47**, 61, 1965.
17. a) Johnson, C.R., Schroeck, C.W., *J. Am. Chem. Soc.*, **95**, 7418, 1973;
    b) Johnson, C.R., Schroeck, C.W., Shanklin, J.R., *J. Am. Chem. Soc.*, **95**, 7424, 1973.
18. Johnson, C.R., *Aldrichimica Acta*, **18**, 3, 1985.
19. Shiner, C.S., Berks, A.H., *J. Org. Chem.*, **53**, 5542, 1988.
20. Allenmark, S., Nielsen, L., Pirkle, W.H., *Acta Chem. Scand., Ser. B*, **B37**, 325, 1983.
21. Wainer, I.W., Alembik, M.C., Johnson, C.R., *J. Chromatogr.*, **361**, 374, 1986.
22. Tamura, Y., Minamikawa, J., Sumoto, K., Fuji, S., Ikeda, M., *J. Org. Chem.*, **38**, 1239, 1973.
23. Johnson, C.R., Kirchoff, R.A., Corkins, H.G., *J. Org. Chem.*, **39**, 2458, 1974.
24. Bolm, C., Müller, P., *Tetrahedron Lett.*, **36**, 1625, 1995.
25. Colonna, S., Stirling, C.J.M., *J. Chem. Soc., Perkin Trans. 1*, 2120, 1974.
26. Drabowicz, J., Łyzwa, P., Dudziński, B., Gawroński, J., unpublished results.
27. Williams, T.R., Booms, R.E., Cram, D.J., *J. Am. Chem. Soc.*, **93**, 7338, 1971.
28. Cram, D.J., Day, J., Rayner, D.R., von Schriltz, D.M., Duchamp, D.J., Garwood, D.C., *J. Am. Chem. Soc.*, **92**, 7369, 1970.
29. Pearson, A.J., Blystone, S.L., Nar, H., Pinkerton, A.A., Roden, B.A., Yoon, J., *J. Am. Chem. Soc.*, **111**, 134, 1989.
30. Akasaka, T., Yoshimura, T., Furukawa, N., Oae, S., *Chem. Lett.*, 417, 1978.
31. a) Annunziata, R., Cinquini, M., Colonna, S., *J. Chem. Soc., Perkin Trans. 1*, 2422, 1979;
    b) Annunziata, R., Cinquini, M., Colonna, S., *J. Chem. Soc., Perkin Trans. 1*, 1685, 1979.
32. Pyne, S.G., *J. Org. Chem.*, **51**, 81, 1986.
33. Pyne, S.G., *Tetrahedron Lett.*, **27**, 1691, 1986.
34. Regellin, M., Weinberger, H., *Tetrahedron Lett.*, **33**, 6959, 1992.
35. Furukawa, N., Yoshimura, T., Ohtsu, M., Akasaka, T., Oae, S., *Tetrahedron*, **36**, 73, 1980.
36. Johnson, C.R., Schroeck, C.W., *J. Am. Chem. Soc.*, **90**, 6852, 1968.
37. Johnson, C.R., Stark, J., Jr., *J. Org. Chem.*, **47**, 1193, 1982.
38. Johnson, C.R., Stark, J., Jr., *J. Org. Chem.*, **47**, 1196, 1982.
39. Johnson, C.R., Stark, J., Jr., *Tetrahedron Lett.*, 4713, 1979.
40. Bolm, C., Felder, C., *Tetrahedron Lett.*, **34**, 6041, 1993.
41. Bolm, C., Seger, A., Felder, C., *Tetrahedron Lett.*, **34**, 8079, 1993.
42. Bolm, C., Müller, J., Schlingloff, G., Zehnder, M., Neuburger, M., *J. Chem. Soc., Chem. Commun.*, 182, 1993.
43. Bolm, C., Felder, M., Müller, J., *Synlett*, 439, 1992.

44. a) Johnson, C.R., Zeller, J.R., *Tetrahedron*, **40**, 1225, 1984.
b) Johnson, C.R., Zeller, J.R., *J. Am. Chem. Soc.*, **104**, 4021, 1982.
45. Johnson, C.R., Penning, T.D., *J. Am. Chem. Soc.*, **110**, 4726, 1988.
46. Salmon, R.G., Sachinvala, N.D., Roy, S., Basu, B., Raychaudhuri, S.R., Miller, D.B., Sharma, R.B., *J. Am. Chem. Soc.*, **113**, 3085, 1991.
47. Johnson, C.R., Barbachyn, M.R., *J. Am. Chem. Soc.*, **104**, 4290, 1982.
48. Johnson, C.R., Barbachyn, M.R., Meanewell, N.A., Stark, Jr., C.J., Zeller, J.R., *Phosphorus Sulfur*, **24**, 151, 1985.
49. Johnson, C.R., Barbachyn, M.R., *J. Am. Chem. Soc.*, **106**, 2459, 1984.
50. Johnson, C.R., Meanwell, N.A., *J. Am. Chem. Soc.*, **103**, 7667, 1981.
51. Barbachyn, M.R., Johnson, C.R., Glick, M.D., *J. Org. Chem.*, **49**, 2746, 1984.
52. Erdelmeier, I., Gais, H.J., Lindner, H.J., *Angew. Chem.*, **98**, 912, (1986): *Angew. Chem. Int. Ed. Engl.*, **25**, 935, 1986.
53. Erdelmeier, I., Gais, H.J., *J. Am. Chem. Soc.*, **111**, 1125, 1989.
54. Trost, B.M., Merlic, C.A., *J. Am. Chem. Soc.*, **110**, 5316, 1988.
55. a) Trost, B.M., Schmuff, N.R., Miller, M.J., *J. Am. Chem. Soc.*, **102**, 5979, 1980;
b) Julia, M., Righini, A., Verpeaux, J.-N., *Tetrahedron*, **39**, 3283, 1983;
c) Masaki, Y., Sakuma, K., Kaji, K., *J. Chem. Soc., Perkin Trans. 1*, 1171, 1985;
d) Cuvigny, T., Julia, M., *J. Organomet. Chem.*, **317**, 383, 1986;
e) Bäckvall, J.-E., Juntunen, S.K., *J. Am. Chem. Soc.*, **109**, 6396, 1988;
f) Gais, H.-J., Ball, W.A., Bund, J., *Tetrahedron Lett.*, **29**, 781, 1988;
g) Trost, B.M., Craig, A.M., *J. Org. Chem.*, **55**, 1127, 1990.
56. Gais, H.J., Müller, H., Bund, J., Scommoda, M., Brandt.J., Raabe, G., *J. Am. Chem. Soc.*, **117**, 2453, 1995.
57. Glass, R.S., Reineke, K., Shanklin, M., *J. Org. Chem.*, **49**, 1527, 1984.
58. Annunziata, R., Cinquini, M., Cozzi, F., *Synthesis*, 767, 1982.

Chapter 5

Chiral "Onium" Derivatives of Sulfur and Chiral Sulfur Ylides

It seems reasonable to discuss in this chapter chiral sulfur ylides together with chiral onium salts **2**. This is because the ylides **1** viewed as zwitterionic structures having a carbanion attached directly to a positively charged sulfur atom are usually generated from the salts **2** by deprotonation (Equation 5.1).

$$R{-}\overset{(O)_n}{\overset{\|}{S}}{}^{\oplus}{-}CH_2R^1 \;\; X^{\ominus} \;(R^2 \text{ on S}) \quad \xrightarrow{N^{\ominus}M^{\oplus}} \quad R{-}\overset{(O)_n}{\overset{\|}{S}}{=}CHR^1 \;(R^2 \text{ on S}) \; + MX + NH$$

2 $\qquad$ **1a**

$n = 0$ or 1

$$\updownarrow$$

$$R{-}\overset{(O)_n}{\overset{\|}{S}}{}^{\oplus}{-}\overset{\ominus}{C}HR^1 \;(R^2 \text{ on S})$$

1b $\qquad$ (5.1)

Considering the stereochemical aspects of these two groups of chiral organosulfur compounds, it should be noted that sulfonium salts constitute the oldest family of organosulfur derivatives isolated in enantiomeric forms.[1] The first members have been known since 1900, when Pope and Peachy[1] and Smiles[2] reported independently the preparation of the chiral sulfonium salts **3** and **4** via resolution of racemic mixtures through the classical formation of diastereomeric salts.

$$Me\cdots\overset{\cdot\cdot}{S}{}^{\oplus}(Et){-}CH_2COOH \qquad\qquad Me\cdots\overset{\cdot\cdot}{S}{}^{\oplus}(Et){-}CH_2C(O)Ph$$

3 $\qquad$ **4**

It should be added that the successful resolution of the sulfonium salts **3** and **4** was of great contemporary importance for the development of the theory of optical activity based on the tetrahedral model, even though in the case of sulfonium salts the role of a fourth substituent is played by the lone pair of electrons on the sulfur atom.

On the other hand, the first chiral sulfur ylide, ethylmethylsulfonium phenacylide **5**, was prepared only in 1968 by deprotonation of optically active methylethylphenacylsulfonium perchlorate **6** by means of sodium methoxide (Equation 5.2).[3]

$$Et{-}\overset{\oplus}{S}(Me){-}CH_2C(O)Ph \;\; ClO_4^{\ominus} \quad \xrightarrow{MeONa} \quad Et{-}\overset{\oplus}{S}(Me){-}\overset{\ominus}{C}HC(O)Ph$$

6 $\qquad$ **5** $\qquad$ (5.2)

Since that time a large number of optically active charged "onium" salts and the corresponding sulfur ylides have been prepared and used as substrates in asymmetric synthesis. Below, a survey of the literature approaches to these two classes of chiral tricoordinate sulfur compounds will be presented, and this compilation will be followed by the presentation of their synthetic potential for the preparation of chiral hydrocarbon systems which, in the majority of cases, do not contain sulfur.

5.1 PREPARATION OF CHIRAL SULFUR ONIUM SALTS AND YLIDES

More than 60 years after the classical works of Pope and Peachy[1] and Smiles,[2] Darwish and co-workers[4] prepared a series of chiral sulfonium perchlorates **7–9** by resolution of the corresponding (–) dibenzoyl hydrogen tartrate salts followed by the anion exchange.

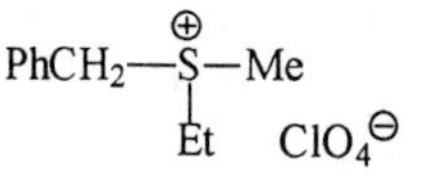

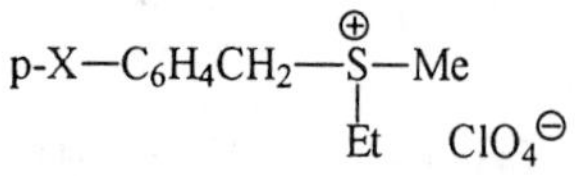

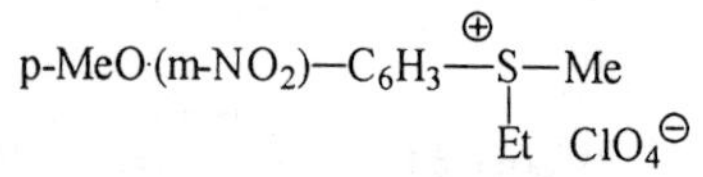

$PhCH_2-S^{\oplus}(Et)-Me$ $ClO_4^{\ominus}$ $p\text{-}X-C_6H_4CH_2-S^{\oplus}(Et)-Me$ $ClO_4^{\ominus}$ $p\text{-}MeO(m\text{-}NO_2)-C_6H_3-S^{\oplus}(Et)-Me$ $ClO_4^{\ominus}$

7, $[\alpha]_D$ = -8.20 (MeOH) **8a**, X = NO_2, $[\alpha]_D$ = +7.8 (MeOH) 9, $[\alpha]_D$ = +10.6 (acetone)

8b, X = Cl, $[\alpha]_D$ = n.r.

At the same time, 1-adamantylethylpropenylsulfonium tetrafluoroborate **10** and 1-adamantylethylmethylsulfonium tetrafluoroborate **11** were resolved via diastereomeric salts with (–) malic acid and (–) dibenzoyltartaric acid, respectively.[5]

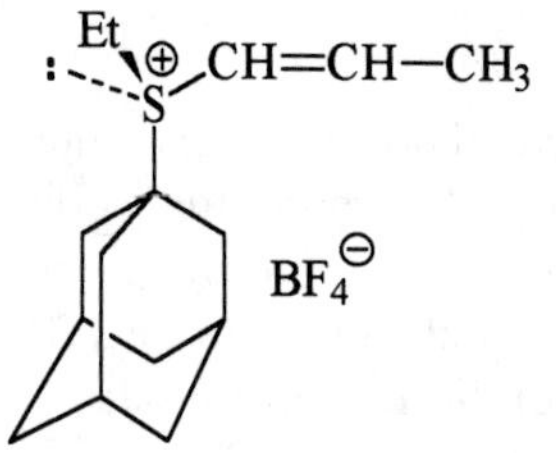

Et CH=CH–CH_3 $BF_4^{\ominus}$ Me Et $BF_4^{\ominus}$

10, $[\alpha]_{365}$ = +22.2 (EtOH) 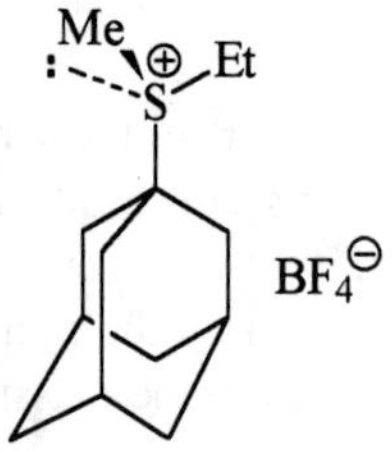 **11**, $[\alpha]_{350}$ = +27.3 (EtOH)

Similarly, the partial resolution of the cyclic sulfonium salts **12–14** was achieved using (–)-di-o-toluoyltartaric acid as the chiral resolving agent.[6]

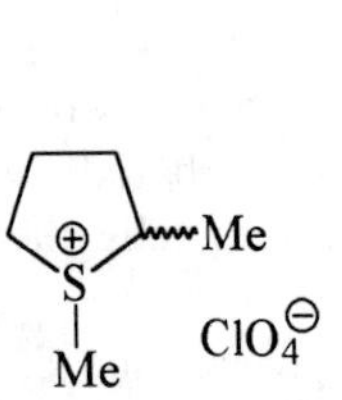

Me $ClO_4^{\ominus}$ Me Me Me $ClO_4^{\ominus}$ Me Me Me $ClO_4^{\ominus}$ Me

12 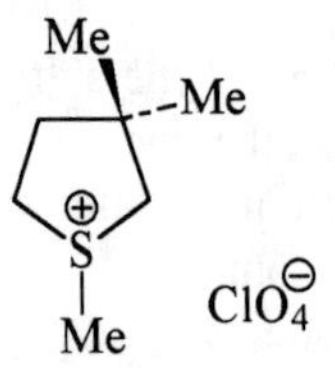 **13** 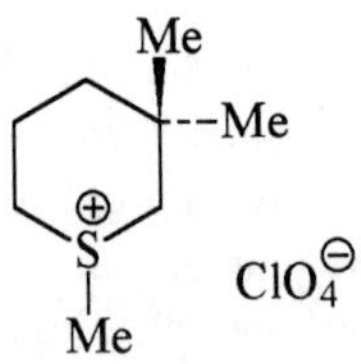 **14**

The enantiomer of the sulfonium salt **15a** was isolated via recrystallization of the corresponding l-thiophanium d-camphorsulfonate.[7]

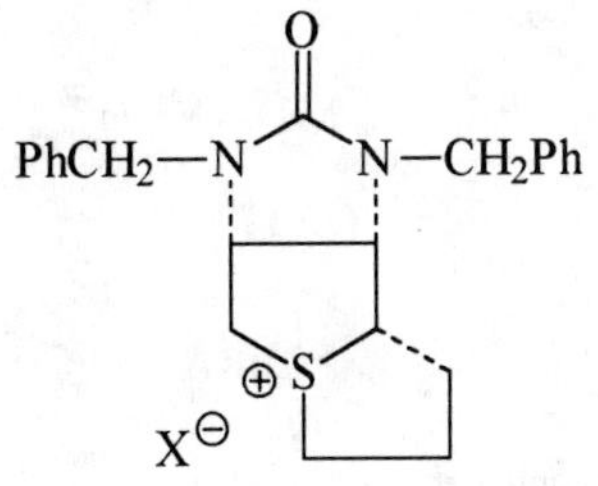

15a, X = camphorosulfonate, $[\alpha]_D = -50.3$ ($CHCl_3$)

More recently, a similar procedure was used[8] for the preparation of optically active 2-chloro-10-(2,5-xylyl)-10-thioxanthenium perchlorate **16b**.

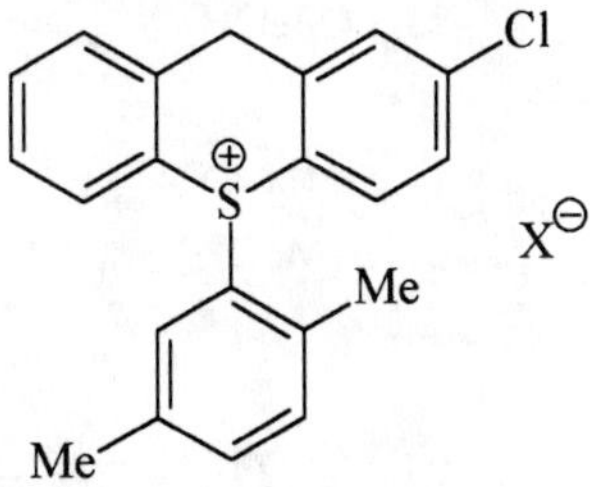

16a, X = α-camphorosulfonate, $[\alpha]_D = +27.8$ ($CHCl_3$)
16b, X = ClO_4^-, $[\alpha]_{400} = +0.71$ (Me_2SO-MeOH 1:4)

A series of chiral aryldialkylsulfonium salts **17a–f** were obtained by fractional crystallization of the diastereomeric d-camphorosulfonates (Scheme 5.1).[9]

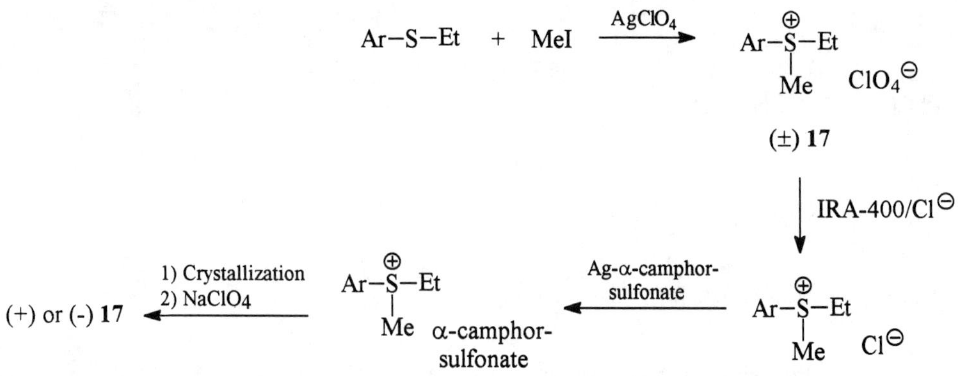

$Ar{-}S{-}Et + MeI \xrightarrow{AgClO_4} Ar{-}S^{\oplus}(Me){-}Et\ ClO_4^{\ominus}$ (±) **17**

IRA-400/$Cl^{\ominus}$ → $Ar{-}S^{\oplus}(Me){-}Et\ Cl^{\ominus}$

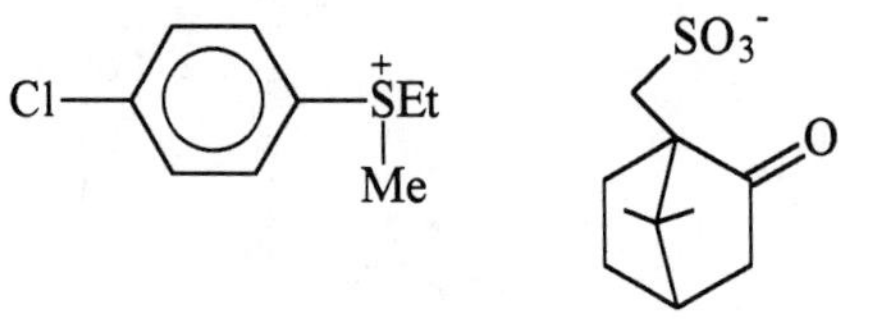

Ag-α-camphor-sulfonate → $Ar{-}S^{\oplus}(Me){-}Et$ α-camphor-sulfonate

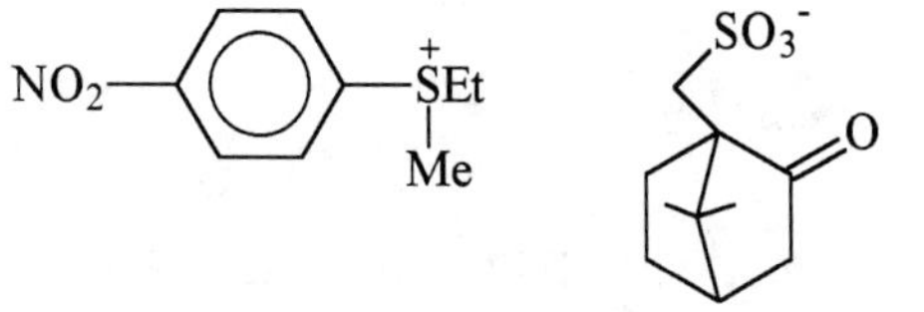

1) Crystallization 2) $NaClO_4$ → (+) or (-) **17**

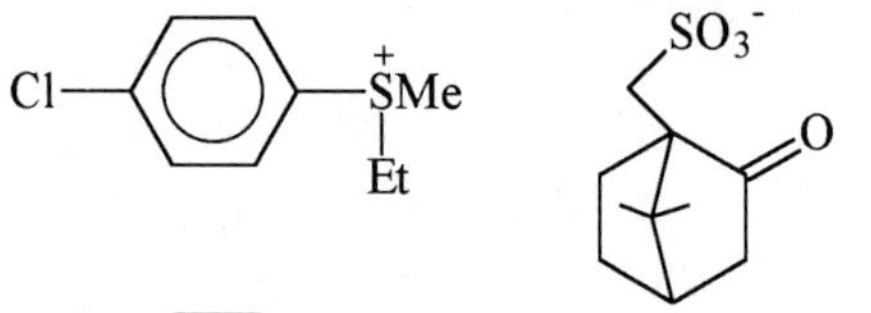

No			$[\alpha]_D$
17a	Cl–C6H4–S+(Me)Et	ClO_4^-	-23.3 (MeOH)
17b	Cl–C6H4–S+(Me)Et	camphorsulfonate (SO_3^-)	+7.52 (MeOH)
17c	NO_2–C6H4–S+(Me)Et	camphorsulfonate (SO_3^-)	+11.3 (MeOH)
17d	Cl–C6H4–S+(Et)Me	camphorsulfonate (SO_3^-)	+46.6 (MeOH)
17e	C6H5–S+(Et)Me	ClO_4^-	+20.3 (acetone)
17f	C6H5–S+(Et)Me	camphorsulfonate (SO_3^-)	+44.57 (MeOH)

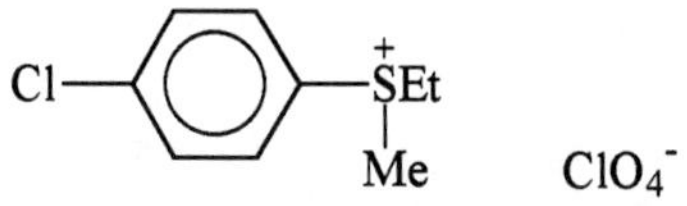

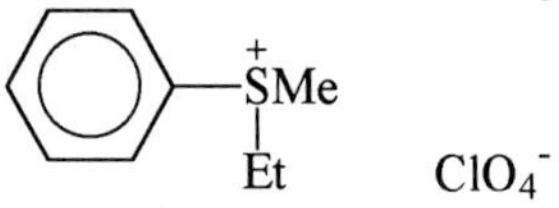

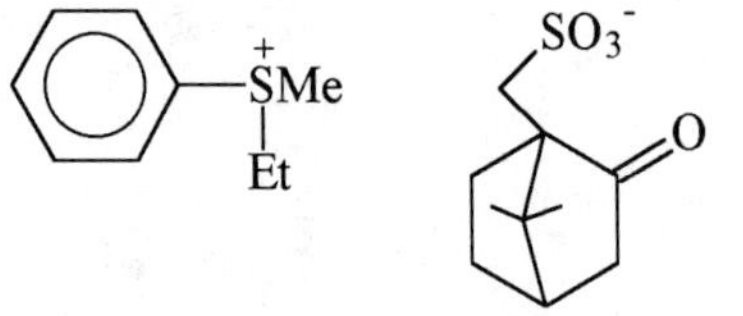

Scheme 5.1

In a similar way, the diastereomerically pure sulfonium salt **18** was obtained.[9a]

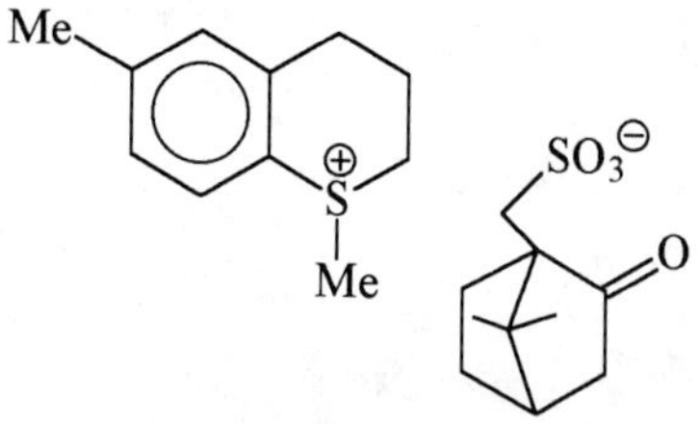

(S)-(+)-**18**
$[\alpha]_D$ +41.2($CHCl_3$)

S-Alkylation of oxathiane **19** with benzyl bromide in the presence of $AgClO_4$ afforded the sulfonium salt **20** as a single diastereomer.[10]

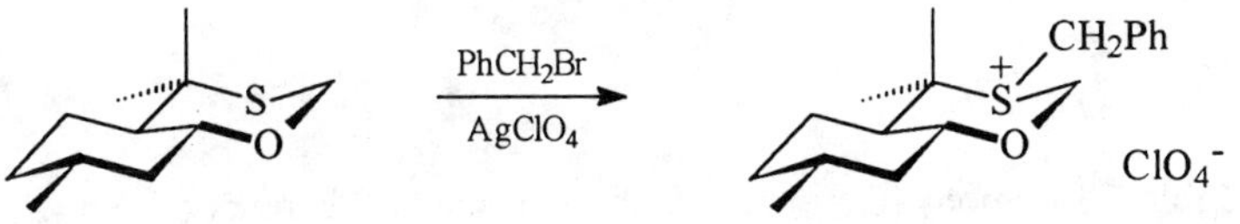

$PhCH_2Br$, $AgClO_4$

(+)-(1R,2S,5R)-**19** **20** ClO_4^-

Diastereomerically pure sulfonium salts **21** and **22** constituted the key intermediates in a multistep stereoselective synthesis of the ethylmethylpropylsulfonium 2,4,6-trinitrobenzene-sulfonate **23** from (–)-(S)-ethyl lactate shown in Scheme 5.2.[11]

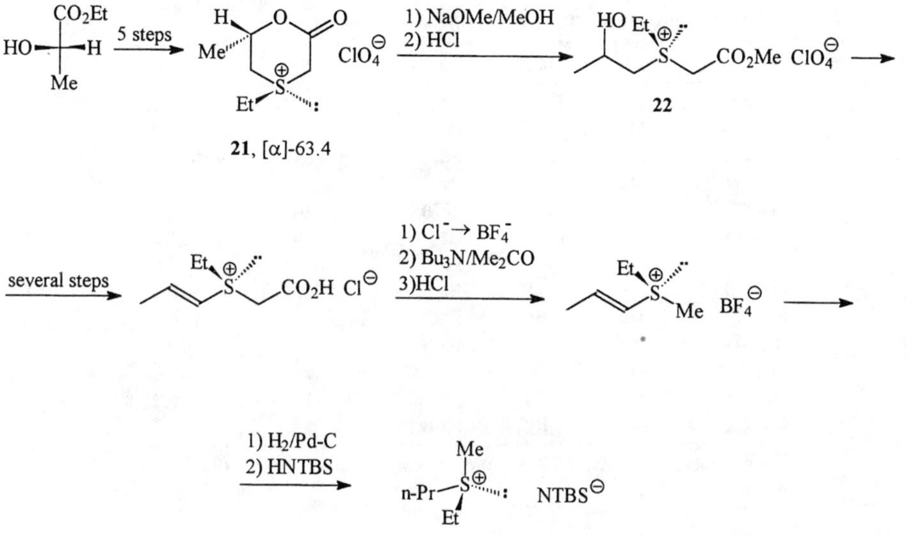

5 steps; 1) NaOMe/MeOH 2) HCl

21, [α]-63.4 **22**

several steps; 1) $Cl^- \rightarrow BF_4^-$ 2) Bu_3N/Me_2CO 3) HCl

1) H_2/Pd-C 2) HNTBS

23, [α]+1.18

Scheme 5.2

In the chemical literature there is a single report on asymmetric synthesis of chiral sulfonium salts. Namely, it was found[12] that, when benzyl ethyl sulfide is reacted with the chiral methoxysulfonium salt (+)-(R)-**24**, the asymmetric methylation takes place, giving the sulfonium salts **7b** with an enantiomeric purity of 60%, as well as chiral methyl p-tolyl sulfoxide **25** (Equation 5.3).

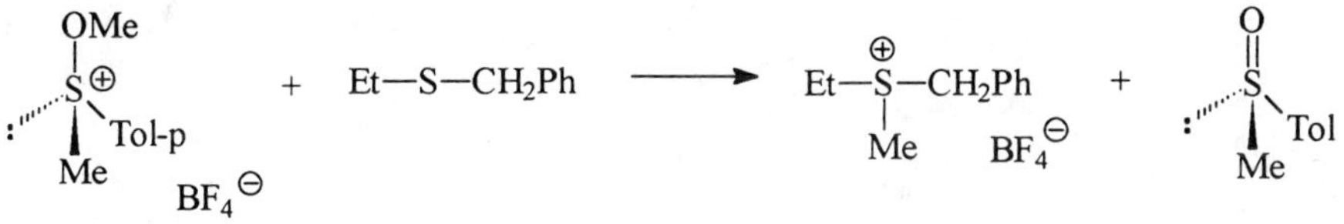

(+)-(R)-**24** + $Et{-}S{-}CH_2Ph$ → **7b** + (+)-(R)-**25a** (5.3)

In addition to resolution and asymmetric synthesis, chiral sulfonium salts have been prepared in a stereoselective way by Andersen and co-workers.[13] This approach is based on the reaction of alkyl Grignard or dialkylcadmium reagents with the chiral alkoxysulfonium salts **26** derived from the corresponding chiral alkyl p-tolyl sulfoxides **25a–c** (Equation 5.4).

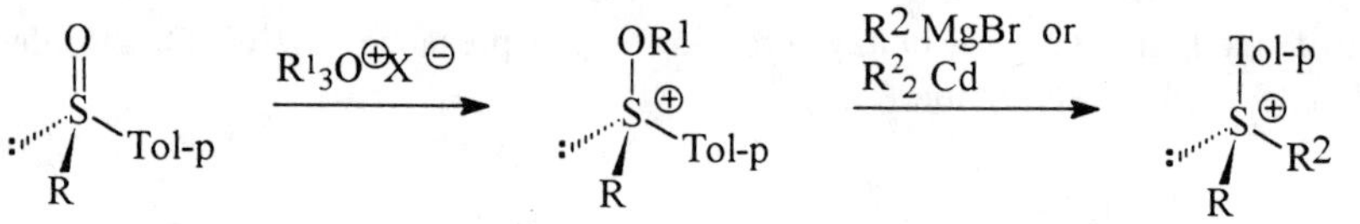

$R^1_3O^{\oplus}X^{\ominus}$; R^2MgBr or R^2_2Cd

(+)-(R)-**25a-c** **26** **27a-g** (5.4)

The data collected in Table 5.1 demonstrate the general character of this reaction, which was found to occur with inversion of configuration at sulfur, although some racemization was observed. Despite its successful application in the synthesis of various dialkylaryl- or diarylalkylsulfonium salts, all attempts to prepare optically active triarylsulfonium salts resulted in the formation of racemic mixtures. It should also be noted that the reaction of O-alkylated dialkyl sulfoxides with alkylmagnesium and alkylcadmium reagents does not produce trialkylsulfonium salts.[13]

Table 5.1 Synthesis of Dialkyl-p-Tolylsulfonium Salts p-TolS$^+$RR^2-X-**27a–g** from (R)-Alkoxyalkyl-p-Tolylsulfonium Salts-**26** (Equation 5.4)

No	R	R^2	X$^-$	Yield [%]	$[\alpha]_D$
(R)-**27a**	Me	Et	TNBS	57	–5.6
(R)-**27b**	Me	Et	BF_4	51	–15.8
(R)-**27c**	Me	Et	Ph_4B	39	–10.5
(R)-**27d**	Me	Et	Br	25	–11.5
(S)-**27a**	Me	Et	TNBS	16	19.2
(S)-**27b**	Me	Et	BF_4	59	21.2
(S)-**27c**	Me	Et	Ph_4B	53	9.0
(S)-**27d**	Me	Et	Br	72	230.0
(R)-**27e**	Me	n-Bu	TNBS	10	–6.6
(R)-**27f**	Me	n-Bu	Ph_4B	10	–10.7
(S)-**27e**	Me	n-Bu	TNBS	75	7.6
(R)-**27g**	Et	n-Bu	TNBS	11	–6.2
(S)-**27g**	Et	n-Bu	Ph_4B	64	10.2

From Andersen, K. K., *J. Chem. Soc., Chem. Commun.*, 1051, 1971. With permission.

Unsymmetrical alkoxysulfonium salts represent another group of tricoordinated sulfur derivatives which are able to exist in enantiomeric forms. A general method for their synthesis is based on the O-alkylation of chiral sulfoxides by triethyloxonium tetrafluoroborate.[14] The synthesis of a few ethoxyalkyl(aryl)-p-tolylsulfonium tetrafluoroborates **28** starting from optically active alkyl(aryl)-p-tolyl sulfoxides **25a–d**[14-16] shown in Scheme 5.3 illustrates this method.

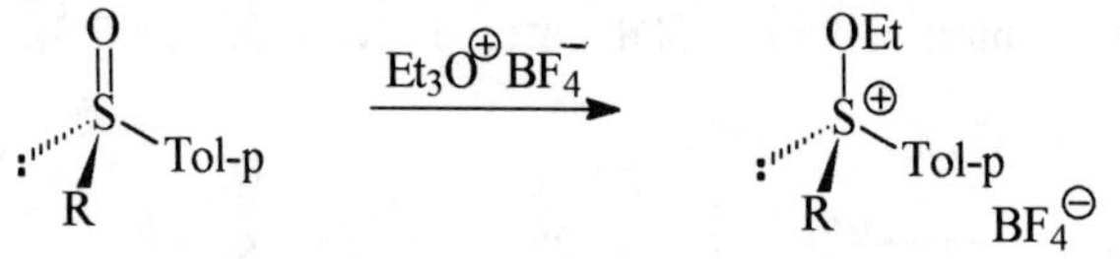

$Et_3O^{\oplus}BF_4^{-}$

(+)-(R)-**25a-d** (R)-**28a-d**

No	R	$[\alpha]_D$ (solvent)	Ref.
a	$PhCH_2$	+202 ($CHCl_3$)	14
b	Me	+149 (acetone)	15
c	Et	n.r.	13c
d	1-Nph	+410 (acetone)	16

Scheme 5.3

It was found later that other "hard" alkylating agents such as FSO_3Me or CF_3SO_3Me may be used for this purpose. Extension of the foregoing approach to optically active sulfinates **29** made it possible to prepare chiral unsymmetrical dialkoxysulfonium salts **30** (Scheme 5.4).[17]

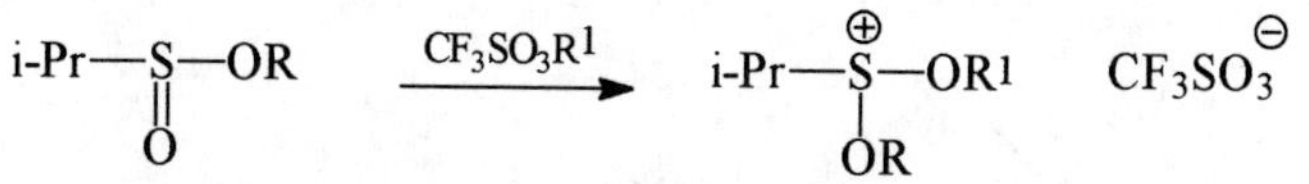

i-Pr—S(=O)—OR —($CF_3SO_3R^1$)→ i-Pr—$S^{\oplus}$(OR)—OR^1 $CF_3SO_3^{\ominus}$

29 **30**

a, R = CD_3, R^1 = Me
b, R = Et, R^1 = Me
c, R = n-Pr, R^1 = Me
d, R = i-Pr, R^1 = Me
e, R = n-Bu, R^1 = Me
f, R = CH_2–Bu-t, R^1 = Me

Scheme 5.4

Similarly, the O-alkylation of chiral sulfinamides leads to chiral alkoxyaminosulfonium salts.[18,19] For instance, methoxypyrrolidino-p-tolyl sulfonium salt **31a** was formed[18] when optically active sulfinamide **32a** was treated with methyl trifluoromethanesulfonate (Equation 5.5).

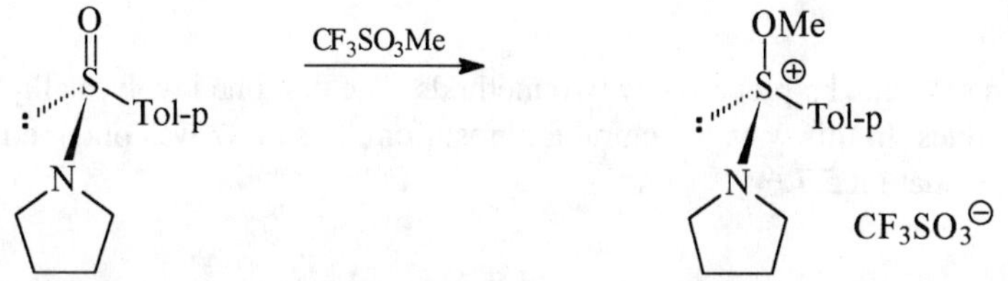

CF_3SO_3Me → $CF_3SO_3^{\ominus}$

(+)-(S)-**32a** (S)-**31a** (5.5)

The reaction of chiral sulfinamide **32b** with the same methylating agent gave the corresponding salt **33**, which was isolated and characterized as a tetraphenylborate (Equation 5.6).[19]

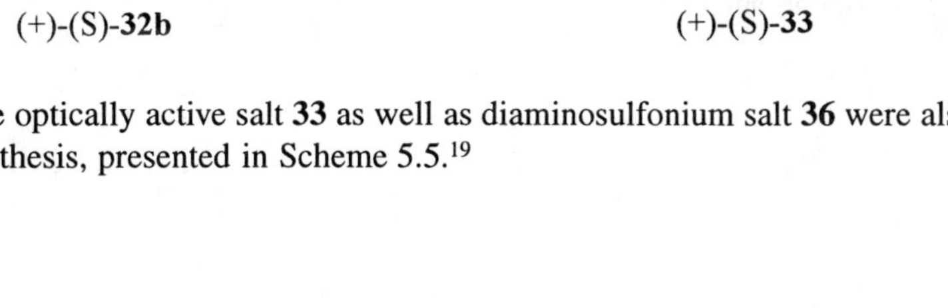

1) CF_3SO_3Me
2) $NaBPh_4$ → $Ph_4B^{\ominus}$

(+)-(S)-**32b** (+)-(S)-**33** (5.6)

The optically active salt **33** as well as diaminosulfonium salt **36** were also prepared by asymmetric synthesis, presented in Scheme 5.5.[19]

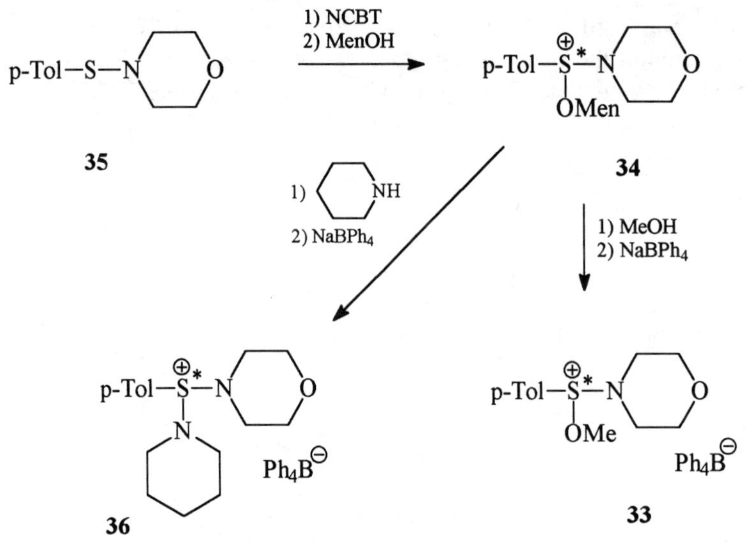

p-Tol—S—N(morpholine) **35** → 1) NCBT 2) MenOH → p-Tol—S⊕*(OMen)—N(morpholine) **34**

34 → 1) piperidine (NH) 2) $NaBPh_4$ → **36** p-Tol—S⊕*(N-piperidinyl)—N(morpholine) $Ph_4B^{\ominus}$

34 → 1) MeOH 2) $NaBPh_4$ → **33** p-Tol—S⊕*(OMe)—N(morpholine) $Ph_4B^{\ominus}$

NCBT = N-Chlorobenzotriazole

Scheme 5.5

Chiral aminosulfonium salts can be prepared by two methods. The first one involves alkylation at nitrogen in chiral sulfilimines. In this way, the chiral aminosulfonium salt **37** was obtained from sulfilimine **38** as shown in Equation 5.7.[20]

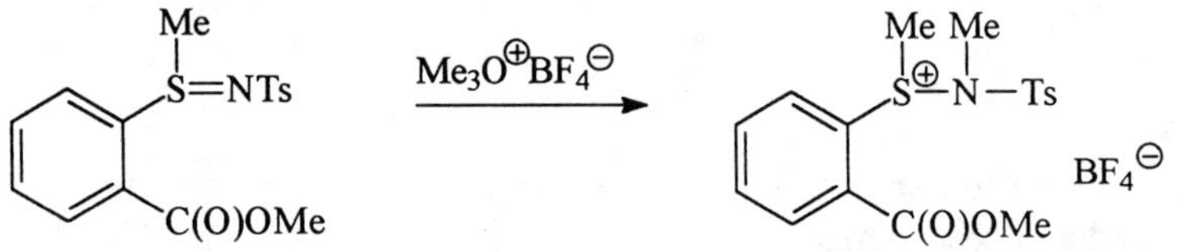

(-)-(S)-**38** (Me, S=NTs, C(O)OMe on benzene) → $Me_3O^{\oplus}BF_4^{\ominus}$ → (-)-(S)-**37** (Me Me, S⊕—N—Ts, C(O)OMe, $BF_4^{\ominus}$) (5.7)

The second approach to chiral aminosulfonium salts consists in the conversion of chiral sulfoxides into aminosulfonium salts by means of N,N-diethylaminosulfinyl tetrafluoroborate.[21] With dialkyl and alkyl aryl sulfoxides the reaction (Scheme 5.6) occurs with predominant retention of configuration at sulfur. However, its stereoselectivity is strongly dependent on the nature of substituents in the starting sulfoxide. In the case of diaryl sulfoxides, this method failed to give optically active aminodiarylsulfonium salts.

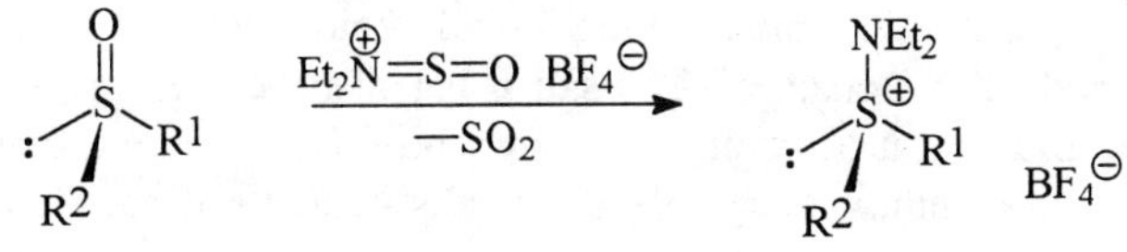

$$\underset{\mathbf{39}}{R^1R^2S{=}O} \xrightarrow[-SO_2]{Et_2\overset{\oplus}{N}{=}S{=}O\ BF_4^{\ominus}} \underset{\mathbf{40}}{R^1R^2\overset{\oplus}{S}NEt_2\ BF_4^{\ominus}}$$

Starting sulfoxide				Aminosulfonium salt	
No	R^1	R^2	$[\alpha]_{589}$ (ee)	No	$[\alpha]_{589}$
a	Me	i-Bu	+64.14 (46.5)	**a**	+32.5
b	Me	n-Bu	−12.25 (11.1)	**b**	−7.93
c	Me	p-Tol	+148.0 (100)	**c**	+10.45
d	Ph	p-Tol	+27.0	**d**	0.00

Scheme 5.6

Oxosulfonium salts of the general formula **41** form another group of chiral "onium" derivatives of sulfur. Thus far, however, methylethylphenyloxosulfonium perchlorate **42a** is the only compound of this type whose enantiomers have been isolated.

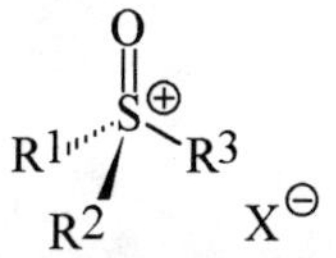

$$\underset{\mathbf{41}}{R^1R^2R^3\overset{\oplus}{S}{=}O\ X^{\ominus}} \qquad \underset{\mathbf{42}}{Me(Et)(Ph)\overset{\oplus}{S}{=}O\ X^{\ominus}}$$

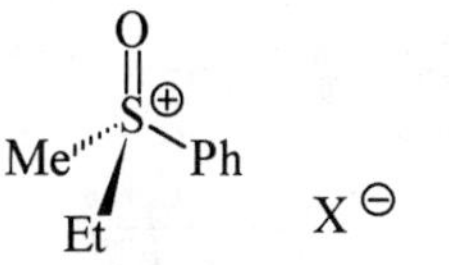

a, X=ClO$_4$
b, X=HgI$_3$

Kobayashi and co-workers[22] have prepared chiral **42a** by three different approaches. The first method is based on the separation of diastereomeric salts of methylethylphenyloxosulfonium mercurytriiodide **42b** with (+)-camphor-10-sulfonic acid. Upon treatment with sodium perchlorate, the enantiomers of **42a** were isolated. Alternatively, the reaction of the chiral (+)-(R)-ethyl phenyl sulfoxide **43** with methyl iodide in the presence of mercuric iodide followed by anion exchange was found to give the chiral salt **42a** (Equation 5.8).[22]

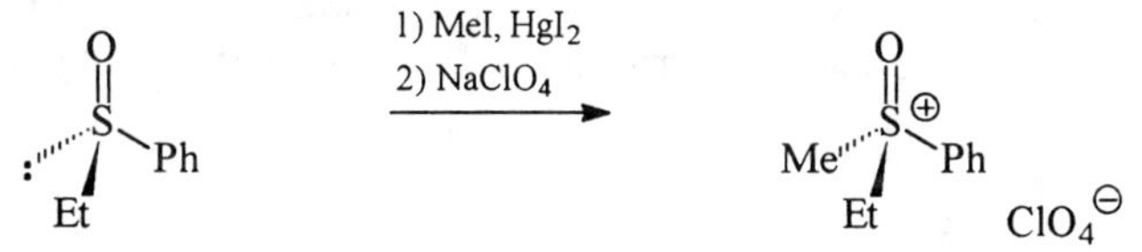

$$\underset{(+)\text{-}(R)\text{-}\mathbf{43}}{Et(Ph)S{=}O} \xrightarrow{\text{1) MeI, HgI}_2\ \text{2) NaClO}_4} \underset{(+)\text{-}(R)\text{-}\mathbf{42a}}{Me(Et)(Ph)\overset{\oplus}{S}{=}O\ ClO_4^{\ominus}} \qquad (5.8)$$

A stereoselective synthesis of (+)-(R)-**42a** consists in the oxidation of (+)-(S)-methylethylphenylsulfonium perchlorate **17e** with sodium perbenzoate (Equation 5.9).[22]

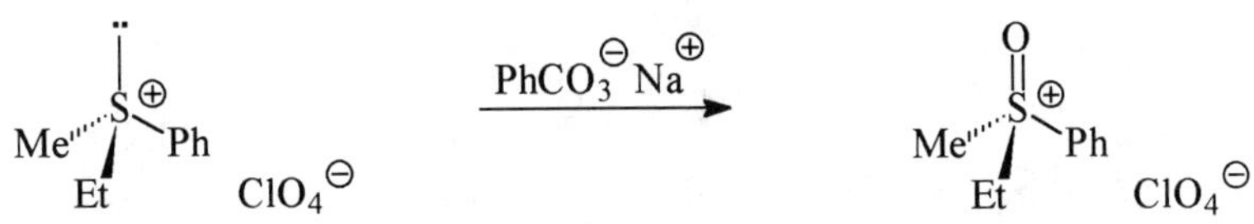

$$\underset{(+)\text{-}(S)\text{-}\mathbf{17e}}{Me(Et)(Ph)\overset{\oplus}{S}\ ClO_4^{\ominus}} \xrightarrow{PhCO_3^{\ominus}\ Na^{\oplus}} \underset{(+)\text{-}(R)\text{-}\mathbf{42a}}{Me(Et)(Ph)\overset{\oplus}{S}{=}O\ ClO_4^{\ominus}} \qquad (5.9)$$

The preparation of chiral methylethyl sulfonium phenacylide **5** by deprotonation of the sulfonium salt **6** has already been mentioned in this chapter.[3] The salt **6** has also been resolved via diastereomeric salts with dibenzoyltartaric acid monohydrate.[4] A few chiral O-substituted diarylsulfonium ylides **44** were obtained[23] from menthoxydiarylsulfonium salts **45** and sodium dimethylmalonate (Equation 5.10).

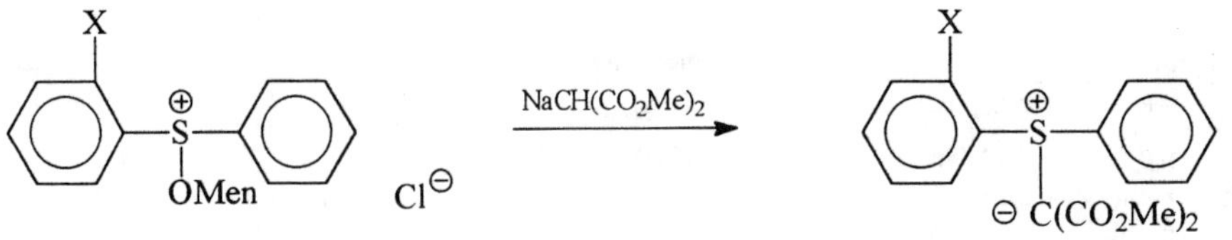

X ⊕ S OMen $Cl^{\ominus}$ **45** — $NaCH(CO_2Me)_2$ → X ⊕ S ⊖ $C(CO_2Me)_2$ **44** (5.10)

From the theoretical point of view, it is interesting to note that optically active thiabenzene structures are best described as chiral cyclic sulfonium ylides. 1-Pentafluorophenyl-2-methyl-2-thianaphtalene **46**, the first example of an optically active thiabenzene, was prepared by Mislow and co-workers[24] by deprotonation of thiochromenium tetrafluoroborate with brucine in anhydrous dimethyl sulfoxide (Equation 5.11).

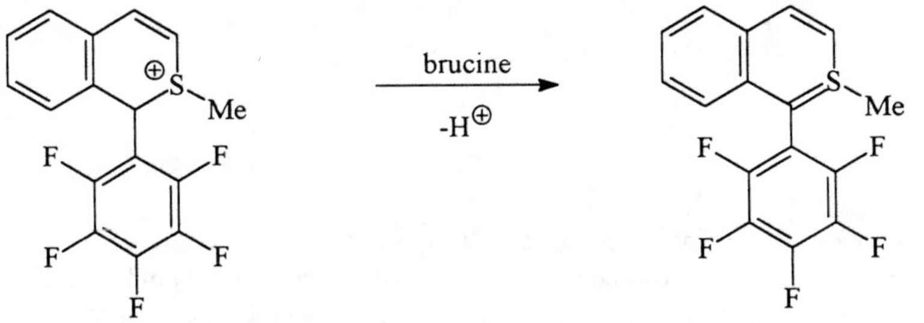

⊕ S Me F F F F F — brucine, $-H^{\oplus}$ → S Me F F F F F **46** (5.11)

Till now, all chiral oxoylides of sulfur have been prepared only *in situ* by deprotonation of the corresponding chiral oxosulfonium salts. The use of a series of oxosulfonium ylides **47–48 and 50–53** derived from N-alkylated sulfoximines and the carbanion **49**, as nucleophilic alkylidene transfer agents for the asymmetric synthesis, will also be discussed below.

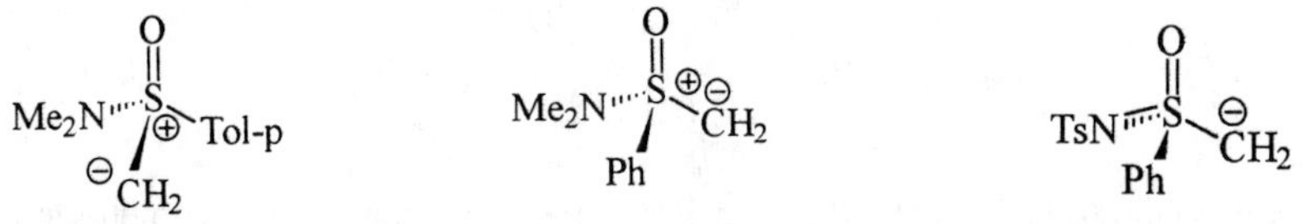

Me_2N–S(O)(⊕)–Tol-p, ⊖CH_2 **(R)-47**; Me_2N–S(O)(⊕)–⊖CH_2, Ph **(S)-48**; TsN=S(O)–⊖CH_2, Ph **(S)-49**

Et_2N⊕–S(O)–Tol-p, ⊖CH_2 **(R)-50**; Me_2N⊕–S(O)–$C_6H_2(Me)_3$, ⊖CH_2 **(R)-51**; Me_2N⊕–S(O)–Tol-p, ⊖CHMe **(R)-52**; Me_2N⊕–S(O)–Tol-p, ⊖cyclopropyl **(R)-53**

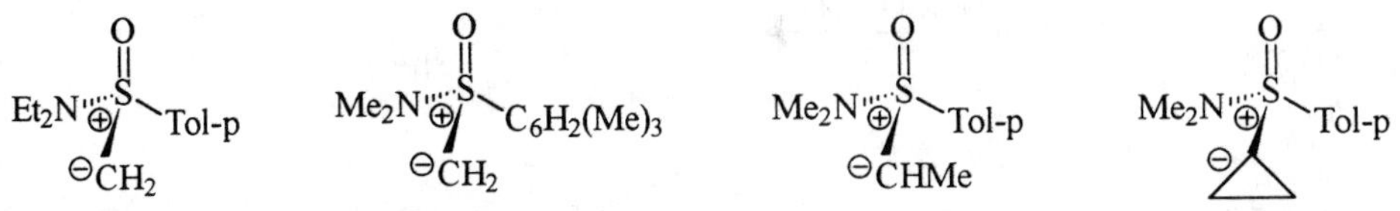

5.2 SYNTHETIC APPLICATION OF CHIRAL SULFUR "ONIUM" SALTS AND YLIDES

The first example of asymmetric induction in the transfer of chirality from the chiral sulfonium sulfur atom to the nitrogen atom was found[12] to occur in the methylation of benzylethylaniline with methoxymethyltolylsulfonium tetrafluoroborate (+)-(R)-**24** (Equation 5.12).

OMe, Me, Tol-p $BF_4^{\ominus}$ + Et, $PhNCH_2Ph$ → Et, $Ph\text{-}N^{\oplus}\text{-}CH_2Ph\ BF_4^{\ominus}$, Me

(+)-(R)-**24** **54**, $[\alpha]_D$ -13.5 (5.12)

(–)-Benzylethylmethylphenylammonium tetrafluoroborate **54** formed in this reaction had optical purity equal to 69%. Asymmetric methylation of benzyl methyl sulfide with this reagent[12] has already been presented in this chapter (see Equation 5.3). Similar asymmetric methylation reactions were observed with the aminoalkoxysulfonium salt **31a**.[25]

Alkylation of the cyclic β-ketoester-2-(methoxycarbonyl)-1-indanone **55** with a series of the chiral sulfonium salts **17a-f** occurs also with asymmetric induction. It was found[9] that the C-alkylation gave always a mixture of optically active 2-alkyl-2-(methoxycarbonyl)-1-indanones **56** and **57** as the main reaction products with ca. 16% ee. The O-alkylated products, 1-alkoxy-2-(methoxycarbonyl)-1-indenes **58** and **59**, were formed as minor components (Scheme 5.7).[26]

CO_2Me

55

K_2CO_3, CH_2Cl_2 Ar, Et $X^{\ominus}$, Me (R)-17a-c

K_2O_3, CH_2Cl_2 Ar, Me $X^{\ominus}$, Et (R)-17d-f

Me, CO_2Me (+)-(R)-56 + CO_2Me, Et (-)-(S)-57 | CO_2Me, Me (-)-(S)-56 + Et, CO_2Me (+)-(R)-57

OMe, CO_2Me 57 | OEt, CO_2Me 58

Scheme 5.7[26]

A very interesting example of asymmetric induction directed by the chiral sulfonium sulfur atom has recently been observed in the electrophilic addition of the enantiopure thiosulfonium salt **60** to nonfunctionalized olefins.[26] The salt **60**, prepared as shown in Scheme 5.8, transfers enantioselectively the MeS^+ group to *trans*-hex-3-ene, leading to the nonracemic thiranium ion **62**. This, in turn, reacts with MeCN–H_2O, allowing the synthesis of chiral difunctionalized alkanes **63** with enantiomeric excesses of up to 86% (Scheme 5.8).[27]

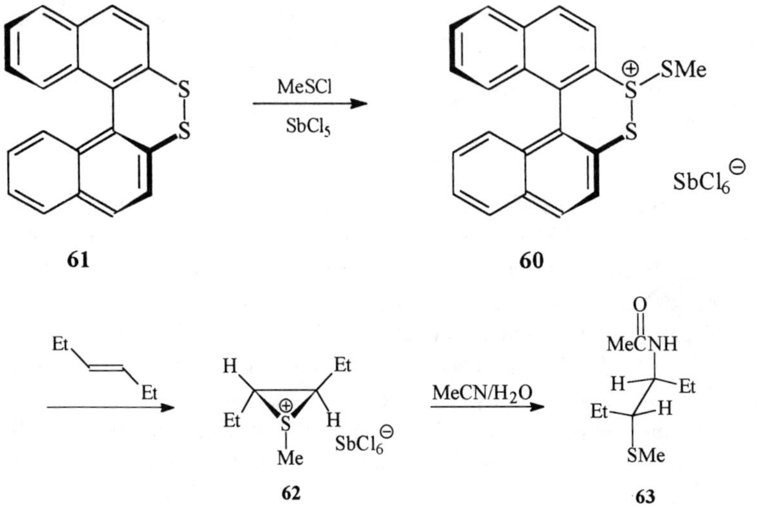

MeSCl, $SbCl_5$ → $SbCl_6^{\ominus}$

61 **60**

MeCN/H_2O

62 **63**

Equiv.	T/°C	Time before MeCN addition [min]	**63**, ee [%]
1.1	-20	10	40
1.1	-20	10	49
1.1	-60	60	56
1.0	-78	120	86

Scheme 5.8[27]

Recently, it was observed that chiral p-tolyl vinyl sulfoxide can be activated for the Diels-Alder reaction by O-alkylation with Meerwein's reagent, allowing cycloadditions to be carried out under mild conditions and with high stereoselectivity.[27] Thus, ethoxy p-tolyl vinylsulfonium tetrafluoroborate **64** was found to react with cyclopentadiene at room temperature over 1.5 h to give the cycloadducts **65–68** (*endo*/*exo* = 92:8) in 50% isolated yield (Scheme 5.9).

EtO, p-Tol, $BF_4^{\ominus}$, H, H, H

64

NaOH

endo major **65**

endo minor **66**

exo minor **67**

exo major **68**

Temp. [°C]	Time [h]	Total yield [%]	*endo* de [%]	*exo* de [%]	*endo*/*exo*
25	1.5	50	86	86	92:8
0	15	46	96	97	91:9
–20	48	32	>99	—	>99:1
–78	36	62	>99	—	>99:1

Scheme 5.9[27]

The diastereomeric excess was high (86% de) for the *exo* and *endo* isomers. When the reaction was carried out at –78°C the formation of the *endo* cycloadduct was only observed and the diastereoisomeric excess was higher than 99%. A high extent of asymmetric induction was also observed in the reaction of the salt **64** with 2,3-dimethylbutadiene and furan (Scheme 5.10).[27]

64 → 69

64 → *endo* minor **70** + *endo* major **71** + *exo* major **72** + *exo* minor **73**

Diene	Temp. [°C]	Time [h]	Total yield [%]	*endo* de [%]	*exo* de [%]	*endo*/*exo*
2,3-diMe-butadiene	25	15	40	>99		—
Furan	25	4.5	70	80	78	48:52
Furan	0	60	90	93	98	55:45
Furan	–20	60	35	>98	>98	59:41
Furan	–10	60	86	91	85	56:44
Furan	–10	60	66	92	87	56:44
Furan	–10	60	69	89	84	55:45

Scheme 5.10[27]

It was also found[27] that trimethylsilyl trifluoromethanesulfonate, $CF_3SO_3SiMe_3$, is able to catalyze the Diels-Alder cycloaddition of optically active p-tolyl vinyl sulfoxide with cyclopentadiene and furan (Scheme 5.11). The reaction was, however, slower in comparison with the sulfonium salt **64**, but its stereochemical outcome (exo/endo and π-facial selectivity) was almost the same. It should be pointed out that the Diels-Alder reactions shown in Schemes 5.9 and 5.11 gave epimeric sulfoxides as the final products because there is an inversion of configuration on conversion of the sulfonium cycloadducts to the corresponding sulfoxides.

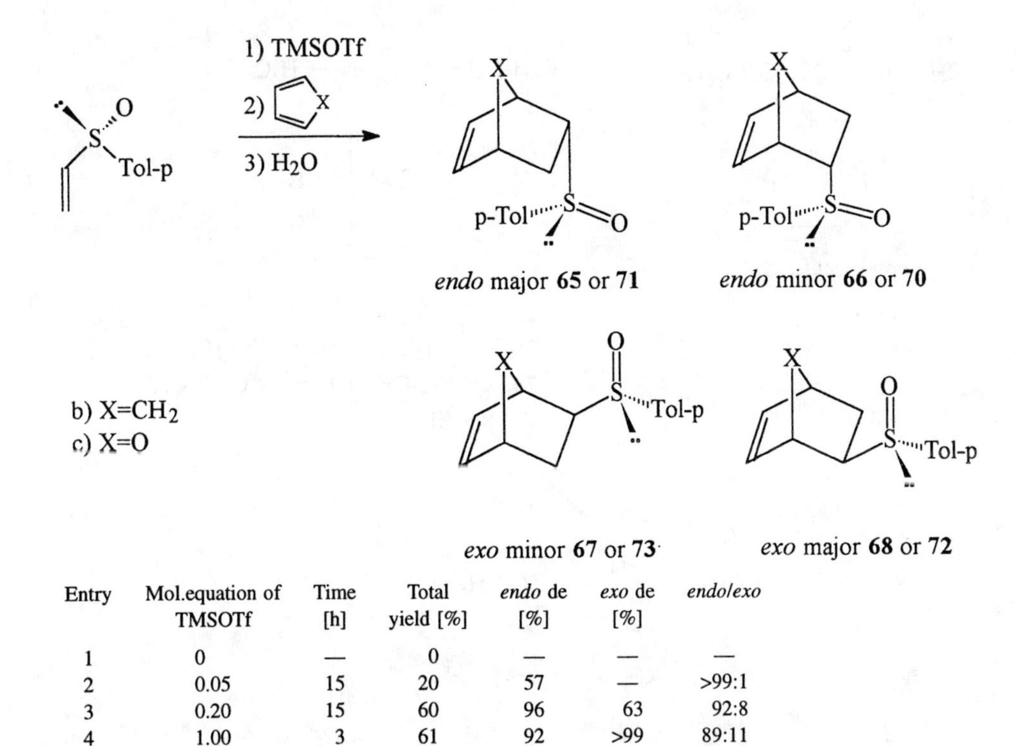

Entry	Mol.equation of TMSOTf	Time [h]	Total yield [%]	*endo* de [%]	*exo* de [%]	*endo*/*exo*
1	0	—	0	—	—	—
2	0.05	15	20	57	—	>99:1
3	0.20	15	60	96	63	92:8
4	1.00	3	61	92	>99	89:11
5	1.00	36	72	71	73	55:45

Entries 1–4: cyclopentadiene; entry 5: furan.

Scheme 5.11

The diastereomerically pure sulfonium salt **15a** is a key substrate in the synthesis of d-α-dehydrobiotin **74**,[7] a natural antimetabolite of the cofactor biotin with antibiotic properties against a number of microorganisms.[28] The sequence of reactions used to achieve this goal is depicted in Scheme 5.12. Thus, the salt **15a**, upon treatment with sodium acetate, gave the acyclic acetate **75**, which was hydrolyzed with alkali to the alcohol **76**. Oxidation of this alcohol to the aldehyde **77** without concomitant oxidation of the sulfenyl sulfur atom was carried out with the Pfizner-Moffat reagent (DCC-DMSO). The additional two carbon atoms were attached by treatment of the aldehyde **77** with sodium salt of triethyl phosphonoacetate to give **78**. The latter, on heating with hydrobromic acid, gave primarily the cyclic sulfonium acid **79** and then the debenzylated acid **80**. This acid was at first esterified to give the unsaturated ester derivative **81**, which was hydrolyzed to d-α-dehydrobiotin **74**.[7]

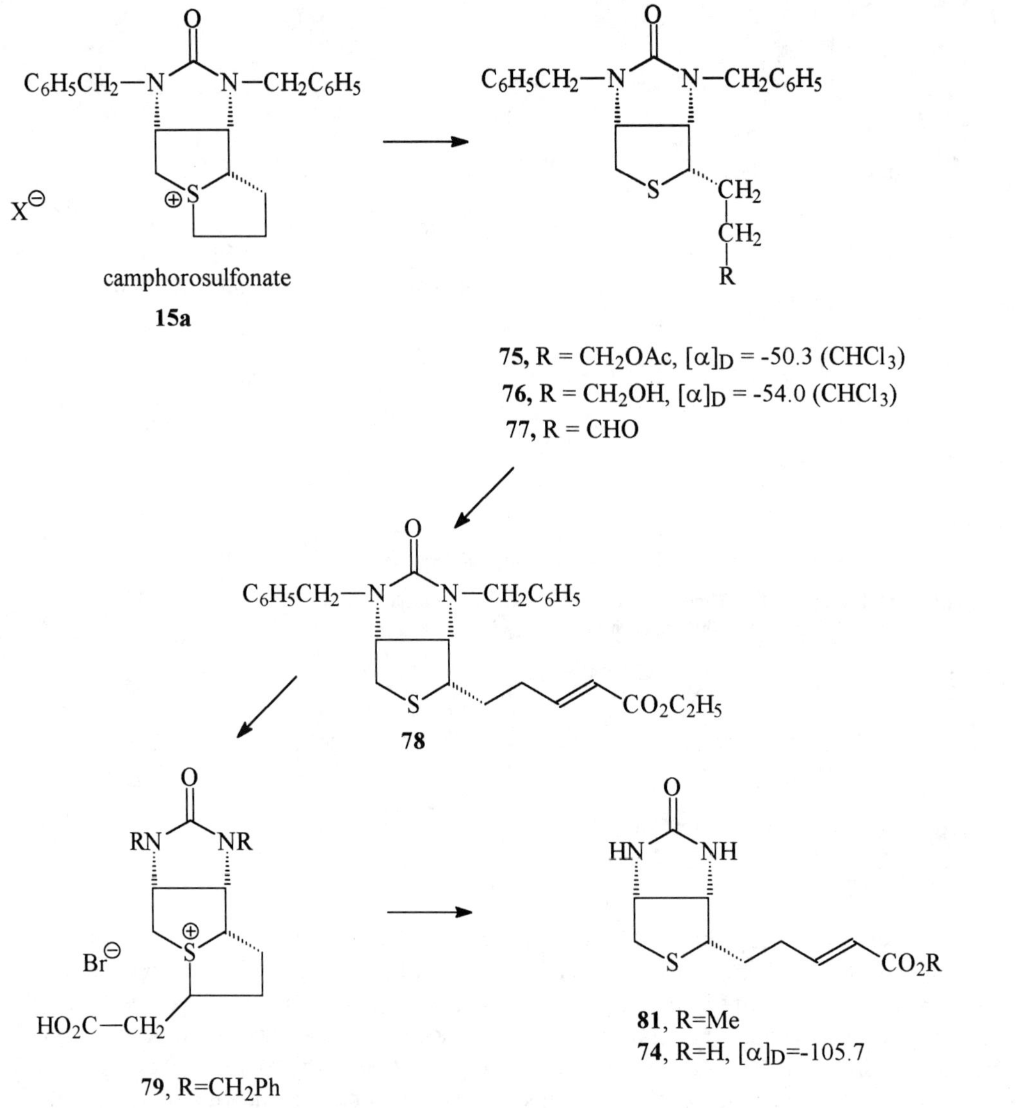

camphorosulfonate

15a

75, R = CH_2OAc, $[\alpha]_D$ = -50.3 ($CHCl_3$)
76, R = CH_2OH, $[\alpha]_D$ = -54.0 ($CHCl_3$)
77, R = CHO

78

79, R=CH_2Ph
80, R=H

81, R=Me
74, R=H, $[\alpha]_D$=-105.7

Scheme 5.12

A general in scope asymmetric synthesis of cyclopropane or epoxide systems developed by Johnson[29] is based on the use of chiral oxosulfur ylides as the agents that induce optical activity. This approach involves asymmetric addition of a chiral ylide to the C=C or C=O bond and subsequent cyclization of the addition product to form a chiral cyclopropane or epoxide system together with chiral sulfinamide (Equation 5.13).

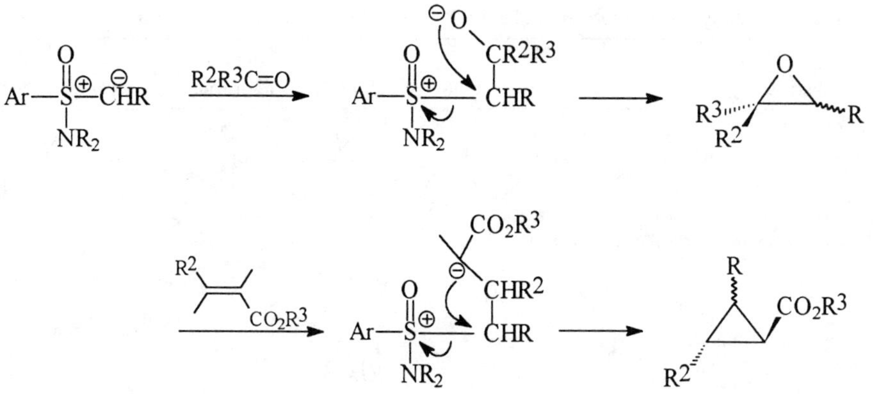

(5.13)

A wide range of chiral oxosulfonium ylides **47–53** used for this purpose have already been presented in this chapter. Chiral cyclopropanes and epoxides obtained by the above asymmetric reaction are listed in Table 5.2.[29]

Generally, this procedure produces chiral cyclopropane and epoxide systems with modest or low enantiomeric purity. Interestingly, the reaction of *trans*-benzalacetophenone with the ylide **47** afforded (1S,2S)-phenylcyclopropane with 35% optical purity, whereas its enantiomer having the (1R,2R)-configuration was formed when *trans*-benzalacetophenone was reacted with the carbanion **49.** It was also found that the enantiomeric excess of the cyclopropanes product increases with increase of steric requirements of the N,N-dialkylamino group.

When benzylideneaniline was allowed to react with the ylide **48**, chiral N-phenyl-2-phenylaziridine **82** of unknown absolute configuration and enantiomeric purity was isolated (Equation 5.14).[30]

Ph-CH=N−Ph + **48** ⟶ Ph-CH—N−Ph

82, $[\alpha]_D = -12.9$

(5.14)

Chiral spiroketone **83** of unknown configuration and enantiomeric purity was also formed when *trans*-benzalacetophenone was treated with the ylide **53**[29a] (Equation 5.15).

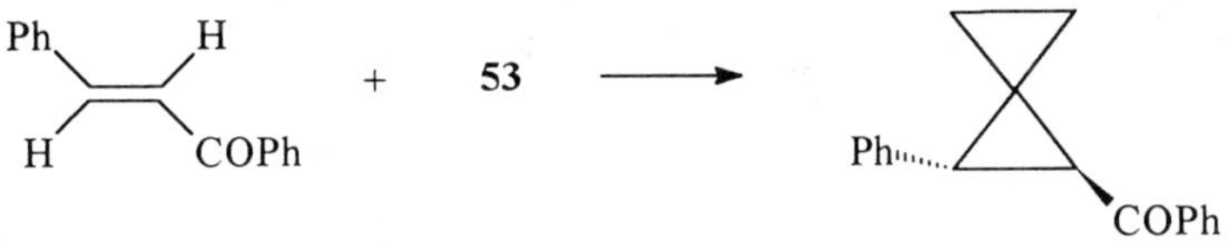

83

(5.15)

The reaction of the sulfonium salt **84** with methyl cyanoacetate in the presence of sodium methoxide afforded exclusively the E-isomer of methyl ester of (1S,2R)-1-cyano-2-phenylcyclopropane-carboxylic acid **85** of 25.5% ee (Equation 5.16).[31]

Table 5.2 Asymmetric Synthesis of Chiral Cyclopropane and Epoxide Derivatives

Substrate	Ylide	Product: Absolute configuration	Product: ee [%]
Methyl *trans*-cinnamate	**47**	Ph, COOMe cyclopropane; 1S,2S	30.4
trans-1,4-Diphenyl-2-butene-1,4-dione	**47**	COPh, COPh cyclopropane; 1S,2S	—
Dimethyl fumarate	**47**	COOMe, COOMe cyclopropane; 1S,2S	15.2
Dimethyl maleate	**47**	COOMe, COOMe cyclopropane; 1S,2S	17.8
Methyl *trans*-crotonate	**48**	Me, COOH cyclopropane; 1R,2R	11.9
trans-Benzalacetophenone	**49**	Ph, COPh cyclopropane; 1R,2R	49
trans-Benzalacetophenone	**47**	Ph, COPh cyclopropane; 1S,2S	35
trans-Benzalacetophenone	**48**	Ph, COPh cyclopropane; 1S,2S	7
Benzaldehyde	**47**	H, Ph epoxide; R	20
p-Chlorobenzaldehyde	**47**	H, p-Cl-C_6H_4 epoxide; R	—
Acetophenone	**47**	H_3C, Ph epoxide; R	—
n-Heptanal	**48**	$CH_3(CH_2)_5$, H epoxide; S	—

$$PhCH{=}CH\text{-}\overset{\oplus}{S}(O)(NMe_2)Ph\ BF_4^{\ominus}\ (\mathbf{84}) \xrightarrow[\text{MeONa}]{N{\equiv}CCH_2CO_2Me} Ph\text{-}CH(CH(CN)CO_2Me)\text{-}\overset{\ominus}{C}H\text{-}\overset{\oplus}{S}(O)(NMe_2)\text{-}Ph \rightleftharpoons MeO_2C(NC)\overset{\ominus}{C}\text{-}CH(Ph)\text{-}CH_2\text{-}\overset{\oplus}{S}(O)(NMe_2)\text{-}Ph \longrightarrow \text{Ph, } CO_2Me\text{, CN cyclopropane}$$

(1S,2R)-**85**, $[\alpha]_D = +64.0$ (5.16)

The carbanions prepared from diastereomerically pure (–)-(S)-neomenthyl-N-tosyl sulfoximine **86a** and **86b** are also efficient asymmetric methylene transfer reagents and give, with prochiral carbonyl compounds, chiral oxiranes with enantiomeric excesses from 50 to 86% (Scheme 5.13).[32]

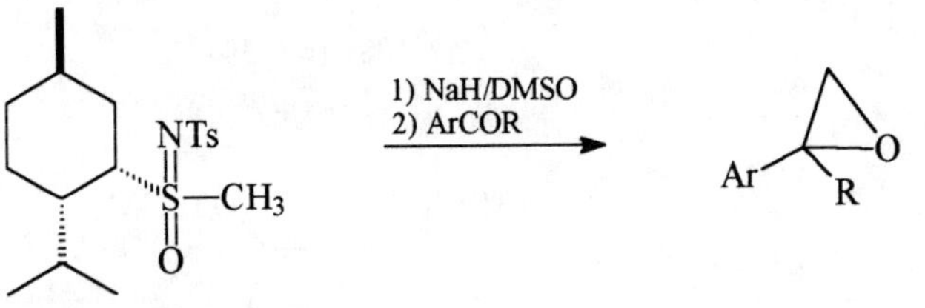

NTs
S—CH_3
O

1) NaH/DMSO
2) ArCOR

Ar O R

86a, $[\alpha]_D$ -60.9
86b, $[\alpha]_D$ +101.9

Ar	R	Yield [%]	ee [%]
Ph	H	42	66.2
p-ClC_6H_4	H	55	55.8
Ph	Me	65	82.3
p-ClC_6H_4	i-Pr	80	86.0

Scheme 5.13

Chiral sulfonium ylides generated *in situ* from cyclic or bicyclic chiral sulfonium salts were recently found to transfer the benzylidene group to some aldehydes with ee approaching 100%. This methodology for the synthesis of optically active epoxides is schematically presented in Scheme 5.14.

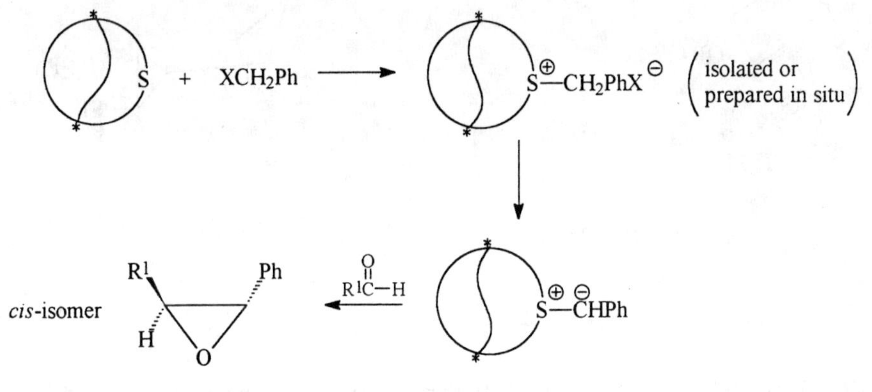

S + XCH_2Ph → $S^{\oplus}$—$CH_2PhX^{\ominus}$ (isolated or prepared in situ)

cis-isomer R1 Ph H O ← $R^1C(O)$—H ← $S^{\oplus}$—$\overset{\ominus}{C}$HPh

Scheme 5.14

The following cyclic sulfonium salts were used for the preparation of the starting ylides:

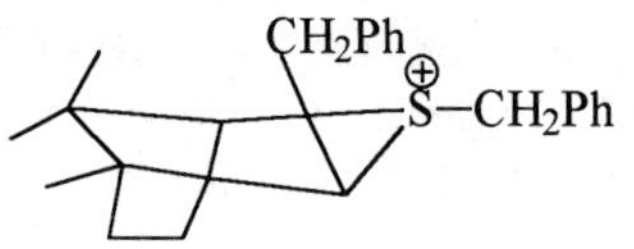

87 (ref. 33)

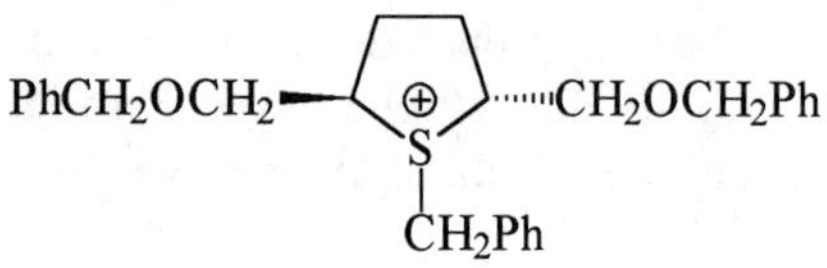

88a (ref. 34)

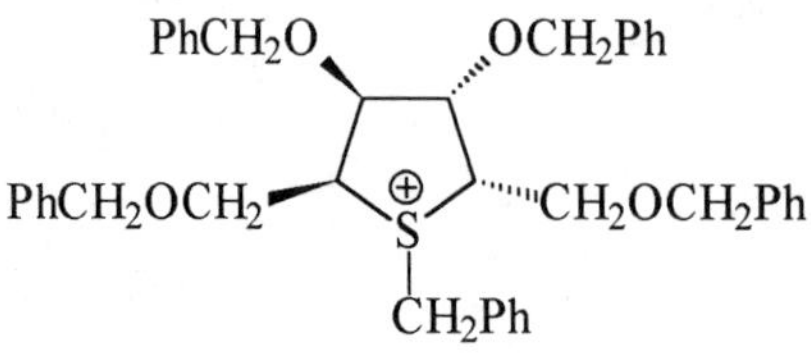

89 (ref.34)

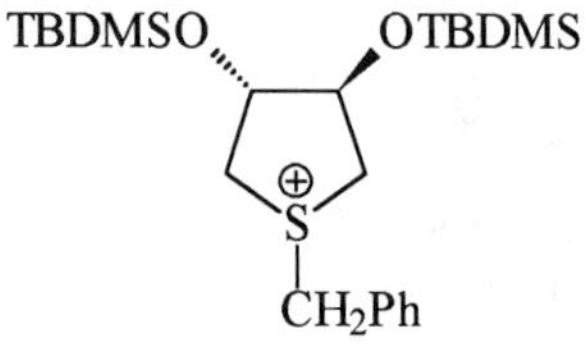

88b (ref. 34)

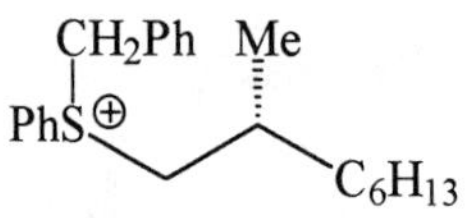

90 (ref. 10)

20 (ref. 10)

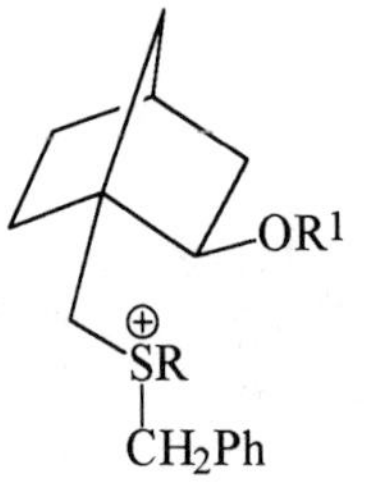

91, R = Me, R^1 = H (ref. 35)

92, R = Et, R^1 = H (ref. 35)

93, R = Me, R^1 = Me (ref. 35)

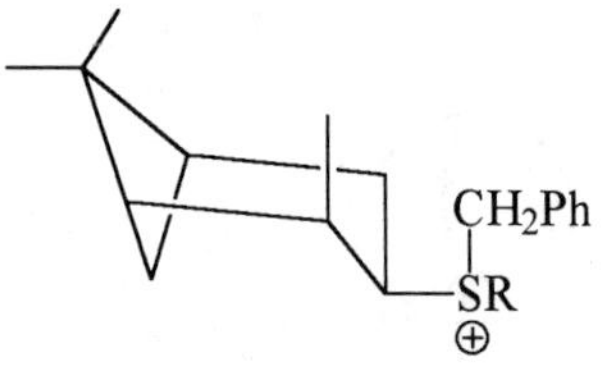

94, R = Me, (ref. 36)

95, R = i-Pr, (ref. 36)

96, R = CH_2Ph, (ref. 36)

Generally, chiral ylides were generated from the sulfonium salts under phase transfer catalytic (PTC) conditions (50% $NaOH/CH_2Cl_2$) or by the addition of sodium salt of DMSO. The results for the reaction of ylides generated in such a way with benzaldehyde and its substituted analogs are shown in Table 5.3.[33-36]

A high extent of asymmetric induction observed in the reaction of the ylide generated from the sulfonium salt **87** was rationalized[33] by assuming that it should adopt preferentially the conformation **97** rather than **98**, since in the latter the aryl ring has severe interactions with the endo hydrogens of the two-carbon bridge. Therefore, electrophilic attack by the carbonyl group on the ylide **97** will occur preferentially from the back of the ylide. This results in chirality of that center as (S). The results shown in the Table 5.3 are in a full agreement with this prediction.[33]

Table 5.3 Preparation of Optically Active Epoxides According to Scheme 5.12

Sulfonium salt	RCHO	Conditions	Yield [%]	ee [%]	Config.	Ref.
87	Ph	PTC	38	>96	S,S	33
87	p-Tol	PTC	32	>96	S,S	33
87	C_6H_{11}	PTC	9	84	(–)	33
88a	Ph	PTC	53	60	S,S	34
89	Ph	PTC	27	64	S,S	34
89	p-$NO_2C_6H_4$	PTC	41	83	S,S	34
88b	Ph	$NaCH_2S(O)Me$	30	15	S,S	34
88b	p-Tol	$NaCH_2S(O)Me$	49	11	S,S	34
20	Ph	PTC	80	72	R,R	10
20	p-Tol	PTC	75	32	R,R	10
20	p-ClC_6H_4	PTC	82	62–100	R,R	10
91*	Ph	PTC	100	47	R,R	35
91*	p-ClC_6H_4	PTC	15	34	R,R	35
92*	p-ClC_6H_4	PTC	90	31	R,R	35
93*	p-ClC_6H_4	PTC	30	28	S,S	35
94	Ph	PTC	67	13	R,R	36
95	Ph	PTC	5	43	R,R	36

* Sulfonium salt generated *in situ* in the reaction between the corresponding sulfide and benzyl bromide.

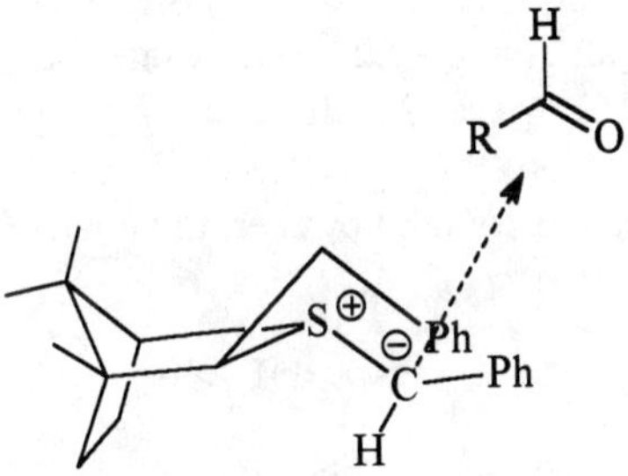

97

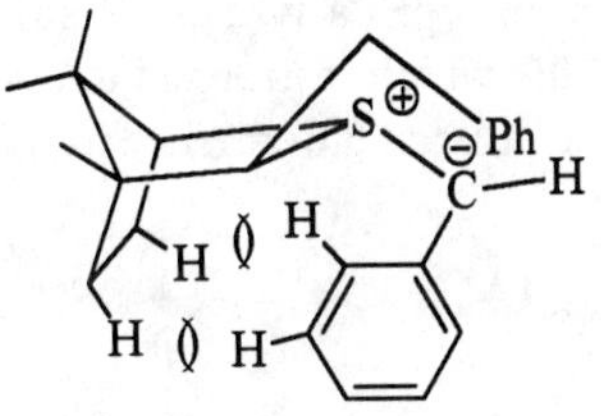

98

A very high asymmetric induction was also observed in the [2,3] sigmatropic rearrangement of the ylide **99** derived from optically active 1-adamantylallylethylsulfonium tetrafluoroborate **10**. It was found that the optically active unsaturated sulfide **100** formed in this process has at least 99% enantiomeric purity (Scheme 5.16).[5] Most probably, the concerted character of the rearrangement is responsible for such an efficient transfer of chirality from the chiral sulfur to the newly created chiral carbon center in the sulfide **100**.[5]

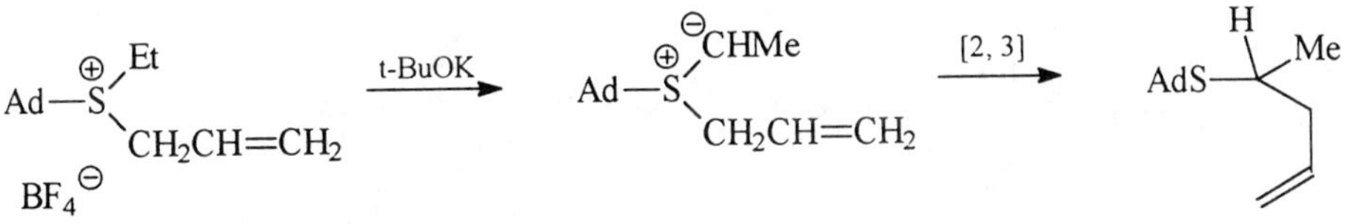

10 —t-BuOK→ **99** —[2, 3]→ **100**

Ad ≡

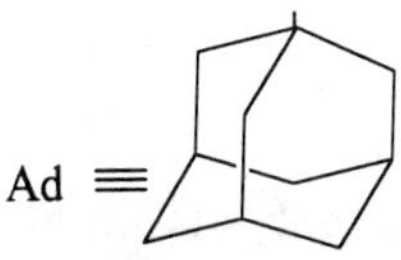

Scheme 5.16

Similarly, very efficient asymmetric induction was observed during [2,3] sigmatropic rearrangement of ylides derived from the diastereomerically pure α-allylated chiral thioxanones **101a–c** (Scheme 5.17).[37] A simple experimental procedure was applied.[37] Thus, dropwise addition of DBU to the allylation reaction mixture completed a one-pot conversion of thioxanone **104** to α-allylated **102** via the intermediacy of a transient ylide.

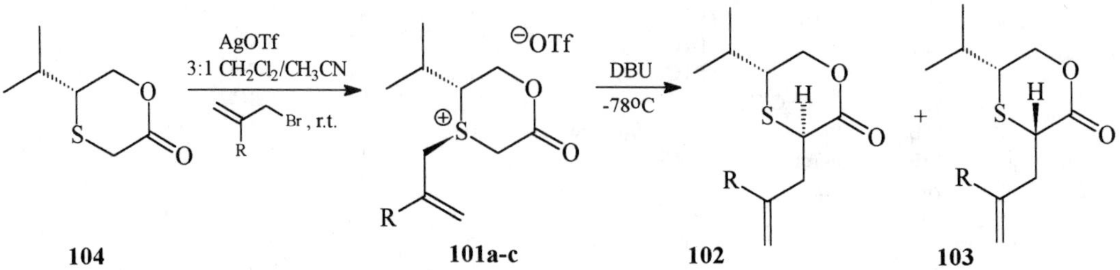

AgOTf, 3:1 CH_2Cl_2/CH_3CN, allyl bromide (R), r.t.; DBU, -78°C

104 → **101a-c** ($^{\ominus}OTf$) → **102** + **103**

	% (102 + 103)	102/103 ratio
a, R = H	82	98:2
b, R = Me	67	98:2
c, R = Et	72	98:2

Scheme 5.17

This is a kinetic product ratio since treating this mixture with DBU resulted in equilibration, by enolization, to a 62:38 diastereomeric mixture. It was also noted that the **102**:**103** ratio is temperature independent over the range from room temperature to –78°C (all giving 98:2 diastereoselectivity).

Deprotonation of optically active sulfonium salt **16b** gave 2-chloro-10-(2,5-xylyl)-10-thiaanthracene **105**, which rearranged to optically active 2-chloro-9-(2,5-xylyl)-10-thioxanthane **106** (Equation 5.17).[9] The enantiomeric excess of this product was found, however, to be only 7%.

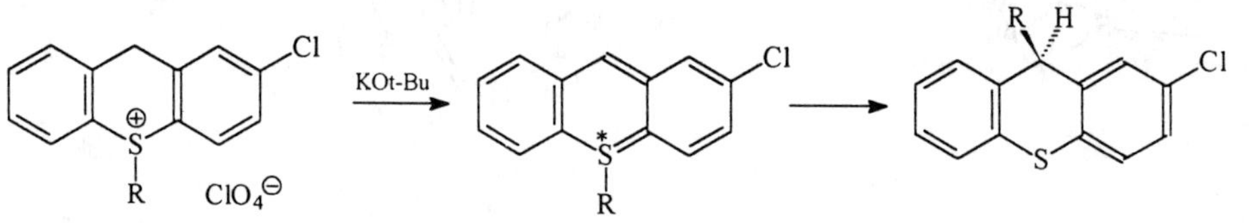

KOt-Bu

16b, $[\alpha]_{400} = 0.71$ → **105** → **106**, $[\alpha]_{400} = -1.64$ (5.17)

It was suggested[9] that this low enantiomeric excess in the produced thioxanthene **106**, which is formed via an intramolecular [1,4] rearrangement, is due to configurational instability of the intermediate thiaanthracene **105**, which racemizes about 10 times as fast as it rearranges at –15°C.

The Sommelet rearrangement of the chiral benzyl sulfonium salts **8a,b** gave α-methylbenzyl sulfides (+)-**107** and (+)-**108** with about 20–25% asymmetric induction (Equation 5.18).[38]

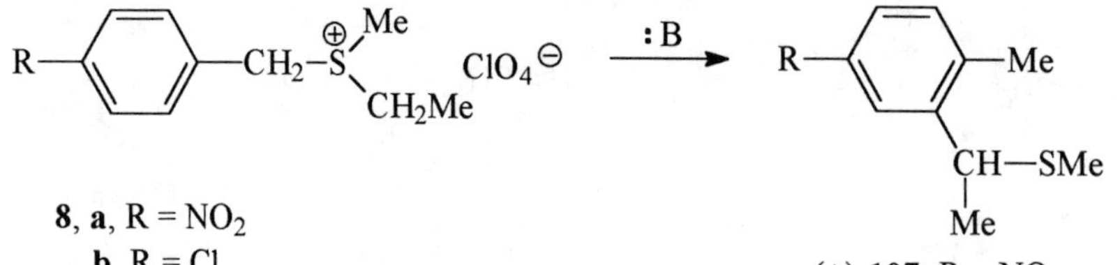

$R-C_6H_4-CH_2-S^{\oplus}(Me)(CH_2Me)\ ClO_4^{\ominus}$ —(:B)→ R–C₆H₃(Me)–CH(Me)–SMe

8, **a**, R = NO_2; **b**, R = Cl

(+)-**107**, R = NO_2

(+)-**108**, R = Cl (5.18)

A stereoselective formation of the lactone of 3,6-anhydro-4,5-O-isopropylidene-2-thiophenoxy-D-allo-heptanic acid **109** was found[39] to occur via 8-membered cyclic sulfonium ylide **110**, then subsequent formation of the oxonium intermediate **111** and stereoselective β-C-glycosidation (Scheme 5.18).

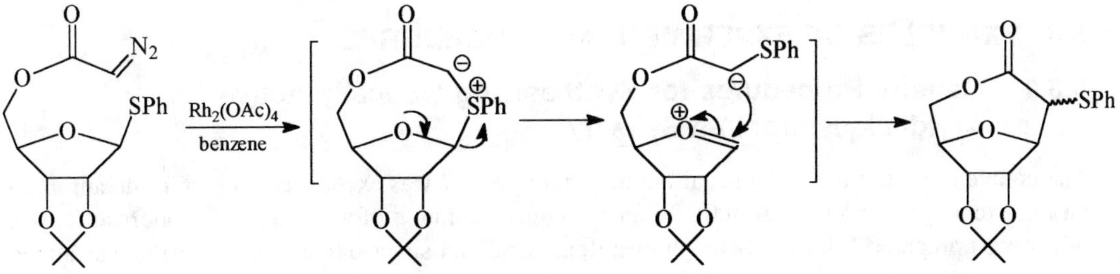

112 → $Rh_2(OAc)_4$, benzene → [110 → 111] → 109a,b

Scheme 5.18

The desired lactone **109** was formed in a refluxing benzene solution of **112** containing 1 mol % of rhodium (II) acetate in 48% yield as a 3:2 mixture of separable epimers. Desulfurization of each isomer of **109** by Raney nickel in acetone converted them to the well-known compound **113** (Equation 5.19).

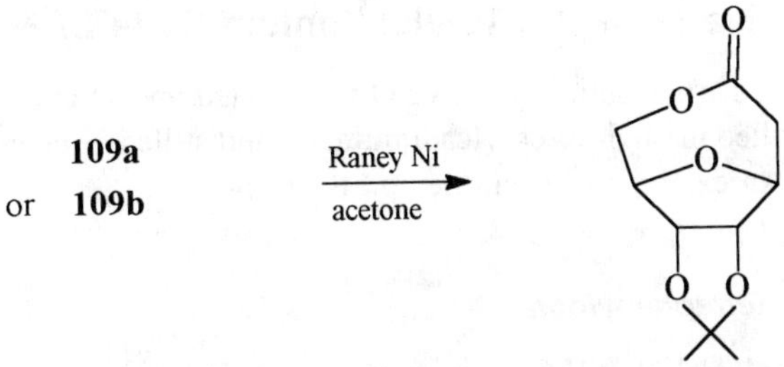

109a or **109b** → Raney Ni, acetone →

113, $[\alpha]_D = +83.4$ (CHCl3) (5.19)

It was also demonstrated[40] that the spiroannulation reaction of the functionalized cyclopentene **114**, under carbenoid generation conditions, proceeded in a stereospecific fashion via a cyclic allylsulfonium ylide **115** to give the spiroketone **116**, which was found to be a key intermediate in the total synthesis of (+) acorenone B-**117** starting from (–) perillaldehyde **118** (Scheme 5.19).

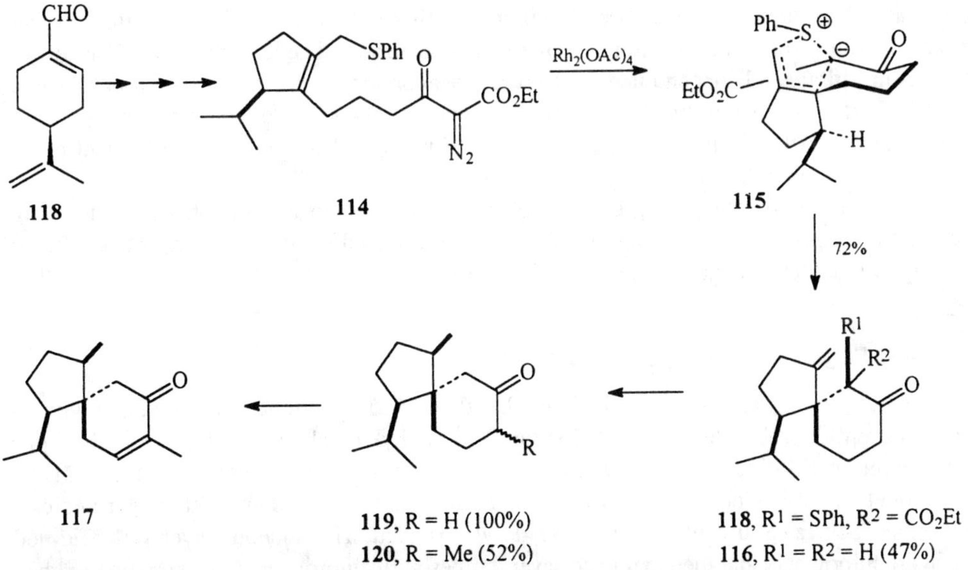

118 → → → **114** → $Rh_2(OAc)_4$ → **115** → 72% →

118, R^1 = SPh, R^2 = CO_2Et

116, $R^1 = R^2$ = H (47%)

119, R = H (100%)

120, R = Me (52%)

117

Scheme 5.19

5.3 EXAMPLES OF EXPERIMENTAL PROCEDURES

5.3.1 General Procedures for Synthesizing Optically Active Aryldialkylsulfonium Salts 17[9a]

The counter anion of the racemic sulfonium perchlorate **17** was exchanged with chloride anion on an ion exchange column (IRA-400). To an acetonitrile solution of silver d-10-camphorosulfonate, which was prepared from d-10-camphorosulfonic acid and silver oxide in acetonitrile, was added an acetonitrile solution of aryldialkylsulfonium chloride at room temperature. Silver chloride was filtered off and the solvent was evaporated to give racemic aryldialkylsulfonium d-10-camphorosulfonate. After fractional recrystallization of diastereoisomers of sulfonium d-10-camphorosulfonate from acetone-ether several times, an optically pure aryldialkylsulfonium d-10-camphorosulfonate was obtained. The sulfonium salt was added to an aqueous solution of sodium perchlorate and the separated oil was extracted with dichloromethane. Evaporation of the solution gave the optically active sulfonium perchlorate, and the salt was recrystallized from acetone-ether.

5.3.2 Synthesis of Optically Active Dialkyl-p-Tolylsulfonium Salts 27[13c]

Dialkyl-p-tolylsulfonium salts were prepared from optically active O-methylated and O-ethylated sulfoxides in three ways: by the use of distilled methyl- or ethylcadmium, by undistilled organocadmium reagents, or by Grignard reagents. An example of each method is given.

1 (+)-(S)-Ethylmethyl-p-Tolylsulfonium Tetraphenylborate

(R)-Ethyl-p-tolyl sulfoxide (1.0 g, 5.9 mmol, $[\alpha]_D^{25}$ + 186.6, acetone) was methylated using trimethyloxonium tetrafluoroborate (0.96 g, 6.5 mmol) in nitromethane. The solution was concentrated and (R)-methoxyethyl-p-tolylsulfonium tetrafluoroborate was precipitated by addition of an excess of ether. The salt was purified by dissolution in methylene chloride, followed by precipitation with ether. After several repetitions, 1.27 g (80%) of the salt were obtained as a thick yellow oil.

Distilled methylcadmium in ether (3 mL, 6.18 mmol, 30% excess, 2.08 M) was added with rapid stirring to a methylene chloride solution of the oil. After 20 min at room temperature, the excess cadmium reagent was hydrolyzed with 5% sulfuric acid, and the entire mixture was extracted with water. The aqueous layers were saturated with ca. 20 g of sodium tetrafluoroborate and extracted with five 25-mL portions of methylene chloride. The organic layer was dried over magnesium sulfate and concentrated on the rotary evaporator to give the tetrafluoroborate as a thick yellow oil. It was purified by dissolution in methylene chloride, followed by precipitation with ether as above. After drying *in vacuo*, 0.88 g (73%) of (+)-(S)-ethylmethyl-p-tolylsulfonium tetrafluoroborate was obtained.

The tetrafluoroborate (0.2 g, 0.79 mmol) was converted to the tetraphenylborate by mixing acetone solutions of the sulfonium salt and sodium tetraphenylborate and adding ether. After several reprecipitations there remained 0.33 g (90% yield).

2 (+)-(S)-Ethylmethyl-p-Tolylsulfonium Salts

(R)-Ethyl-p-tolyl sulfoxide (1.0 g, 5.94 mmol, $[\alpha]_D^{25}$ +185.7· acetone) was ethylated with triethyloxonium tetrafluoroborate (1.19 g, 6.26 mmol) in methylene chloride. Methylcadmium prepared from cadmium chloride (2.0 g, 10.9 mmol) and methylmagnesium bromide (7.4 mL, 21.8 mmol, 3.0 M in ether) was added at 0°C. After 15 min, the mixture was poured into water and extracted with ether. Several grams of sodium bromide were added. The aqueous layer was acidified with 5% hydrochloric acid and then extracted several times with chloroform. Concentration on the rotary evaporator without external warming gave 1.06 g (72.3%) of crude bromide.

The 2,4,6-trinitrobenzenesulfonate salt was obtained from the crude bromide (0.53 g) in acetone-ether using 2,4,6-trinitrobenzenesulfonic acid to yield 0.22 g (22%).

The tetraphenylborate salt was prepared in a similar way from sodium tetraphenylborate (0.74 g) and the crude bromide (0.53 g) in acetone, yielding 0.10 g (9.6%).

3 (+)-(S)-Ethylmethyl-p-Tolylsulfonium Tetraphenylborate

(R)-Ethyl-p-tolyl sulfoxide (1.0 g, 5.9 mmol, $[\alpha]_D^{25}$ + 185.7, acetone) was ethylated with triethyloxonium tetrafluoroborate (1.32 g, 6.6 mmol) in methylene chloride. Methylmagnesium bromide (2.0 mL, 6 mmol, 3.0 M) was added slowly at –78°C. After hydrolysis with 5% sulfuric acid, the entire mixture was extracted with an equal volume of ether and the two layers were separated. The aqueous layer was saturated with sodium bromide (ca. 35 g) and extracted with chloroform. Concentration gave 0.52 g (40%) of the crude sulfonium bromide as a yellow oil. The bromide was converted to the tetraphenylborate as above.

5.3.3 Preparation of the Ethoxysulfonium Salts 28a–d[14] (General)

Triethyloxonium fluoroborate (0.190 g, 1 mmol) was added to a solution of the sulfoxide (1 mmol) in 2–5 mL of methylene chloride and stirred for 30 min at room temperature. Addition of anhydrous ethyl ether at 0°C effected precipitation of the white, crystalline solid. The product was purified by recrystallization from methylene chloride-ether mixtures.

The (R)-benzyl-p-tolylethoxysulfonium fluoroborate **28a** had mp 115–117°C, $[\alpha]_D^{22}$ +203 (chloroform).

The (S)-benzyl-p-tolylethoxysulfonium fluoroborate **28a** had mp 115–117°C, but $[\alpha]_D^{22}$ –202.

5.3.4 Synthesis of (Diethylamino)Isobutylmethylsulfonium Tetrafluoroborate 40a[21]

To a solution of 1.55 g (0.0075 mol) of N,N-diethylaminosulfinyl tetrafluoroborate in 3.75 mL of acetonitrile (2 M solution) at –25°C was added 0.8 g (0.0075 mol) of sulfoxide (+)-(S)-**39a**; $[\alpha]_D$ +64.14 (46.5% ee). The reaction mixture was stirred for 1 h and warmed slowly to room temperature and then stirred for an additional 1 h at room temperature. Evaporation of the solvent afforded the crude product which was dissolved in methylene chloride (25 mL). After filtration the solvent was evaporated to give the salt **40a** as a slightly yellow liquid. It was washed with ether and dried *in vacuo* to afford 1.1 g (94%) of **40a**: $[\alpha]_D$ + 32.5.

According to the procedure described above, other aminosulfonium tetrafluoroborates **40** were prepared.

5.3.5 Diels-Alder Reaction between (S)-64 and Cyclopentadiene[27]

The following standard procedure was followed at –78°C: 3.11 g of (S)-**64** (11 mmol) dissolved in 15 mL dichloromethane under argon was cooled to –78°C. After standing at –78°C for 36 h the reaction was quenched by addition of NaOH 0.2 N (11 mmol) and stirring 30 min. A dichloromethane extraction, washing by saturated NaCl solution, drying on $MgSO_4$ and concentration *in vacuo* yielded 2.66 g of oily **65**. Isolated yield: 62%; *endo*/*exo* >100:1; de >99% (measured by ^1H NMR). Purification by flash chromatography on silica gel (ether/ArOEt = 4:1) gave (1R,2R,4R,R_S)-(+)-**65** 1.6 g as a yellow oil (62% yield); $[\alpha]_D$ +207.6 (c = 1.18, acetone).

REFERENCES

1. Pope, W. J., Peachy, S. J., *J. Chem. Soc.,* 1072, 1900.
2. Smiles, S. J., *J. Chem. Soc.,* 1174, 1900.
3. Darwish, D., Tomilson, R. L., *J. Am. Chem. Soc.,* **90**, 5938, 1968.
4. Darwish, D., Hui, S. H., Tomilson, R. L., *J. Am. Chem. Soc.,* **90**, 5631, 1968.
5. Trost, B. M., Hammen, R. F., *J. Am. Chem. Soc.,* **95**, 962, 1973.
6. Garbesi, A., Corsi, N., Fava, A., *Helv. Chim. Acta,* **53**, 1499, 1970.
7. Field, G. F., Zally, W. J., Sternbach, *J. Am. Chem. Soc.,* **92**, 3520, 1970.
8. Maryanoff, C. A., Hayes, K. S., Mislow, K., *J. Am. Chem. Soc.,* **99**, 4412, 1977.
9. a) Umemura, K., Matsuyama, H., Watanabe, N., Kobayashi, M., Kamigata, N., *J. Org. Chem.,* **54**, 2374, 1989;
   b) Kobayashi, M., Umemura, K., Watanabe, N., Matsuyama, H., *Chem. Lett.,* 1067, 1985;
   c) Kobayashi, M., Umemura, K., Matsuyama, H., *Chem. Lett.,* 327, 1987.
10. Solladie-Cavallo, A., Adib, A., *Tetrahedron,* **48**, 2453, 1992.
11. Kelstrup, E., Kjaer, A., *J. Chem. Soc., Chem. Commun.,* 629, 1975.
12. Tsumori, K., Minato, H., Kobayashi, M., *Bull. Chem. Soc., Jpn.,* **46**, 3503, 1973.
13. a) Andersen, K. K., Caret, R. L., Ladd, D. L., *J. Org. Chem.,* **41**, 3096, 1976;
    b) Andersen, K. K., Caret, R. L., Ladd, D. L., *Int. J. Sulfur Chem.,* A, **2**, 196, 1972;
    c) Andersen, K. K., *J. Chem. Soc., Chem. Commun.,* 1051, 1971.
14. Johnson, C. R., McCants, D., Jr., *J. Am. Chem. Soc.,* **87**, 5404, 1965.
15. Annunziata, R., Cinquini, M., Colonna, S., *J. Chem. Soc., Perkin Trans. 1,* 1231, 1973.
16. Annunziata, R., Cinquini, M., Colonna, S., *J. Chem. Soc., Perkin Trans. 1,* 404, 1975.
17. Mikołajczyk, M., Bujnicki, B., Drabowicz, J., unpublished results.
18. Mikołajczyk, M., Bujnicki, B., Drabowicz, J., *Bull.Acad.Pol.Sci. Ser. Sci.Chim.,* **25**, 267, 1977.
19. Okuma, K., Minato, H., Kobayashi, M., *Bull. Chem. Soc. Jpn.,* **53**, 435, 1980.
20. Bohman, O., Allenmark, S., *Chem. Scrip.,* **4**, 202, 1973.
21. Drabowicz, J., Bujnicki, B., Mikołajczyk, M., *J. Org. Chem.,* **46**, 2788, 1981.
22. Kamiyama, K., Minato, H., Kobayashi, M., *Bull. Chem. Soc. Jpn.,* **46**, 3895, 1973.
23. Moriyama, M., Oae, S., Numata, T., Furukawa, N., *Chem. Ind. (London),* 163, 1976.
24. Maryanoff, B. E., Stackhouse, J., Senkler, G. H., Jr., Mislow, K., *J. Am. Chem. Soc.,* **97**, 2718, 1975.
25. Minato, H., Yamaguchi, K., Okuma, K., Kobayashi, M., *Bull. Soc. Chem. Jpn.,* **49**, 2590, 1976.
26. Lucchini, V., Modena, G., Pasquato, L., *J. Chem. Soc., Chem. Commun.,* 1565, 1994.
27. Ronan, B., Kagan, H. B., *Tetrahedron: Asymmetry,* **2**, 75, 1991.
28. a) Hanka, L. J., Bergy, M. B., Kelly, R. B., *Science,* **154**, 1667, 1967;
    b) Hanka, L. J., Reineke, L. M., Martin, D. G., *J. Bacteriol.,* **100**, 42, 1969;
    c) Rubin, S. H., Scheiner, J., *Arch. Biochem.,* **23**, 400, 1949.
29. a) Johnson, C. R., Schroeck, C. W., *J. Am. Chem. Soc.,* **90**, 6852, 1968;
    b) Johnson, C. R., Schroeck, C. W., *J. Am. Chem. Soc.,* **93**, 5303, 1971;
    c) Johnson, C. R., Schroeck, C. W., *J. Am. Chem. Soc.,* **95**, 7418, 1973;
    d) Johnson, C. R., Schroeck, C. W., Shanklin, J. R., *J. Am. Chem. Soc.,* **95**, 7424, 1973;
    e) Johnson, C. R., Janiga, R. J., *J. Am. Chem. Soc.,* **95**, 7692, 1973.
30. Johnson, C. R., Kirchoff, R. A., Reischer, R. J., Katekar, G. F., *J. Am. Chem. Soc.,* **95**, 4287, 1973.
31. Johnson, C. R., Lockhard, J. P., *Tetrahedron Lett.,* 4589, 1971.
32. Taj, S. S., Soman, R., *Tetrahedron: Asymmetry,* **5**, 1513, 1994.
33. Breau, L., Durst, T., *Tetrahedron: Asymmetry,* **2**, 367, 1991.
34. Breau, L., Ogilvie, W. W., Durst, T., *Tetrahedron Lett.,* **31**, 35, 1990.
35. Furukawa, N., Sugihara, Y., Fujihara, H., *J. Org. Chem.,* **54**, 4222, 1989.
36. Aggarwal, V. K., Kalomiri, M., Thomas, A. P., *Tetrahedron: Asymmetry,* **5**, 723, 1994.
37. Tahir, S. H., Olmstead, M. M., Kurth, M. J., *Tetrahedron Lett.,* **32**, 335, 1991.
38. Campell, S. J., Darwish, D., *Can. J. Chem.,* **54**, 193, 1976.
39. Kim, G., Kang, S., Kim, S. N., *Tetrahedron Lett.,* **34**, 7627, 1993.
40. Kido, F., Abiko, T., Kato, M., *J. Chem. Soc., Perkin Trans. 1,* 229, 1992.

Index

A

Acetate, 255–256
α-Acetoxy-α-(dimethoxyphosphoryl)methyl p-tolyl sulfide, 124
2-Acyl dithiolane sulfoxide, 138
Acorenone, 263
Acylation
  carbanion, 73–77
  dithioacetal monoxides, 94
α-Acylaziridines, 206
N-Acyliminocyclization, 168
1-Adamantylallylethylsulfonium tetrafluoroborate, 261
1-Adamantylethylpropenylsulfonium tetrafluoroborate, 242
(+)-A Factor 25, 188
Alanine ethyl ester, 38
Alcohols
  epoxy sulfoxide reactions, 111
  α-hydroxy, 113
  onium derivative reactions, 255–256
  SPAC reaction, 133–135
  sulfinic ester preparation, 7–15
  sulfoximine synthesis, 206–210, 236–237
  synthesis from allyl sulfoxides, 80–82
Alcoholysis, α-phosphoryl sulfoxide, 118
Aldehyde fluoride, 32
Aldehydes
  bis-sulfoxide synthesis, 99–100
  carbanion reactions, 61–70
  α-halogeno sulfoxide reactions, 113
  hydrolysis and isolation, 96
  lithium reactions, 95
  onium derivative reactions, 255–256
  sulfoximine reactions, 210–211, 236–237
  vinyl sulfoxide synthesis, 154
Alkaloids, synthesis, 168–169
Alkanes, 252
Alkanesulfinates, 10, 11, 16–26, 27
Alkenes, 218, 231
1-Alkenylsulfoximines, 228–229
Alkoxyaminosulfonium salts, 247
1-Alkoxy-2-(methoxycarbonyl)-1-indenes, 251
Alkoxysulfonium salts, 245–246
β-Alkoxyvinyl sulfoxides, 174
3-Alkylalkanoic acids, 224
Alkyl(aryl)sulfenylmethyl sulfoxides, 91–96
Alkyl aryl sulfoxides, see Dialkyl and alkyl aryl sulfoxides
Alkyl(aryl)-p-tolyl sulfoxides, 246–247
Alkylation
  dithioacetal monoxides, 92–93
  α-halogeno sulfoxides, 109–111
  α-sulfinylcarboxylate reactions, 131
  α-sulfonyl sulfoxides, 102
  sulfoxides, 57–59
N-Alkylidenesulfinamides, 43
O-Alkyl O-isopropyl phosphorothioic acid, 122
Alkyllithium, 31, 225
2-Alkyl-2-(methoxycarbonyl)-1-indanones, 251
N-Alkyl sulfinamides, 43
N-(Alkylthio)oxazolidinones, 29–30
Alkyl p-tolyl sulfoxides
  halogenation, 106–107, 114–115
  O-menthyl p-toluenesulfinate in synthesis, 28
1-Alkynylmagnesium bromides, 21
Alkynylmethyl sulfoxides, 20–21
1-Alkynyl p-tolyl sulfoxide, 21
Allenes, synthesis, 26
Allenic sulfoxides, 20–21
Allyl alcohols
  allyl sulfoxides in synthesis, 80–82
  epoxy sulfoxide reactions, 111
  SPAC reaction, 133–135
Allylamines, synthesis, 36–37
Allylmagnesium bromide, 35–36, 39
Allyl sulfinates, 23–24
Allyl sulfoxides, 122
  carbanion reactions, 82–89
  experimental procedures, 89–90
  racemization, 78
  sigmatropic rearrangement, 78–82
  sulfinates in synthesis, 20
Allyl sulfoximines, 229–232, 238
cis-Allyl p-toluenesulfinates, 24
Allyl p-tolyl sulfoxides
  carbanion reactions, 84–85
  racemization, 78
  synthesis, 78, 89
Allymagnesium bromide, 174
α,β-unsaturated sulfones, 98–99, 102, 105
α,β-unsaturated sulfoxides
  conjugate additions
    nitrogen nucleophiles, 152–157
    silicon nucleophiles, 157–158
  cyclopropanation, 173–174
  Diels-Alder cycloaddition, 158–170
  electrophilic addition, 144–142
  experimental procedures, 175–178
  Michael additions
    carbon nucleophiles, 145–148
    oxygen nucleophiles, 148–152
  Pummerer rearrangement, 170–173

radical cyclizations, 174
sulfinyl diene cycloaddition, 174–175
Aluminum, 30, 218–219, 222
Amides, 168
α-hydroxy, 113
sulfilimine synthesis, 233
Amidosulfites, 3
Amidothiosulfites, 3
Amines
sulfinic ester preparation, 15
synthesis, 112
Amino acids, 36–39
Aminoalcohols, 14
α-Amino aldehydes, 110
Aminoalkoxysulfonium salt, 251
Aminoalkylation, carbanion, 70–73
α-Amino ketones, 110
2-Amino-2-methyl-4-butenoic acid, 39
α-Aminonitriles, 37, 38
β-Aminophosphonic acid esters, 40
Aminosulfonium salts, 248–249
Amphotericin B, 66
Andersen method, 16–25, 78
Aphidicolin, 188–189
Apicophilicity, 4
Arenesulfinates, 10, 11, 16–26, 27
N-Arylaziridines, 112–113
Aryl cycloalkenyl methyl sulfoxides, 81, 82
Aryldialkylsulfonium salts, 243–244, 264
N-Arylimines, 112–113
Aryl phenyl N-(unsubstituted) sulfilimines, 195–198, 234
N-(Arylthio)oxazolidinones, 29–30
α-Arylthio-β-oxosulfoxides, 96
Aziridine-2-carboxylates, 41–42
Aziridines, 112
2,2′-Azobisisobutyronitrile, 106

B

Benzalacetophenone, 257
Benzaldehyde, 63, 83, 121, 211
Benzene, 19, 28, 234
Benzene-sulfinyl chloride, 8
Benzonitrile, 31
3-Benzoyloxy-2-butanone, 138
Benzyl t-butyl sulfoxide, 49–50
Benzylethylmethylphenylammonium tetrafluoroborate, 251
N-Benzylidene-p-toluenesulfinamide, 40
Benzylmagnesium chloride, 19
Benzyl methyl sulfoxide, 48, 49–50, 61, 108
Benzyl methyl (S)-2-(p-toluenesulfinyl)maleate, 162
1-Benzyl 4-methyl 2-p-tolylsulfinylmaleate, 176
Benzyl sulfoxides, 52–56
Bipyramidal sulfur compounds, 4
Bis-sulfoxide, 98–100
9-Borabicyclo nonane, 38
Borane, 208–209
Boron trifluoride, 13
Boschnialactone, 166–167
Bromination, sulfoxides, 106–108, 114–115
Bromine, 106–107, 114
α-Bromobenzyl methyl sulfoxide, 108
Bromomethyl p-tolyl sulfoxide, 95
2-Bromo-1,4,5-trimethoxynapthalene, 23
t-Butanesulfinates, 10, 14
t-Butyl benzyl sulfoxide, 71
Butylcyclohexanone, 238
Butyllithium, 23, 49, 123
t-Butylmagnesium bromide, 131
t-Butylmagnesium chloride, 14
n-Butylmagnesium iodides, 19
1-τ-Butylsulfonyl-1-p-tolylsulfinylethene, 103
t-Butyl p-tolylsulfinylacetate, 141
t-Butyl α-(p-tolylsulfinyl)-β-hydroxymyristate, 142
Butyraldehyde, 94
γ-Butyrolactones, 69–70

C

Canadine, 156–157
Carbanions
alkyl and arylsulfenylmethyl sulfoxides, 91–96
allyl sulfoxides, 78–90
dialkyl and alkyl aryl sulfoxide generation
acylation, 73–74
alkylation, 57–59
aminoalkylation, 70–73
experimental procedures, 74–77
hydroxyalkylation, 61–70
Michael addition to carbonyl compounds, 60–61
halogeno sulfoxides, 106–115
oxosulfoxides and derivatives, 129–142
phosphoryl sulfoxides and sulfinyl phosphonium ylides, 116–127
properties and generation, 47–56
sulfinyl-, sulfonyl-, and sulfoximino sulfoxides, 98–105
Carbocyclic nucleosides, 168
Carbon nucleophiles, 145–148
α-Carboxy-ω-hydroxyalkenyl sulfoxides, 180–181
Carboxylic esters, 60–61
Carnegine, 71–72, 154–156, 175–176
Cesium fluoride, 32
Chalcone, 210
Chelation, 183–184, 190–191, 193
Chiral compounds, see specific compounds
Chirality-structure relationship, 1–5
Chlorination
alkyl p-tolyl sulfoxides, 115
α-phosphoryl sulfoxide, 126–127
sulfoxides, 106–108, 115
α-Chloroalkyl aryl sulfoxides, 109
1-Chloroalkyl p-tolyl sulfoxides, 112–113
α-Chloroalkyl p-tolyl sulfoxides, 106, 109
Chloroamines, 112
Chlorohydrines, 109–110, 113
2-Chloromethylcyclohexenyl sulfoxide, 174
α-Chloromethyl p-tolyl sulfoxide, 109
m-Chloroperbenzoic acid, 29–30
α-Chloro-α-phosphoryl sulfoxide, 126–127
N-Chlorosuccinimide, 115
α-Chloro sulfide, 125

Chlorosulfites, 11–12
α-Chloro sulfoxide, 109–110
Chlorosulfurane, 5
O-Cholesteryl methanesulfinates, 19
Chroman-4-ones, 189–190
Chroman ring, 150–151
Chromatography, 10–11, 35, 237
Chromene, 150
O-Cinnamyl p-toluenesulfinate, 25
δ-Coniceine, 168
Conjugate addition, see Michael addition
Copper, 226, 230–233
Copper-lithium reagents, 19
Crotyl sulfoxides, 78–79, 89
Cuparenone, 186–187
N-β-Cyanoethylsulfilimine, 197
Cyanohydrins, 210, 211
1-Cyano-2-phenylcyclopropane-carboxylic acid, 257–258
Cycloaddition, see Diels-Alder cycloaddition
2-Cycloalkenolides, 186
Cycloalkenone sulfoxides
  asymmetric radical addition, 190–191
  conjugate addition of nucleophiles
    applications, 186–190
    sulfinyl alkenolides, 185–186
    sulfinyl cycloalkenones, 182–185
  experimental procedures, 191–193
  synthesis, 180–182
Cyclobutanones, 69
Cyclohexanol, 14
Cyclopentadienes, 126, 158–170, 177, 254, 265
Cyclopentenes, 263
Cyclopentenones, 83–87, 89–90, 92–93, 170
1-Cyclopenten-1-ylmethyl-N-methyl-S-phenylsulfoximine, 238
Cyclopropanation, 173–174, 214, 221
Cyclopropanes, 173, 256–258
Cyclopropyl ketones, 214, 215

D

Dane's diene, 169–170
Debenzylated acid, 255–256
trans-Decalyl-like transition state, 83
α-Dehydrobiotin, 255–256
Deoxy-*ent*-prostaglandin, 92
Deuterium, 47–48
Diacetone-d-glucose, 10, 27
Dialkylaminomagnesium halides, 30
Dialkyl and alkyl aryl sulfoxides
  acylation, 73–74
  alkylation, 57–59
  aminoalkylation, 70–73
  experimental procedures, 74–77
  hydroxyalkylation, 61–70
  Michael addition to carbonyl compounds, 60–61
Dialkylcopperlithium, 225
N,N-Dialkyl p-toluenesulfinamides, 43
Dialkyl-p-tolylsulfonium salts, 245–246, 264–265
Dialkylzinc reagents, 11
Diaminosulfonium salt, 247–248
1,5-Dicarbonyl adducts, 185
α,α-Dichloro-γ-arylthio-γ-butyrolactones, 170
2,2-Dichloro-4-(3,4-dimethoxyphenyl)-3-methyl-4-p-tolylthiobutyrolactone, 177
Dichloroiodobenzene, 106–107
Dichlorolactone, 172–173
Dicyclohexylidene-D-glucose, 10
α,α-Dideuteriobenzyl α′-chlorobenzyl sulfoxide, 108
α,α-Dideuteriodibenzyl sulfoxide, 108
Diels-Alder cycloaddition, 100, 122
  acyclic α,β-unsaturated sulfoxide, 158–170, 174–175
  cyclopentadienes, 177
  onium derivatives, 252–255, 265
  vinyl sulfoximines, 231
Dienone, 188
Dienophile, 159
Diethoxyphosphorylvinyl p-tolyl sulfoxide, 125–126
trans-2-N,N-Diethylacetamide-1,3-dithiolane S-oxide, 94
Diethylaluminum cyanide, 37–38
(Diethylamino)isobutylmethylsulfonium tetrafluoroborate, 265
Diethyl malonate, 145 N,N-Diethyl p-toluenesulfinamide, 13–14, 27–28, 33–34
Diethylzinc, 210
Difluoro compounds, 139
α,β-Dihalosulfoxides, 144–145
3,4-Dihydro-6-methoxy-1-vinylnaphthalene, 169
Dihydroxycycloalkanones, 216–218
Dihydroxyketones, 216
Diisobutylaluminumhydride, 32, 136–138, 142
N,N-Di-isopropyl p-toluenesulfinamide, 14
1,4-Dimethoxybenzene, 22–23
3,4-Dimethoxyphenyl-3-methyl-γ-butyrolactone, 178
3,4-Dimethoxyphenyl-3-methyl-4-p-tolyl-thio-γ-butyrolactone, 178
Dimethoxyphosphorylmethyllithium, 118
Dimethoxyphosphorylmethyl p-tolyl sulfoxide, 117, 120–121, 123, 124, 127
α,γ-Dimethylallyl 2,4-dimethylbenzenesulfinate, 26
Dimethylamino-methyl p-tolueneoxosulfonium tetrafluoroborate, 31
Dimethylamino-methyl-p-tolyloxosulfonium fluoroborate, 235
2,3-Dimethylbutadiene, 253
Dimethyl ethanephosphonate, 125
Dimethyloxosulfonium methylide, 42
Dimethyl-S-phenylsulfoximine, 235, 236–238
Dimethylsulfonium methylide, 42
Dimethyltetrahydrofuran, 184
N,N-Dimethyl p-toluenesulfinamide, 31
1,3-Diols, 138
1,7-Dioxaspiro [5,5,]undecane, 148–150
β,δ-Dioxosulfoxides, 138
Diphenyl N-(substituted) sulfilimines, 233
Disparlure, 109–110
1,3-Dithiane, 49
trans-1,3-Dithiane S,S-dioxide, 99–100
Dithioacetal monoxides, 91–94, 138
DMSO, carbanion, 55
γ-n-Dodecanolactone, 131–132

E

Electronegativity, apicophilicity and, 4
Electrophiles
  acyclic α,β-unsaturated sulfoxide reactions, 144–145
  carbanion reactions, 49–56
  α-sulfinylcarboxylate reactions, 131–135
Enolates, 186–188, 221–222
Enones, 83–87, 212–216
Ephedrine, 203
Epoxides, 59
  onium derviative reactions, 256–261
  synthesis, 137
Epoxy silane, 111
Epoxy sulfoxides, 110–111
Epoxytrichothec-9-ene, 88
Erythronolides, 65–67
Ester enolate anions, 186
Esters
  β-aminophosphonic acid, 40
  carboxylic, 60–61
  α-hydroxy, 113
  onium derivative reactions, 255–256
  SPAC procedures, 132–135, 142
  structure-chirality relationship, 3
  sulfoximinyl, 222
  synthesis, 33–34
Estrone methyl ether, 187–188
Ethanethiol, 126–127
Ethoxyalkyl(aryl)-p-tolylsulfonium tetrafluoroborates, 246–247
β-Ethoxycarbonyl-vinyl sulfoxides, 160–161
α-Ethoxycarbonyl-vinyl tolyl sulfoxide, 160–161
Ethoxysulfonium salts, 265
1-Ethoxythoxybutyl-2-methylenecyclopentane, 238
Ethoxy p-tolyl vinylsulfonium tetrafluoroborate, 252
Ethyl ether, 28
Ethyl ethylthiomethyl sulfoxide, 91
N-Ethyl-(1)-hydroxycyclopentenyl methyl phenyl sulfoximine, 210
Ethylmagnesium iodide, 16
Ethylmethylpropylsulfonium 2,4,6-trinitrobenzene-sulfonate, 245
Ethylmethylsulfonium phenacylide, 241
Ethylmethyl-p-tolylsulfonium salts, 264–265
Ethylmethyl-p-tolylsulfonium tetraphenylborate, 264, 265
Ethyl p-tolyl sulfoxide, 107

F

3-Fluorinated 2-oxopropyl sulfoxides, 76–77
β-Fluoro-alkyl-β-aminosulfoxides, 141
Fluoroalkyl-oxosulfoxides, 139
Formylchroman, 150–151
2-Formyl pyrrole, 126
Furan, 253–254

G

Ginseng sesquiterpene panasinsene, 218–219
Grignard reagents, 25, 95
  N-alkylidenesulfinamide synthesis, 31
  cycloalkenone sulfoxide reactions, 183
  sulfoxide synthesis, 16, 18–20, 34–35

H

α-Halogenoalkyl p-tolyl sulfoxides, 106
N-Halogenosuccinimides, 106–107
α-Halogeno sulfoxides, 106–115
α-Halomethyl sulfoxides, 91
N-Halosuccinimides, 114–115
Heptane lactones, 166–167
α-Heteroatom-substituted sulfoxides, 21
Hexacoordinate sulfur compounds, 5
δ-n-Hexadecanolactone, 131–132
Hirsutene, 87–88
Horner-Wittig reaction, 119–123, 126, 181
Hydrides, 135–139
Hydrocarbons
  sulfinic esters in synthesis, 24–25
  sulfoximine reactions, 225, 229–230
Hydrogen, α-sulfinyl carbanions, 47–53
Hydrogen cyanide, 211
Hydrogenolysis, cycloalkenone sulfoxide, 188
Hydrolysis, sulfinylcarboxylates, 129–130
α-Hydroxyaldehydes, 138
Hydroxyalkenyl sulfoxides, 180–181
α-Hydroxyalkylacrylates, 132
Hydroxyalkylation
  carbanion, 61–70
  dithioacetal monoxides, 93–94, 95–96
  α-sulfinylcarboxylates, 142
exo-2-Hydroxy-10-bornyl, 163–165
3-Hydroxybutyrates, 138
β-Hydroxy carboxylic acids, 131
4-Hydroxy-2-cyclohexanones, 137
4-Hydroxy-2-cyclohexenone, 137
[2-Hydroxy-7,7-dimethylbicyclo-[2.2.1]heptan-1-y]methylsulfinylmaleate, 177
Hydroxylation, bis-sulfoxide anion, 98–99
2-Hydroxy-2-phenylethyl, 163
3-Hydroxy-3-phenyl-propyne, 26
β-Hydroxysulfinates, 18
Hydroxysulfoxides, 137, 139
  alkylation, 58–59
  hydroxyalkylation, 61, 63
β-Hydroxy sulfoximines, 207–222, 236–238
β-Hydroxythioacetamides, 139
α-Hydroxyvinyl p-tolyl sulfoxides, 67–68

I

N-Iodosuccinimide, 60
Imines, carbanion reactions, 76
β-Iminosulfoxides and analogs, 140–141
1-Iododecane, 109
Isocarbacyclins, 230
Isopropanol, 14
O-Isopropyl methanephosphonothioic acid, 122

J

Juvabiols, 64–65

K

Ketene, 100 β-Ketoester-2-(methoxycarbonyl)-1-indanone, 251
Ketones, 110, 257
carbanion reactions, 61–70
sulfoximine reactions, 207, 210–219, 236, 237–238
β-Ketosulfoxides, see β-Oxo sulfoxides

L

Laburnine, 168
Lactam, 168
Lactones, 69–70, 166–167, 172–173, 177–178, 262–263
Leucine, 37–38
Lewis acid, 190
Ligand number, in compound classification, 1, 3, 4
α-Lithiobenzyl sulfoxides, 48
α-Lithio-α-methylbenzyl phenyl sulfoxide, 53–55
α-Lithiosulfide, 48–49
α-Lithiosulfone, 48–49
Lithiosulfoxides, 48–49, 75, 151
α-Lithiosulfoximine, 48–49
α-Lithiovinyl sulfoxide, 150–151
Lithium
aldehyde reactions, 95
allyl sulfoxide reactions, 83
carbanion, 53–56
cycloalkenone sulfoxide reactions, 186–190
hydroxyalkylation, 63–64
sulfoximine reactions, 224–228, 230
Lithium alkoxide, 228
Lithium amides, 30, 57
Lithium bis(dimethylphenylsilyl)cuprate, 157, 176
Lithium bis(trimethylsilyl)amide, 32
Lithium α-bromoacrylate, 58
Lithium-copper reagents, 19
Lithium dienolate, 188
Lithium diisopropylamide, 41–42
Lithium dimethylcuprate, 190
Lithium enolate, 40
Lithium methanephosphonate, 40
Lithium methyl acetate, 37
Lithium salts, 219
Lithium sulfoxides, 75

M

Magnesium, 94, 98, 129, 183, 227
Maleimides, 164, 174
Menthoxydiarylsulfonium salts, 250
1-Menthoxytrimethylsilane, 8
Menthyl arenesulfinates, 16, 30–31, 180
O-Menthyl benzenesulfinate, 8
Menthyl p-toluenesulfinate, 7–8, 16, 26–27, 28
alkyl(aryl)sulfenylmethyl sulfoxide synthesis, 91, 95
allyl sulfoxide synthesis, 78 α-phosphoryl sulfoxide synthesis, 118, 127
in sulfinamide synthesis, 30
α-sulfinyl phosphonium ylide synthesis, 119
α-sulfinyl sulfoxide synthesis, 98
α-sulfonyl sulfoxide synthesis, 101
10-Mercaptoisoborneol, 164
Mercury, 218
Mesitylenesulfinate, 10
Mesityl methyl sulfoximine, 199
O-Mesitylsulfonyl hydroxylamine, 201
Mesylates, 147–148
Metallated alkyl(aryl) methyl sulfides, 91, 95
2-Metallated 2-cycloalkenone ketals, 180
α-Metallated methyl sulfones, 101
Methanesulfinates, 25–26
Methionine sulfoximine, 198
α-Methoxyaldehydes, 94–95
Methoxycarbonylmethylcyclopentanone, 192–193
α-Methoxyphenylacetaldehyde, 95–96, 138
Methoxypyrrolidino-p-tolyl sulfonium salt, 247
Methoxysulfonium salt, 245
Methylallyl p-toluenesulfenate, 78–79
γMethylallyl p-tolyl sulfone, 26
Methylation, α-phosphoryl sulfoxides, 125–126
α-Methylbenzylamine, 29
N-α-Methylbenzyl p-toluenesulfinamide, 29
Methyl bromoacetate, 40
Methylcarbinols, 157–158
3-Methylcarboxylic acids, 147
Methylcyclohexanone, 193
N-Methyl-3,4-dihydro-6,7-dimethoxy-1-[(p-tolylsulfinyl)]isoquinoline, 176
α-Methylene butyrolactones, 58
2-Methylene-1,3-dithiolane 1,3-dioxide, 100–101
α-Methylene hydrogens, 47, 116
N-Methylephedrine, 10
Methylethylphenyloxosulfonium perchlorate, 4, 249
Methylethylphenylsulfonium perchlorate, 249
Methyl glyoxalate, 122
6-Methylheptanal, 109–110
Methyl γ-hydroxy-α,β-unsaturated esters, 132–135, 142
N-Methyl-2-imidazolyl sulfoxides, 63–64
Methyl iodide, 41–42, 56
Methyl jasmonate, 187
Methyllithium, 19, 32, 49
N-Methyllithium anilide, 30
Methylmagnesium bromide, 198
Methyl methylthiomethyl sulfoxide, 91
Methyl-p-nitrophenylsulfoximine, 198–199
Methyl phenyl sulfoxide, 53
Methylphenylsulfoximine, 30, 199, 203–205, 234
α-Methylpropargyl p-toluenesulfinate, 26
Methyl sulfinylcarboxylates, 142
N-Methylsulfoximine, 231
(*cis*-6-Methyltetrahydropyran-2-yl)acetic acids, 152
Methylthiomethyl Grignard reagent, 95
Methylthiomethyl p-tolyl sulfoxide, 95
Methyltitanium triisopropoxide, 193
O-Methyl p-toluenesulfinate, 15
Methyl 3-p-tolylsulfinylpropenoate, 122
Methyl p-tolyl sulfoxide, 75, 189
aminoalkylation, 70–71
halogenation, 107
imine reactions, 76
α-sulfinyl sulfoxide synthesis, 98
Methyl p-tolyl sulfoxides, 245–246
S-Methyl-S-(p-tolyl)sulfoximine, 201–202, 235

Methyltriphenylphosphonium iodide, 127
Methyl vinyl p-tolyl sulfoxide, 153
Michael additions
  acyclic α,β-unsaturated sulfoxide reactions, 145–148, 152–158
  allyl sulfoxides to carbanyl compounds, 83–84, 89–90
  cycloalkenone sulfoxides, 182–193
  α-phosphoryl sulfoxide reactions, 125–126
  α-sulfinylcarboxylate reactions, 131
  sulfoxides, to carboxylic esters, 60–61
  vinyl sulfoximines, 222–234
Monoterpenes, 220

N

Naphthalene, 22–23
Neomenthyl-N-tosyl sulfoximine, 259
Nickel, 207, 219–220, 226–228, 227, 237, 263
Nickel acetylacetonate, 210
Nitriles, 207
Nitrogen, 152–157, 225
Nitrones, 72–73
Norephedrine-derived oxazolidin-2-one, 42–43
Norvaline, 37–38
Nucleophiles, 148–158, 182–190
Nucleophilic substitution reaction, 3

O

Ohno's intermediate, 168
Onium derivatives, 137
  applications, 251–263
  experimental procedures, 264–265
  stability, 1–2
  structure, 241
  synthesis, 241–250
Organometallic reagents, 222–227, 230–233
  N-alkylidene sulfinamide reactions, 32
  cycloalkenone sulfoxides, 182–185, 190–191
  sulfinamide synthesis, 30–31
  sulfinic ester preparation, 11, 12
  sulfoxide synthesis, 16, 17–18
Oxathiane, 245
Oxaziridine, 33
Oxazolidinones, 29, 42–43
3-Oxocyclopentanecarboxylic acid, 92
11-Oxoequilenin, 186
δ-Oxo-β-hydroxysulfoxides, 138
Oxosulfonium salts, 4, 60
β-Oxosulfoxides
  cycloalkenone sulfoxide reactions, 189
  reduction, 142
  synthesis, 73–75, 76, 96, 129, 135–139
Oxosulfur ylides, 256–257
Oxygen nucleophiles, 148–152
Ozonolysis, 113

P

Palladium complexes, 24
Panasinsene, 218–219
Pentacoordinate sulfur compounds, 4
1-Pentafluorophenyl-2-methyl-2-thianaphtalene, 250
Pentalene, 89
Perchlorate, 249
Perhydro-cyclopenta[α]phenanthrene, 170
Perillaldehyde, 263
Phenylalanines, 41
3-Phenylbutanoic acid, 224
2-Phenylbutylamine, 36
3-Phenylbutyric acid, 145
trans-2-Phenylcyclohexanol, 11–12, 18–19
Phenylglycine, 37–38
2-Phenylhexane, 224
N-Phenyl imines, 71
Phenyliodonium bis-trifluoroacetate, 60–61
N-Phenyl-2-phenylaziridine, 257
Phenylserine, 41
N-Phenyl sulfinamides, 43
Phenylsulfonimidates, 204
N-(Phenylsulfonyl)(3,3-dichlorocamphoryl)oxaziridine, 33
N-Phosphate, 198
Phosphonate α-carbanions, 118
α-Phosphonium substituted sulfoxides, 116
Phosphonothionates, 122
α-Phosphoryl sulfoxides, 154, 181
  asymmetric reactions, 124–127
  experimental procedures, 127
  synthesis, 116–118
  α,β-unsaturated sulfoxide synthesis, 116–118
N-Phthalimidosulfoximines, 202
Piperidine, 80, 98–99, 102
  SPAC procedure, 132–135, 142
Potassiomethyl phenyl sulfoxide, 53
Potassium t-butoxide, 112
Potassium carbonate, 115
Propargylic alcohol, 114–115
Propenyl p-tolyl sulfoxide, 153
pro-R hydrogen, 48, 50, 51
Protonic acids, 13
Pseudomonas, 142
Pseudorotation, 4
Pummerer reaction, 124–125, 156–157, 170–173
Pyramidal sulfur compounds, stability, 1
Pyridine, 108

Q

Quinine, 117

R

Radical reactions, 174, 190–191
Raney nickel desulfurization, 207, 219–220, 237, 263
Rothrockene, 220

S

Salicylate, 189
Sedamine, 154–156
Sigmatropic rearrangement
  allyl sulfoxides, 78–82
  SPAC reactions, 132–135
Silanes, 157–158

Silica gel chromatography, 35–36
Silicon nucleophiles, 157–158
Silver perchlorate, 14
Silylation, carbanion, 57
ω-Silyloxyalkenyl-sulfoxides, 181
ω-Silyloxycarbonyl compounds, 181
Simmons-Smith cyclopropanation, 214
Sodium, 222
Sodium bis-(trimethyl-silyl)amide, 98–100
Sodium borohydride, 209
Sodium dimethyl malonate, 24–25
Sodium methoxide, 188
Sodium thiolates, 91
Sodium p-toluenesulfinate, 27
SPAC procedure, 132–135, 142
Spiroketone, 257, 263
Steroids, 169–170
Structure-chirality relationship, 1–5
trans-Styryl p-tolyl sulfoxide, 121, 145
Sulfenamide, 1
Sulfenate, 78–82
Sulfides, 233
Sulfilimines
  applications, 206
  experimental procedures, 233–234
  synthesis, 195–198
Sulfinamides, 247
  structure, 2
  structure-chirality relationship, 3
  synthesis, 43
Sulfinate esters, see Sulfinic esters
Sulfinates, structure, 2
Sulfinic acid derivatives
  sulfinamides and N-alkylidenesulfinamides
    diastereomeric, 28–30
    enantiomeric, 30–33
    experimental procedures, 42–43
    sulfoxide synthesis, 33–42
  sulfinic esters, 7–28
    diastereomeric, 7–12
    enantiomeric, 13–15
    experimental procedures, 26–28
    sulfoxide synthesis, 16–26
Sulfinic esters
  structure-chirality relationship, 3
  synthesis, 33–34
Sulfinylacetates, 133
Sulfinyl alkenolides
  applications, 186–190
  conjugate addition of nucleophiles, 185–186
  experimental procedures, 191–193
  synthesis, 180–182
Sulfinylaziridines, 112–113
Sulfinylbutadienes, 67–68
2-Sulfinyl-2-butenolide, 188
α-Sulfinyl carbanions, see Carbanions
α-Sulfinylcarboxylates
  hydroxyalkylation, 142
  synthesis, 129–135
Sulfinyl chlorides
  sulfinamide synthesis, 29
  sulfinic ester preparation, 7–10, 15
Sulfinyl compounds, structure, 1–2
Sulfinyl cycloalkenones
  applications, 186–190
  conjugate addition of nucleophiles, 182–185, 192–193
  experimental procedures, 191–193
  synthesis, 180
Sulfinyl cyclohexanones, 23
Sulfinyl cyclopentenones, 184, 190–193
Sulfinyldiacetates, 129–131
Sulfinyl dienes, 174–175
Sulfinyl dienophiles, 169–170
3-Sulfinyl-4,5-dihydroisoxazoles, 140
α-Sulfinyl hydrazones, 140
α-Sulfinyl maleates, 163–165
α-Sulfinyl maleimides, 163–165, 168
α-Sulfinyl-N-methoxyacetimidates, 140
N-Sulfinyloxazolidinones, 29–30, 33–34, 35
2-Sulfinyl oxazolines, 140
α-Sulfinyl phosphonium iodide, 119
α-Sulfinyl phosphonium salts and ylides
  experimental procedures, 127
  sulfoxide synthesis, 120–123
  synthesis, 119–120
α-Sulfinyl sulfoxides, 98–101
Sulfinyl sulfur, 31
Sulfites, 3, 12, 14
Sulfones, 3, 24
Sulfonic acids, 3
Sulfonimidoyl chlorides, 3
Sulfonium acid, 255–256
Sulfonium salts, see Onium derivatives
Sulfonyl chloride, 198–199
α-Sulfonyl sulfoxides, 101–104
N-Sulfonylsulfoximines, 231, 232
Sulfoxide Piperidine Aldehyde Condensation (SPAC)
    procedure, 132–135, 142
Sulfoxides
  acyclic α,β-unsaturated
    α,β-unsaturated reactions, 170–174
    conjugate addition of silicon, 157–158
    Diels-Alder cycloaddition, 158–170
    electrophilic addition, 144–145
    experimental procedures, 175–178
    Michael addition of nitrogen, 152–157
    Michael addition of oxygen, 145–152
    sulfinyl diene cycloadditions, 174–175
  alkyl and arylsulfenylmethyl, 91–96
  allyl, 78–90
  cycloalkenone, 180–193
  dialkyl and alkyl aryl
    acylation, 73–74
    alkylation, 57–59
    aminoalkylation, 70–73
    experimental procedures, 74–77
    hydroxyalkylation, 61–70
    Michael additions, 60–61
  α-halogeno, 106–115
  O-menthyl p-toluenesulfinate in synthesis, 28
  β-oxo and related derivatives
    experimental procedures, 141–142
    iminosulfoxides and analogs, 140–141
    oxosulfoxides, 135–139

sulfinylcarboxylates, 129–135
thioxosulfoxides, 139
phosphoryl and sulfinyl phosphonium, 116–127
structure, 2
sulfinic esters in synthesis, 16–21
sulfinyl-, sulfonyl-, and sulfoximino, 98–105
α-sulfinyl carbanions, 47–56
synthesis, 34–35
Sulfoximine nitrogen, 225
Sulfoximines, 250
applications, 206–233
experimental procedures, 235–239
structure-chirality relationship, 3
synthesis, 198–205
α-Sulfoximino sulfoxides, 104–105
Sulfoxonium salts, structure, 2
Sulfurane, structure-chirality relationship, 4–5
Sulfur compounds, structure-chirality relationship, 1–5
Sulfur onium derivatives, see Onium derivatives
Sulfurous acid, structure-chirality relationship, 3

T

Talaromycines, 148–150
Temperature, sulfinamide synthesis, 29
Tert-butyl phenylmethyl sulfoxide, 75
Tert-butyl sulfoxides, hydroxyalkylation, 62–63
Tetracoordinate sulfur compounds, 3–4
Tetracyclic amides, 168
Tetrahedral, tetracoordinate sulfur compounds, 3–4
Tetrahydrofurans, 24, 212
carbanion chelation, 49, 50
stereoselective synthesis, 65, 66
Tetrahydropalmatine, 71–72
Tetrahydropyranylacetic acid, 152
Tetramethylammonium salts, 117
Tetramethylethylenediamine, 53
Thermolysis, sulfoximines, 237
THF, 74–76, 104–105
Thiaanthracenes, 262
Thiane oxide, 57, 73, 74–75
Thionyl chloride, 11
Thiosulfinates, 2
Thiosulfonium salt, 252
Thioxanones, 262
Thioxanthenes, 262
Thioxanthenium perchlorate, 243
β-Thioxosulfoxides, 139–141
Thiranium, 252
Thujopsene, 24
Titanium, 162–163, 183, 210–211
Titanium tetrachloride, 125
α-Tocopherol, 150–151
p-Toluenesulfinamides, 13–14, 27–28, 43
amino acid synthesis, 38
β-aminophosphonic acid ester synthesis, 40
sulfinic ester synthesis, 33–34
p-Toluenesulfinates, 7–8, 11, 13, 15, 16, 24–27, 28, 34
p-Toluenesulfinylation, 42–43
p-Toluenesulfinyl chloride, 26–27, 29
3-(p-Toluenesulfinyl)chromone, 189
p-Toluenesulfinyl cycloalkenones, 183–184
2-p-Toluenesulfinyl-2-cyclopentenone, 192
2-p-Toluenesulfinyl-2-cyclopentenone ethylene ketal, 191–192
2-p-Toluenesulfinyl-1,4-dimethoxybenzene, 22–23
cis-n-(p-Toluenesulfinyl)-2-methoxycarbonylaziridines, 40
p-Toluenesulfonyl chloride, 27
N-(p-Toluenesulfonyl)-S-methyl-S-(p-tolyl)sulfoximine, 235
N-p-Toluenesulfonylsulfilimines, 196–198, 234
Toluenethiosulfonic acid, 3
N-p--Tolylsulfenylpiperidine, 80
Tolylsulfinyl-1,3-butadienes, 76
p-Tolyl sulfinyl-N,N-dimethylthioacetamide, 139
Tolylsulfinylmethane, 98, 104
p-Tolylsulfinylmethyltriphenylphosphonium iodide, 119
p-Tolylsulfinylmethyltriphenylphosphonium ylide, 119
p-Tolylsulfinyl-2-propen-1-yl-cyclopentanone, 89–90
p-Tolyl sulfoxides, 28
Tolylthiomethyl p-tolyl sulfoxide, 95
p-Tolyl vinyl N-phthalimidosulfoximine, 231
Tolyl vinyl sulfoxide, 159
Trans-fused chairlike transition state, 83
Trialkylboranes, 191
Tricoordinate sulfur compounds, 1–3
1-(Triethoxymethyl)ethylidene-p-toluenesulfinamide, 38
Trifluoroacetic acid, 14, 27–28, 40
Trifluoro-2-phenyl-(p-tolylsulfinyl)propan-2-ol, 75
3,3,3-Trifluoroprop-1-enyl p tolyl sulfoxides, 146
Trihydroxysulfoximine derivatives, 216
Trimethylsilyl chloride, 209

V

Vinyl chlorides, 114
Vinyl sulfoxides, 83, 121–123
carbon nucleophile addition, 147–148
Diels-Alder reaction, 158–166
nitrogen nucleophile addition, 152–157
oxygen nucleophile addition, 148–152
synthesis, 20, 175–176
Vinyl sulfoximines, 204, 222–232, 236
Vitamin E, 150–151

W

Wittig reaction, see Horner-Wittig reaction

Y

Ylides, see Onium derivatives; α-Sulfinyl phosphonium salts and ylides

Z

Zinc, 227
Zinc bromide, 177, 182
Zinc chloride, 126, 135–136, 137, 162–167, 177
Zinc sulfoxides, 75